Advanced Structured Materials

Volume 88

More information about this series at http://www.springer.com/series/8611

Magomed F. Mekhtiev

Vibrations of Hollow Elastic Bodies

Springer

Magomed F. Mekhtiev
Institute of Mathematics and Mechanics
of National Academy of Sciences
of Azerbaijan
Baku State University
Baku
Azerbaijan

ISSN 1869-8433 ISSN 1869-8441 (electronic)
Advanced Structured Materials
ISBN 978-3-319-74353-0 ISBN 978-3-319-74354-7 (eBook)
https://doi.org/10.1007/978-3-319-74354-7

Library of Congress Control Number: 2017963979

Printed on acid-free paper

This Springer imprint is published by the registered company Springer International Publishing AG part of Springer Nature
The registered company address is: Gewerbestrasse 11, 6330 Cham, Switzerland

Preface

Shell theory is one of the most important fields of modern mechanics. It develops methods for calculating thin-walled structures which are widely used in modern plants and machinery. Requirements of strength, lightness and efficiency of a modern design make thin shells indispensable constructional elements. However, calculation of shells (calculation of stress-strained state of shells) on the basis of three-dimensional equations of elasticity theory involves considerable mathematical difficulties. That is why we have to resort to various approximate methods simplifying the calculation of shells. Here, a fundamental geometrical assumption, i.e. the smallness of the thickness of the shell to its remaining two dimensions, is taken into account. Namely, the problem of reduction of a three-dimensional problem of elasticity theory to a two-dimensional problem is the main content of shell theory. Obviously, there are many possible ways of transforming the problems of elasticity theory to the problems of shell theory for thin-walled structures such as shells. The main relevant results concerning mostly static problems of shell theory are consecrated in Vorovich's (1966, 1975), Goldenveyzer's (1969, 1975), Koiter and Simmonds (1973), Sayir and Mitropoulos (1980), Petraszkievicz (1992) reviews. The problem of reduction of a three-dimensional dynamic problem of elasticity theory to a two-dimensional problem of shell theory is described in particular in the works of Aynola and Nigul (1965) and Grigolyuk and Selezov (1973), Achenbach (1969), Berdichevskii and Khan'Chau (1980), Goldenveizer et al. (1993). Here, we also note the monographs by Kilchevsky's (1963), Berdichevskii (1983), Kaplunov et al. (1998), Le (2012), Aghalovyan (2015) where these issues are rather thoroughly discussed and where a bibliography on the discussed topics can also be found. As it is hardly possible to find a detailed literature survey on the topics mentioned above, it is both practically impossible and there is no need to give an overview of the results on shell theory carried out at different times by means of different methods.

Many of the major achievements in the field of plates and shells are obtained by scientists from the CIS (the Commonwealth of Independent States, or former USSR). The most significant contributions to the development of the foundations of the classical shell theory were made by S. A. Ambartsumyan, V. V. Bolotin, I. N. Vekua,

V. Z. Vlasov, I. I. Vorovich, B. G. Galerkin, K. Z. Galimov, A. L. Goldenveyzer, E. I. Grigolyuk, N. A. Kilchevsky, A. I. Lure, H. M. Mushtari, V. V. Novozhilov, P. M. Ogibalov, Yu. N. Rabotnov, V. V. Sokolovsky, S. P. Timoshenko, K. F. Chernyh, P. M. Nahdi, E. Reissner and others.

It is clear from the ongoing discussion that the modern theory is a well-developed branch of solid mechanics. However, the process of the development of the theory of shells cannot be regarded as a complete one. First of all, there constantly appear new technology designs the calculation of which is impossible in the framework of the existing versions of the theory of shells.

Therefore, the emergence of new variants of the theory of shells is inevitable. One can surmise that the number of such nontraditional structures will increase. It is connected with the modern technology achievement of an extremely high level of exploitational parameters.

On the other hand, the theory of shells as a theory must be internally consistent. As is known, the classical theory of shells is not deprived of contradictions.

Therefore, the major problem of the theory of shells is the formulation of different variants of boundary value problems and evaluation of their applicability scale. The problem, of course, is not only of theoretical interest but of great practical importance. Over many years of development in this particular field of elasticity theory, a considerable amount of material accumulated both on formulation of different variants of applied theories together with the development of solution (calculation) methods. At various stages of its development, the state of the theory of shells was subjected to some critical analysis. A closer look at the current state of the theory reveals a need for additional research on the comparative analysis of various applied theories established in the domain of their applicability.

Below we will consider the ways of bringing the two-dimensional problems to the three-dimensional ones using the smallness of shell thickness in their constructions compared to its other dimensions. Among these methods, the asymptotic method takes a special place. Asymptotic approach, probably for the first time, was applied to the problems of shell theory by Shterman (1924) and later—by Krauss (1929). Asymptotic integration of the equations of two-dimensional theory of shells undergone a great development in the works of A. L. Goldenveyzer (since 1939). The combination of complex transformation of equations of V. V. Novozhilov's shell theory with the asymptotic methods is represented in Chernykh's works (1962, 1964).

Thus, the asymptotic method developed by A. L. Goldenveyzer, I. I. Vorovich and their students made a significant contribution to the development of the theory of plates and shells. This method proved to be very effective in the study of problem of the limiting transition from three-dimensional elasticity problems into two-dimensional ones. Due to works of authors mentioned above, it became possible to solve such important issues as the establishment of the range of applicability of the applied theories of plates and shells, in particular the classical Kirchhoff–Love's theory. Further development of this method allowed to create effective methods for calculating the three-dimensional stress state and to solve practically important problems of stress concentration at the holes in plates and shells of constant thickness in the static case.

However, the question of the relationship between two-dimensional theories and the corresponding three-dimensional problems of elasticity theory for plates and shells of variable thickness has not practically been studied.

The problem of passage to the limit in dynamic problems of elasticity theory is particularly critical. Since there are now a number of dynamic applied theories of shells based on various hypotheses, lack of data on the comparative analysis raises the question of establishing the range of applicability of each of them on the basis of three-dimensional dynamic theory of elasticity. Here, the most important point is the question of determining the natural frequencies and the forms of vibrations of shells from the position of three-dimensional theory of elasticity.

Many of the difficulties associated with the study of these problems are caused by the presence of several parameters in the original problems. For example, the interplay of such parameters as relative thickness, curvature, oscillation frequency, curvature of the holes even for sufficiently smooth external loads can generate a fairly complex stress–strain state which is impossible to calculate correctly not only within the classical theory but also in such approved revised theories as the theory of S. P. Timoshenko, V. Z. Vlasov and others.

In addition, the asymptotic method enables to effectively solve boundary value problems for such elastic bodies that cannot be attributed to shells (e.g. thick cylindrical or conical rings) and for which at the same time the spatial theory is powerless because of the proximity of the parameters describing the boundary surfaces.

This monograph is devoted to these range of questions.

The book consists of four chapters. In the first chapter, an axisymmetric dynamic problem of elasticity theory for a hollow cylinder is investigated by the method of homogeneous solutions. Homogeneous solutions depending on the roots of the dispersion equation are constructed. The classification of the roots of the dispersion equation is presented.

The basis of the classification procedure is at the order of a root with respect to a small parameter ε characterizing the thinness of the shell depending on the frequency of the driving forces. It is shown that in the high-frequency domain (in terms of spatial problems) in the first term of the asymptotic behaviour, the dispersion equation coincides with the well-known Rayleigh–Lamb equation for an elastic strip.

The classification of homogeneous solutions has been conducted and it is shown that each group of roots of the dispersion equation corresponds to its type of homogeneous solutions. Asymptotic expansions of the homogeneous solutions, allowing the calculation of the stress–strain state at different values of the frequency of the driving forces, are obtained.

A generalized condition of orthogonality of homogeneous solutions for a hollow cylinder which allows the accurate solution of the problem of forced vibrations of a hollow cylinder for certain end boundary conditions is proved. In the general case of the end loading, the original boundary value problem is reduced to the solution of infinite system of linear algebraic equations by means of the Lagrange variational principle.

Also presented is a method for constructing applied theories intended to relieve stresses from cylindrical boundaries of a shell. Together with homogeneous solutions, they permit the solution of the inhomogeneous problem.

The importance of the evaluation of boundary conditions in the formation of the spectrum and vibration shapes of a shell is of great interest in dynamic problems of elasticity theory. Therefore, practically all boundary conditions which can be stated in three-dimensional theory of elasticity are studied. In particular, some solutions of the problem of forced vibrations of a hollow cylinder with a clamped side surface are presented. It is shown that the solution of this problem in the first term of its asymptotic expansion coincides with the known solution for the elasticity theory for an elastic strip. We also discuss the problem of torsional vibrations and the vibrations under mixed boundary conditions on the side surface of a cylinder. These problems appeared to be very simple both from a physical and a mathematical point of view. In fact, these problems, in mathematical terms, are reduced to solving boundary value problems for Helmholtz equation.

The second chapter deals with a three-dimensional dynamic problem of elasticity theory for a spherical layer. Homogeneous solutions are derived for the case of axisymmetric vibrations. In the case of axisymmetric vibrations, homogeneous solutions are constructed. One way of constructing of the heterogeneous solutions is pointed out. An asymptotic analysis of homogeneous solutions for a spherical shell corresponding to different groups of roots of the dispersion equation is performed.

It is shown that in contrast to a cylindrical shell, ultra-low frequency vibrations for a spherical shell are not available. A generalized condition of orthogonality of systems of homogeneous solutions is proved. One class of boundary conditions on the side surface, admitting an exact solution of the problem of forced vibrations of a spherical shell is identified. In the case of a general loading by Hamilton's variational principle, the boundary value problem is reduced to solving an infinite system of linear algebraic equations. Matrices of such systems are known for a spherical shell in the static case, and for an elastic strip in the dynamic elasticity problems. The problem of torsional vibrations of a spherical shell is solved analytically. The need for a detailed discussion of this problem is revealed when considering non-axisymmetric elasticity problems.

Non-axisymmetric dynamic problem of elasticity theory for a spherical layer is considered. Due to spherical symmetry, the general boundary value problem is divided into two problems one of which coincides with the boundary value problem for axisymmetric vibrations of a hollow sphere, and the second one describes the vortex motion of a hollow sphere and coincides with the boundary value problem for purely torsional vibrations of a hollow sphere.

The third chapter provides an asymptotic process for finding the frequencies of free axisymmetric vibrations of an isotropic hollow cylinder and a closed hollow sphere based on the dynamic equations of elasticity theory. An asymptotic process is thoroughly built for a cylinder with free side surfaces and with simply supported at the ends and for a closed hollow sphere with free facial surfaces. These problems are considered to be a model, since the study of the asymptotic processes for other boundary conditions has no fundamental difficulties.

A comparison of the results obtained in Kirchhoff–Love theory with the results obtained by the three-dimensional elasticity theory is given. For a cylinder and a sphere, there are obtained two frequencies in the first term of the asymptotic expansions, coinciding with the frequencies determined by the application of shell theory, and a countable set of frequencies which are not available in the applied theory of shells. The frequencies of the thickness vibrations of cylindrical and spherical shells are determined.

The fourth chapter is dedicated to the development of the asymptotic method of integrating three-dimensional equations of elasticity theory for a conical shell and a plate of variable thickness and the analysis of three-dimensional stress–strain state on the basis of this method.

In the first part of the fourth chapter, the solution of the problem of elasticity theory for a truncated hollow cone of variable thickness is obtained by the method of homogeneous solutions.

An asymptotic analysis of the characteristic equation establishes the existence of three groups of zeros with the following asymptotic properties: $\lambda = O(1)$, $\lambda = O\left(\varepsilon^{-1/2}\right)$ $\lambda = O(\varepsilon^{-1})$ (ε is a parameter of thin-walledness) each of which corresponds to its type of stress–strain state.

The first group of zeros corresponds to a penetrating solution coinciding with the known Mitchell-Neuber's solution. The stress state, defined by this solution, is equivalent to the resultant vector of forces applied to one end of the shell.

The second group of zeros corresponds to the solution of end-effect type similar to the end effect of the applied theory of shells. The first terms of the asymptotic expansions of the stress state obtained through this solution are equivalent to the bending moment and shearing forces.

The third group of zeros corresponds to the solution of boundary layer type which in the first term of the asymptotic expansion coincides with Saint-Venant's end effect in the theory of thick plates. Using the principle of Lagrange virtual displacements, the boundary value problem is reduced to solving an infinite system of linear algebraic equations. Matrices of these systems are known in the theory of thick plates of constant thickness. Their inversion can be achieved by using the reduction method. A method for constructing applied theories intended to relieve stresses from conical shell boundaries is shown.

The second part of the fourth chapter investigates the asymptotic behaviour of the axisymmetric stress–strain state of the plate, the thickness of which is $h = \varepsilon r$, where r is the distance from the centre of the plate, and ε is still a small parameter. Here, we are not talking about an arbitrary plate but the particular form of the conical shell discussed in the first part of the fourth chapter which it takes during the degeneration of its midsurface into a flat one. Since this is a special case of degeneration, all the arguments of the previous chapters have to be repeated.

When constructing refined applied theories for the plates of variable thickness, instead of the traditional linear-independent solutions $P_\nu(\cos\theta)$, $Q_\nu(\cos\theta)$ of Legendre's equation, we introduce, for convenience, another set of linearly independent solutions of Legendre equation $T_\nu(\theta) = P_\nu(\cos\theta) + P_\nu(-\cos\theta)$ and

$F_\nu(\theta) = P_\nu(\cos\theta) - P_\nu(-\cos\theta)$ which are respectively odd and even functions with respect to the midplane of the plate. The chosen form of solutions makes it possible to divide the general problem into two independent ones: the problem of the tension–compression of a plate and the plate bending problem. Such a division greatly simplifies the process of building refined applied theories for plates of variable thickness.

Also considered are the problems of equilibrium of an elastic hollow cone with a fixed side surface and with mixed boundary conditions on the side surface.

It is shown that in the case of fixed side surface, the solution of this problem in the first term of the asymptotic expansion coincides with the known solution for the elasticity theory for an elastic strip. A generalized orthogonality condition for a hollow cone is proved.

Proceeding from Papkovich–Neuber's general solution, a stress–strain state of a plate with variable thickness subject to the action of non-axisymmetric loads is studied.

The behaviour of the solution as the parameter of thin-walledness tends to zero is investigated. It is shown that the stress–strain state of a plate consists of the stress–strain state penetrating deep into the plate and the end effect similar to Saint-Venant's end effect. Kirsch's problem for plates of variable thickness is solved.

In the last section of the fourth chapter, torsional vibrations of a conical shell and a plate of variable thickness are discussed. At first, the problem is solved exactly. Then an asymptotic analysis of the problem of harmonic torsional waves spreading in a conical shell and plate of variable thickness is given. Depending on the frequency of the driving forces, a form of wave formation is studied. Asymptotic formulas determining the frequency of torsional vibrations of a conical shell and a plate of variable thickness are obtained.

The derived homogeneous and inhomogeneous solutions not only reveal the qualitative features of three-dimensional solutions in shell theory but they can serve as an effective device of solving specific boundary value problems, as well as a basis for the assessment of simplified theories. As in the general case of loading, the solutions of dynamic and static elasticity problems are reduced to solving infinite systems of linear algebraic equations. So, in the appendix, a solution of the axisymmetric problem on stress concentration around a circular hole in the plate whose boundary is loaded with normal forces of the form $\sigma_r = \chi(\eta^p - k_p\eta)$ is presented. Here χ is a constant, k_p is a parameter that has been selected so that the load is self-balanced, η is a transverse coordinate. Numerical results for the solution of the problem is presented. This problem can be regarded as a model for the corresponding problems in the theory of shells; it is relatively simple and at the same time contains all the characteristic features of problems in three-dimensional elasticity theory. The exact solution of the problem of axisymmetric vibrations of a cylindrical shell under given mixed boundary conditions at the ends is derived by means of homogeneous solutions. Numerical analysis is carried out for the parabolic distribution of normal stress on the end surface and zero end radial displacement. Some numerical analysis of the problem of dynamic torsion of a spherical layer by forces distributed on the surface of a tapered cut is conducted.

Summarizing the results, we note that the following results of the author are obtained in the monograph:

1. The forced vibrations of a cylinder and spherical layers are investigated by the method of homogeneous solutions. A possible form of wave formation is studied depending on the frequency of the driving forces. A complete asymptotic analysis of solutions of three-dimensional dynamic problems of elasticity theory is conducted as the parameter of thin-walledness tends to zero. A comparison of the asymptotic solution with the solutions obtained by the applied theories of shells is given. A generalized orthogonality condition of homogeneous solutions which allows accurate solutions to the problem of forced vibrations of a hollow cylinder and a spherical layer under specific conditions of shell end bearings are derived. In the general case of loading, the boundary value problem is reduced to solving an infinite system of linear algebraic equations employing Lagrange and Hamilton variational principles.
2. A qualitative study of some applied theories is considered; the limits of their applicability are determined. In particular, it is shown that all existing applied theories of shells inadequately describe the stress–strain state in the vicinity of concentrates and are not suited for the study of high-frequency vibrations of thin and thick shells.
   The detailed study of the properties of homogeneous solutions, in fact, of independent wave motion types, which may be realized in the considered elastic bodies, provides the basis for stating the correct methodological problem of constructing refined theories for thin-walled elements. In this connection, refined applied theories, more accurately describing the processes occurring in thin shells rather than the classical two-dimensional theory of shells, are constructed and these allow obtaining the solutions of inhomogeneous problems to a given degree of accuracy.
3. An asymptotic process for finding the frequencies of free axisymmetric vibrations of a hollow cylinder and a closed hollow sphere is derived. As is well known, even in relatively simple cases the analysis of the frequency equations is quite challenging. It is, therefore, essential to determine all the frequencies in a certain frequency range.
   The author's approaches allow to create algorithms that are able to capture all natural frequencies in a given interval, and this undoubtedly represents a scientific and practical value.
   Comparisons of the results obtained by Kirchhoff–Love and Timoshenko's theories with the results obtained by the three-dimensional elasticity theory are presented. It is shown that in the problems on free vibrations, the applied theory of shells approximately approximates only the lowest part of the frequency spectrum, but is unable to describe the phenomenon of end resonance.
4. An asymptotic method of integrating three-dimensional equations of elasticity theory for a conical shell and plates of variable thickness is developed. Homogeneous and inhomogeneous solutions are established and a generalized orthogonality condition for a cone is proved. Asymptotic analysis of the

problems of harmonic torsional waves spreading in a conical shell and in a plate of variable thickness is conducted through which asymptotic formulas are obtained that allow the determination of the frequencies of the mentioned bodies.

5. It is proved that the derived homogeneous and inhomogeneous solutions not only reveal the qualitative features of the three-dimensional solutions in the theory of shells, but can serve as an effective technique of solving specific boundary value problems, as well as a basis for assessing simplified theories.

Baku, Azerbaijan Magomed F. Mekhtiev

References

Achenbach, J.D.: An asymptotic method to analyze the vibrations of an elastic layer. Trans. ASME, Ser. E, J. Appl. Mech. **36**(1) (1969)

Aghalovyan, L.: Asymptotic theory of anisotropic plates and shells (2015)

Ainola L.A., Nigul W.K.: Wave processes of deformation of elastic plates and shells// News of USSR Academy of Sciences. **14**(1), 3–63 (1965)

Berdichevskii, V.L., Khan'Chau, L.: High-frequency long-wave shell vibration. J. Appl. Math. Mech. **44**(4), 520–525 (1980)

Berdichevskii, V.L.: Variational Principles in Mechanics of Continuous Media. Nauk, Moscow (1983)

Chernykh, K.F.: The linear theory of shells. Part 1. L.: LSU, p. 274 (1962)

Chernykh, K.F.: The linear theory of shells. Part 2. L: LSU, p. 395 (1964)

Goldenveizer, A.L.: Some questions of general linear theory of shells// Proc. VII All-Union conf. on the theory of shells and plates. Moscow: Nauka, 1970, pp. 749–754. (1969)

Goldenveizer, A.L., Kaplunov, J.D., Nolde, E.V.: Int. J. Solids Structures. **30**, 675–694 (1993)

Grigolyuk, E.N., Selezov I.T.: Non-classical theory of vibrations of rods, plates and shells. M: VINITI, p. 272 (1973)

Kaplunov, J.D., Kossovitch, L.Y., Nolde, E.V.: Dynamics of thin walled elastic bodies. Academic Press (1998)

Kilchevsky, N.A.: Fundamentals of analytical mechanics of shells, vol. 1. Kiev, Ukrainian Academy of Sciences, p. 354 (1963)

Koiter, W.T., Simmonds, J.G.: Foundations of shell theory. In: Theoretical and Applied Mechanics, pp. 150–176. Springer, Berlin, Heidelberg (1973)

Krauss, F.: Uber aie gerundgleichungen der Elasitatstheorie Schwachdeformieten Schalen// Math. Ann. Bd. **101**(1) (1929)

Le, K.C.: Vibrations of shells and rods. Springer Science & Business Media (2012)

Petraszkievicz, W.: Appl. Mech. Rev. **45**, 249–250 (1992)

Sayir, M., Mitropoulos, C.: On elementary theories of linear elastic beams, plates and shells. Zeitschrift für angewandte Mathematik und Physik ZAMP, **31**(1), 1–55 (1980)

Vorovich, I.I.: Some mathematical problems in the theory of plates and shells// Proc. II All-Union Congress on the theor. and appl. Mechanics. Review reports, 3, Nauka, 1966 (1964)

Vorovich, I.I.: Some results and problems of the asymptotic theory of plates and shells// Proceedings of the I All-Union school on the theory and numerical methods of calculation of plates and shells. Tbilisi, pp. 51–150 (1975)

About the Book

The book is inspired by justification and refinement of highly diverse approximate dynamic models for engineering structures arising in modern technology, including high-tech domains dealing with nano and metamaterials. The dynamic equations of 3D elasticity are applied to the analysis of harmonic vibrations of hollow bodies of canonical shapes. New exact homogeneous and inhomogeneous solutions are derived for cylinders, spheres, cones, including spherical and conical layers, as well as for plates of variable thicknesses. Associated dispersion relations are subject to detailed asymptotic treatment in case of a small thickness. A classification for vibration spectra is suggested over a broad frequency domain. The range of validity of various existing 2D theories for thin-walled shells is evaluated by comparison with 3D benchmark solutions. A number of numerical examples are presented. Refined versions of 2D dynamic formulations are developed. Boundary value problems for hollow bodies are also considered.

Contents

About the Author

Magomed F. Mekhtiev graduated from the Department of Mechanics-Mathematics of Baku State University. He defended his Ph.D. thesis at the chair of elasticity theory of Rostov State University and there. In 1989, he defended his Doctoral thesis in Leningrad (St. Petersburg) State University. During 1966–1991, he occupied various positions in the Institute of Mechanics and Mathematics of National Academy of Sciences of Azerbaijan. He has worked in Baku State University since 1991. In 1994, he became professor. Scientific research direction of Prof. M. F. Mekhtiyev is mathematical methods of solid mechanics and qualitative questions of optimal control. He has published over 120 scientific papers and two monographs in this field. M. F. Mekhtiyev is awarded with gold medal by Scientific-Industrial Chamber of European Union. At present, he is Dean of the Department of Applied Mathematics and Cybernetics and heads the chair of Mathematical Methods of Applied Analysis.

Chapter 1
Asymptotic Analysis of Dynamic Elasticity Problems for a Hollow Cylinder of Finite Length

Abstract In this chapter, we investigate forced vibrations of an isotropic hollow cylinder under the action of axisymmetric loads by the method of homogeneous solutions. Depending on the frequency of the driving forces a possible form of wave formation in a hollow cylinder is explored. The asymptotic behaviour of the solutions of three-dimensional dynamic problems of elasticity theory is studied as the wall-thickness parameter tends to zero hence corresponding to a thin walled structure. The comparison of the asymptotic solutions with the solutions obtained by the applied theories is given. A generalized orthogonality condition of homogeneous solutions is proved which allows an accurate solution of the problem of forced vibrations of a hollow cylinder with mixed end conditions. In the general case of loading of a cylinder by means of the Lagrange variational principle the boundary value problem is reduced to the solution of a system of linear algebraic equations. A method of constructing applied theories designed for stress relief from cylindrical boundaries of the shell is suggested. Together with the construction of the homogeneous solutions the solution of the inhomogeneous problem follows. The problems of torsional vibrations of a hollow cylinder with mixed boundary conditions on the side surface are solved exactly. The vibrations of a cylinder with a fixed side surface are considered as well. It is shown that in the first term of the asymptotic expansion the solution of this problem coincides with the solution of a similar problem in the theory of elasticity for an elastic strip.

1.1 Construction of Homogeneous Solutions

Consider the axisymmetric problem of elasticity theory for a hollow cylinder. The position of the points on the cylinder in space is defined by the cylindrical coordinates r, φ, z varying within (Fig. 1.1).

$$R_1 \leq r \leq R_2,\ 0 \leq \varphi \leq 2\pi,\ -l \leq z \leq l \tag{1.1.1}$$

It is assumed that the lateral surface of the cylinder is free from stresses, i.e.

M. F. Mekhtiev, *Vibrations of Hollow Elastic Bodies*,
Advanced Structured Materials 88,
https://doi.org/10.1007/978-3-319-74354-7_1

Fig. 1.1 Hollow cylinder

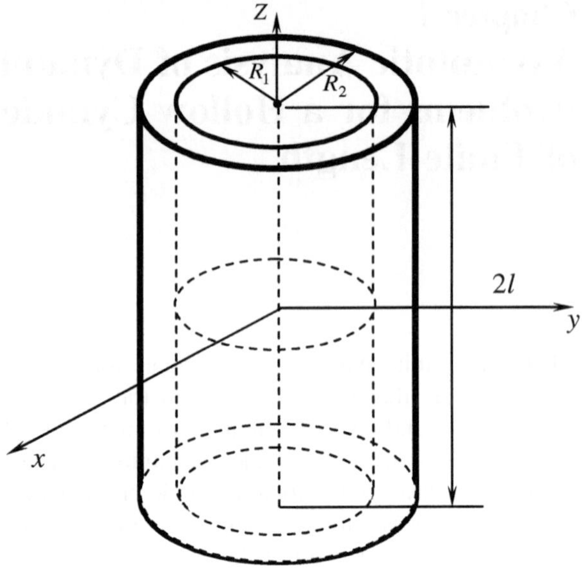

$$\sigma_r = 0,\ \tau_{rz} = 0$$

at

$$r = R_n, \quad -l \le z \le l, \quad (n = 1, 2) \tag{1.1.2}$$

while the rest of the boundary conditions are given as follows

$$\sigma_z = Q^{\pm}(r)e^{i\omega t}, \tau_{rz} = T^{\pm}(r)e^{i\omega t}$$

at

$$z = \pm l\,(k = 1, 2). \tag{1.1.3}$$

The equations of motion in terms of displacements in a cylindrical coordinate system have the form:

$$\begin{gathered}\frac{1}{1-2v}\frac{\partial x}{\partial \rho} + \Delta U_\rho - \frac{1}{\rho^2}U_\rho = \frac{2g(1+v)R_0^2}{E}\frac{\partial^2 U_\rho}{\partial t^2}\\ \frac{1}{1-2v}\frac{\partial x}{\partial \xi} + \Delta U_\xi = \frac{2g(1+v)R_0^2}{E}\frac{\partial^2 U_\xi}{\partial t^2}\\ x = \frac{\partial U_\rho}{\partial \rho} + \frac{U_\rho}{\rho} + \frac{\partial U_\xi}{\partial \xi}.\end{gathered} \tag{1.1.4}$$

Here $\rho = R_0^{-1}r$, $\xi = R_0^{-1}z$ are dimensionless coordinates, $R_0 = 1/2(R_1 + R_2)$ is the radius of the mid-surface of the shell, E is Young's modulus, v is Poisson's ratio, g is the density of the shell material, Δ is Laplace operator, and $U_\rho = R_0^{-1}U_z$, $U_\xi = R_0^{-1}U_z$ are the nondimensional displacements.

Stress tensor components are expressed by means of the displacements as follows:

$$\begin{aligned} \sigma_r &= 2G\left(\frac{\partial U_\rho}{\partial \rho} + \frac{\nu}{1+2\nu}x\right), \quad \sigma_\varphi = 2G\left(\frac{U_\rho}{\rho} + \frac{\nu}{1-2\nu}x\right) \\ \sigma_z &= 2G\left(\frac{\partial U_\xi}{\partial \xi} + \frac{\nu}{1-2\nu}x\right), \quad \tau_{rz} = 2G\left(\frac{\partial U_\xi}{\partial \rho} + \frac{\partial U_\rho}{\partial \xi}\right) \end{aligned} \tag{1.1.5}$$

where G is the shear modulus.

The solutions of (1.1.4) will be sought in the form:

$$U_\rho = U(\rho)\frac{dm}{d\xi}\, e^{i\omega t}, \; U_\xi = W(\rho)m(\xi)\, e^{i\omega t} \tag{1.1.6}$$

where the function $m(\xi)$ is subject to the condition

$$\frac{d^2m}{d\xi^2} - \mu^2 m(\xi) = 0 \tag{1.1.7}$$

for which the parameter μ is determined through the fulfilment of the boundary conditions on the surface.

Substituting (1.1.6) into (1.1.4) and separating variables relative to a pair of functions (U, W) we obtain the following system of ordinary differential equations:

$$\begin{aligned} &L_1(\mu, \lambda)(U, W) = U'' + \tfrac{1}{\rho}U' \\ &+ \left(\alpha^2 - \tfrac{1}{\rho^2}\right)U + \tfrac{1}{2(1-\nu)}(W' - \mu^2 U) = 0 \\ &L_2(\mu, \lambda)(U, W) = \tfrac{1}{1-2\nu}\mu^2\left(U' + \tfrac{U}{\rho} + W\right) \\ &+ W'' + \tfrac{1}{\rho}W' + \gamma^2 W = 0 \end{aligned} \tag{1.1.8}$$

$$\lambda^2 = \frac{2g(1+\nu)R_0^2\omega^2}{E}, \; \alpha^2 = \mu^2 + \frac{1-2\nu}{2(1-\nu)}\lambda^2, \; \gamma^2 = \mu^2 + \lambda^2.$$

here primes denote derivatives with respect to ρ and λ is a frequency parameter.

Taking into account (1.1.6), the formulas (1.1.5) take the form:

$$\begin{aligned} \sigma_r &= 2G\left[U' + \frac{\nu}{1-2\nu}\left(U' + \frac{U}{\rho} + W\right)\right]\frac{dm}{d\xi}e^{i\omega t} \\ \sigma_\varphi &= 2G\left[\frac{U}{\rho} + \frac{\nu}{1-2\nu}\left(U' + \frac{U}{\rho} + W\right)\right]\frac{dm}{d\xi}e^{i\omega t} \\ \sigma_z &= 2G\left[W + \frac{\nu}{1-2\nu}\left(U' + \frac{U}{\rho} + W\right)\right]\frac{dm}{d\xi}e^{i\omega t} \\ \tau_{rz} &= G(\mu^2 U + W')m(\xi)e^{i\omega t}. \end{aligned} \tag{1.1.9}$$

Substituting (1.1.9) into (1.1.2) we obtain the following homogeneous boundary conditions for the functions $U(\rho,\mu,\lambda)$, and $W(\rho,\mu,\lambda)$:

$$M_1(\mu,\lambda)(U,W)|_{\rho=\rho_n} = \left[U' + \tfrac{\nu}{1-2\nu}\left(U' + \tfrac{U}{\rho} + W\right)\right]_{\rho=\rho_n} = 0,$$

$$M_2(\mu,\lambda)(U,W)|_{\rho=\rho_n} = \left[W' + \left(W' + \mu^2 U\right)\right]_{\rho=\rho_n} = 0. \tag{1.1.10}$$

Thus, the system of Eq. (1.1.8) together with boundary conditions (1.1.10) generates a spectral problem for a pair of functions $(U,\ W)$. Let us undertake the analysis of the suggested spectral problem. Without going into details, we give the final solution of Eq. (1.1.8) in the following form:

$$\begin{aligned} U(\rho,\mu,\lambda) &= -\alpha z_1(\alpha\rho) - z_1(\gamma\rho) \\ W(\rho,\mu,\lambda) &= \mu^2 z_0(\alpha\rho) + \gamma z_0(\gamma\rho). \end{aligned} \tag{1.1.11}$$

Here $z_k(x) = C_{1k}J_k(x) + C_{2k}Y_k(x)$, $J_0(x)$, $J_1(x)$, $Y_0(x)$ are Bessel functions of the first and second kind respectively; $C_i(i = 1,2,3,4)$ are arbitrary constants.

On satisfying homogeneous boundary conditions (1.1.10) we obtain a linear system of algebraic equations in the unknowns C_i:

$$\begin{aligned} &\left[\tfrac{\alpha}{\rho}Z_1(\alpha\rho) - \delta^2 Z_0(\alpha\rho) + \tfrac{1}{\rho}Z_1(\gamma\rho) - \gamma Z_0(\gamma\rho)\right]_{\rho=\rho_n} = 0 \\ &\left[2\mu^2\alpha Z_1(\alpha\rho) + (2\mu^2+\lambda^2)Z_1(\gamma\rho)\right]_{\rho=\rho_n} = 0 \end{aligned} \tag{1.1.12}$$

$$\delta^2 = \mu^2 + 1/2\lambda^2$$

The condition for the existence of nontrivial solutions of system (1.1.12) leads to the following dispersion equation:

$$\begin{aligned} \Delta(\mu,\lambda) = {} & 8\pi^{-2}\rho_1^{-1}\rho_2^{-1}\mu^2\left(2\mu^2+\lambda^2\right)^2 - \lambda^4\alpha^2\rho_1^{-1}\rho_2^{-1} \\ & \times L_{11}(\alpha)L_{11}(\gamma) + 2^{-1}\alpha\lambda^2\left(2\mu^2+\lambda^2\right)^2\rho_2^{-1}L_{01}(\alpha)L_{11}(\gamma) \\ & + 2^{-1}\alpha\lambda^2\left(2\mu^2+\lambda^2\right)^2\rho_1^{-1}L_{10}(\alpha)L_{11}(\gamma) - 2\lambda^2\gamma\mu^2\alpha^2 \\ & \times \rho_1^{-1}L_{10}(\gamma)L_{11}(\alpha) - 2\lambda^2\gamma\mu^2\alpha^2\rho_2^{-1}L_{01}(\gamma)L_{11}(\alpha) \\ & - 4^{-1}\left(2\mu^2+\lambda^2\right)^4 L_{00}(\alpha)L_{11}(\gamma) - 4\mu^4\alpha^2\gamma^2 \\ & \times L_{00}(\gamma)L_{11}(\alpha) + \alpha\gamma\mu^2\left(2\mu^2+\lambda^2\right)^2 \\ & \times \left[L_{01}(\gamma)L_{10}(\alpha) + L_{01}(\alpha)L_{10}(\gamma)\right] = 0 \end{aligned} \tag{1.1.13}$$

$$\begin{aligned} L_{ii}(x) &= J_i(x\rho_1)Y_i(x\rho_2) - J_i(x\rho_2)Y_i(x\rho_1) \\ L_{ij}(x) &= J_i(x\rho_1)Y_i(x\rho_2) - J_j(x\rho_2)Y_i(x\rho_1) \\ i,j &= 0,\ 1 \end{aligned}$$

The transcendental Eq. (1.1.13) defines a countable set of roots μ_k and the corresponding constants $C_1\mu_k$, $C_2\mu_k$, $C_3\mu_k$, $C_4\mu_k$ are proportional to the cofactors of the elements of any row of the determinant of the system. If the matrix of the linear system (1.1.13) is expanded in terms of cofactors in the first row we obtain:

$$\begin{aligned}
C_1\mu_k &= C_k\big[4\pi^{-1}\rho_2^{-1}\alpha_k\mu_k^2(2\mu_k^2+\lambda^2)Y_1(\alpha_k\rho_1)\\
&\quad-\alpha_k\lambda^2(2\mu_k^2+\lambda^2)\rho_2^{-1}Y_1(\alpha_k\rho_2)L_{11}(\gamma_k)+2^{-1}(2\mu_k^2+\lambda^2)^3\\
&\quad\times Y_0(\alpha_k\rho_2)L_{11}(\gamma_k)-2\alpha_k\gamma_k\mu_k^2(2\mu_k^2+\lambda^2)Y_1(\alpha_k\rho_2)L_{10}(\gamma_k)\big]\\
C_2\mu_k &= C_k\big[4\pi^{-1}\rho_2^{-1}\alpha_k\mu_k^2(2\mu_k^2+\lambda^2)J_1(\alpha_k\rho_1)\\
&\quad-\alpha_k\lambda^2(2\mu_k^2+\lambda^2)\rho_2^{-1}J_1(\alpha_k\rho_2)L_{11}(\gamma_k)+2^{-1}(2\mu_k^2+\lambda^2)^3\\
&\quad\times J_0(\alpha_k\rho_2)L_{11}(\gamma_k)-2\alpha_k\gamma_k\mu_k^2(2\mu_k^2+\lambda^2)J_1(\alpha_k\rho_2)L_{10}(\gamma_k)\big]\\
C_3\mu_k &= C_k\big[2\pi^{-1}\rho_2^{-1}\mu_k^2(2\mu_k^2+\lambda^2)Y_1(\gamma_k\rho_1)\\
&\quad+2\rho_2^{-1}\lambda^2\mu_k^2\alpha_kY_1(\gamma_k\rho_2)L_{11}(\alpha_k)+4\gamma_k\mu_k^4\alpha_k^2Y_0(\gamma_k\rho_2)L_{11}(\alpha_k)\\
&\quad-\alpha_k\mu_k^2(2\mu_k^2+\lambda^2)^2Y_1(\gamma_k\rho_2)L_{10}(\alpha_k)\big]\\
C_4\mu_k &= C_k\big[2\pi^{-1}\rho_2^{-1}\mu_k^2(2\mu_k^2+\lambda^2)J_1(\gamma_k\rho_1)\\
&\quad+2\rho_2^{-1}\lambda^2\mu_k^2\alpha_kJ_1(\gamma_k\rho_2)L_{11}(\alpha_k)+4\gamma_k\mu_k^4\alpha_k^2J_0(\gamma_k\rho_2)L_{11}(\alpha_k)\\
&\quad-\alpha_k\mu_k^2(2\mu_k^2+\lambda^2)^2J_1(\gamma_k\rho_2)L_{10}(\alpha_k)\big]
\end{aligned} \tag{1.1.14}$$

Substituting (1.1.14) into (1.1.11), summing through all roots and taking into account formulas (1.1.6) and (1.1.9) we obtain the homogeneous solutions in the following form:

$$\begin{aligned}
U_\rho &= \sum_{k=1}^{\infty} C_kU_k(\rho)\frac{dm_k}{d\xi}e^{i\omega t}\\
U_\xi &= \sum_{k=1}^{\infty} C_kW_k(\rho)m_k(\xi)e^{i\omega t}\\
\sigma_r &= 2G\sum_{k=1}^{\infty} C_kQ_{rk}(\rho)\frac{dm_k}{d\xi}e^{i\omega t}\\
\sigma_\varphi &= 2G\sum_{k=1}^{\infty} C_kQ_{\phi k}(\rho)\frac{dm_k}{d\xi}e^{i\omega t}\\
\sigma_z &= 2G\sum_{k=1}^{\infty} C_kQ_{zk}(\rho)\frac{dm_k}{d\xi}e^{i\omega t}\\
\tau_{rz} &= G\sum_{k=1}^{\infty} C_k\tau_k(\rho)m_k(\xi)e^{i\omega t}
\end{aligned} \tag{1.1.15}$$

Here C_k are arbitrary constants.

$$
\begin{aligned}
U_k(\rho) &= \alpha_k Z_1(\alpha_k\rho) - Z_1(\gamma_k\rho),\\
W_k(\rho) &= \mu_k^2 Z_0(\alpha_k\rho) - \gamma_k Z_0(\gamma_k\rho),\\
Q_{rk}(\rho) &= \frac{\alpha_k}{\rho} Z_1(\alpha_k\rho) - \delta_k^2 Z_0(\alpha_k\rho) + \frac{1}{\rho} Z_1(\gamma_k\rho) - \gamma_k Z_0(\gamma_k\rho),\\
Q_{\varphi k}(\rho) &= -\frac{\alpha_k}{\rho} Z_1(\alpha_k\rho) - \frac{\nu}{2(1-\nu)}\lambda^2 Z_0(\alpha_k\rho) - \frac{1}{\rho} Z_1(\gamma_k\rho),\\
Q_{zk}(\rho) &= \left[\mu_k^2 - \frac{\nu}{2(1-\nu)}\lambda^2\right] Z_0(\alpha_k\rho) + \gamma_k Z_0(\gamma_k\rho),\\
\tau_k(\rho) &= -2\mu_k^2\alpha_k Z_1(\alpha_k\rho) - \left(2\mu_k^2 + \lambda^2\right) Z_1(\gamma_k\rho).
\end{aligned}
\tag{1.1.16}
$$

1.2 Analysis of the Roots of the Dispersion Equation

Let us undertake the analysis of the roots of the dispersion Eq. (1.1.13). As it is clearly seen from formula (1.1.12), the dispersion equation has a very complicated structure. For an effective study of the location of the roots of (1.1.12) we make some assumptions on the geometric parameters of the cylinder. Suppose:

$$
\rho_1 = 1 - \varepsilon,\ \rho_2 = 1 + \varepsilon,\ 2\varepsilon = \frac{R_2 - R_1}{R_0} = \frac{2h}{R_0}. \tag{1.2.1}
$$

Let us suggest that ε is a small parameter. Substituting (1.2.1) into (1.1.13), we obtain

$$
D(\mu, \lambda, \varepsilon) = \Delta(\mu, \lambda, \rho_1, \rho_2) = 0 \tag{1.2.2}
$$

One can show that $D(\mu, \lambda, \varepsilon)$ is an even function of its arguments.

The case $\lambda_0^2 = gR_0^2\omega^2/E = 1,\ \lambda_0^2 = \frac{1}{1-\nu^2}$ and $\mu = 0$ is a particular case and is treated separately.

The following statement can be formulated with respect to the zeros of the function $D(\mu, \lambda, \varepsilon)$: the function $D(\mu, \lambda, \varepsilon)$ has three groups of zeros for finite $\lambda[\lambda = 0(1)\ as\ \varepsilon \to 0]$:

(a) The first group consists of two zeros $\mu_k = O(1)(k = 1, 2)$;
(b) The second group consists of four zeros at the order $O(\varepsilon^{-1/2})$;
(c) The third group contains a countable set of zeros which are of the order $O(\varepsilon^{-1})$.

Let us prove this assertion. To do this we expand $D(\mu, \lambda, \varepsilon)$ into a series in powers of ε. On doing so we obtain:

$$\begin{aligned}D(\mu,\lambda,\varepsilon) &= 64(1+\nu)^2(1-\nu)^{-2}\lambda_0^4\varepsilon^2\langle(1-\nu^2)(1-\lambda_0^2)\mu^2\\&+(1-\nu^2)\lambda_0^2[1-(1-\nu^2)\lambda_0^2]+3^{-1}\{\mu^6+2(1+\nu)\\&\times[(3-2\nu)\lambda_0^2-2(1-\nu)]\mu^4+(1+\nu)[(1+\nu)(4\nu^2-16\nu+11)\lambda_0^4\\&+2(2\nu^2+9\nu-9)\lambda_0^2+9(1-\nu)]\mu^2+2(1-\nu^2)\\&\times(1+\nu)^2(3-4\nu)\lambda_0^6-2(1+\nu)^2(6\nu^2-14\nu+7)\lambda_0^4\\&+9(1-\nu^2)\lambda_0^2\}\varepsilon^2+\frac{1}{45}(-8\mu^8+\cdots)\varepsilon^4+\cdots\rangle=0\end{aligned} \tag{1.2.3}$$

We are seeking μ_k in the form of the following expansion

$$\mu_k = \mu_{k0}+\mu_{k2}\varepsilon^2+\cdots\,(k=1,2). \tag{1.2.4}$$

After substituting (1.2.4) into (1.2.3) we obtain

$$(\lambda_0^2-1)\mu_{k0}^2+\lambda_0^2[(1-\nu^2)\lambda_0^2-1]=0$$

$$\begin{aligned}\mu_{k2} &= -6^{-1}(1-\nu^2)^{-1}\left(\frac{\lambda_0}{\lambda_0^2-1}\right)^4\nu^2\mu_{k0}^{-1}\left[(1-\nu^2)(1+\nu)^2\lambda_0^8\right.\\&+(1+\nu)^2(6\nu-7)\lambda_0^6-(1+\nu)(8\nu^2+4\nu-14)\lambda_0^4\\&\left.+(4\nu^3+10\nu^2-4\nu-11)\lambda_0^2+3(1-\nu^2)\right].\end{aligned} \tag{1.2.5}$$

We will prove that all the other zeros of the function $D(\mu,\lambda,\varepsilon)$ increase unboundedly when $\varepsilon\to 0$. Let us proceed by contradiction, assuming that $\mu_k\to\mu_k^*\neq\infty$ as $\varepsilon\to 0$. Then the limit relation $D(\mu,\lambda,\varepsilon)=\varepsilon^2 D_0(\mu^*,\lambda_0)$ holds true as $\varepsilon\to 0$. Thus, the limit points of the set of roots μ_k in the limit $\varepsilon\to 0$ can be determined from the equation

$$D_0(\mu_k^*,\lambda_0)=0.$$

In this case

$$D_0(\mu_k^*,\lambda_0)=(\lambda_0^2-1)\mu_k^*+\lambda_0^2[(1-\nu^2)\lambda_0^2-1]. \tag{1.2.6}$$

The last equality implies that no other root, other than the one found above, which are bounded as $\varepsilon\to 0$ do not exist.

Thus, we proved that all the remaining zeros of the function $D(\mu_k,\lambda_0,\varepsilon)$ tend to infinity as $\varepsilon\to 0$. They can be divided into two groups according to their behaviour for $\varepsilon\to 0$:

(1) $\varepsilon\mu_k\to 0$, $\varepsilon\to 0$;
(2) $\varepsilon\mu_k\to const$, $\varepsilon\to 0$.

We first define those μ_k for which $\varepsilon\mu_k \to 0$, $\varepsilon \to 0$. For this we use the expansion (1.2.3). Let's assume that the leading term of the asymptotics expansion has the form

$$\mu_k = \chi_{k0}\varepsilon^{-\beta}, \chi_{k0} = O(1) \quad as \quad \varepsilon \to 0, 0<\beta<1 \tag{1.2.7}$$

Substituting (1.2.7) into (1.2.3) and keeping only the leading order terms we obtain the following limiting equation for χ_{k0}:

$$\begin{aligned}&\left[(1-\nu^2)\left(1-\lambda_0^2\right)\chi_{k0}^2 + O\left(\varepsilon^{2\beta}\right)\right]\varepsilon^{-2\beta} + 3^{-1}\left[\chi_{k0}^6 + O\left(\varepsilon^{2\beta}\right)\right]\varepsilon^{2-6\beta}\\ &\quad + O\left[\max\left(\varepsilon^{4-8\beta}, \varepsilon^{2-4\beta}\right)\right] = 0.\end{aligned} \tag{1.2.8}$$

We consider three cases:

$$(a)\, 0<\beta<\frac{1}{2} \quad (b)\, \beta = \frac{1}{2} \quad (c)\, \frac{1}{2}<\beta<1$$

In case (a) passing in (1.2.8) to the limit $\varepsilon \to 0$ we find that $\chi_{k0} = 0$, which is contrary to the assumption (1.2.7). Similarly, in case (c) we arrive at a contradiction. Finally, in case (b) we have

$$\chi_{k0}^2\left[3\left(1-\nu^2\right)\left(1-\lambda_0^2\right) + \chi_{k0}^4\right] = 0 \tag{1.2.9}$$

Seeking now $\mu_k (k = 3, 4, 5, 6)$ in the form of an expansion as

$$\mu_k = \varepsilon^{-1/2}\left[\mu_{k0} + \mu_{k1}\sqrt{\varepsilon} + \mu_{k2}\varepsilon + \cdots\right] \tag{1.2.10}$$

And substituting the proposed expansion into (1.2.3) we obtain

$$\begin{aligned}\mu_{k0} &= \chi_{k0}, \mu_{k1} = 0, \mu_{k2} = \frac{1}{20}\frac{1}{\mu_{k0}}\frac{1}{1-\lambda_0^2}\\ &\quad\times\left[(1+\nu)(17-7\nu)\lambda_0^4 + \left(24\nu^2 - 10\nu - 29\right)\lambda_0^2 + 12\left(1-\nu^2\right)\right]\end{aligned} \tag{1.2.11}$$

As can be seen from formula (1.2.9), in the case for $\lambda_0^2 < 1$ we get four complex roots and for $\lambda_0^2 > 1$ we get two real and two purely imaginary roots which correspond to penetrating solutions.

To construct the asymptotics of zeros of the third group we seek $\mu_n (n = k - 6,\ \ k = 7, 8)$ as

$$\mu_n = \frac{\delta_n}{\varepsilon} + \lambda_0^2 O(\varepsilon) \tag{1.2.12}$$

On substituting (1.2.12) into the dispersion Eq. (1.1.13) and converting it using the asymptotic expansions of the functions $J_\nu(x)$, $Y_\nu(x)$ for δ_n we obtain the following equation:

$$\sin^2 2\delta_n - 4\delta_n^2 = 0. \tag{1.2.13}$$

It should be noted that Eq. (1.2.13) coincides with the equation determining the indicators of St. Venant's boundary effects in shell statics [1]. Since Eq. (1.2.13) has a countable set of roots such that $\varepsilon\mu_k \to const$ for $\varepsilon \to 0$ then Eq. (1.2.12) corresponds to a boundary layer solution of shell statics.

In principle, the case $\mu\varepsilon \to \infty$ is possible at the limit $\varepsilon \to 0$.

We show that this case is impossible. In fact, denote $\mu_k\varepsilon$ by x_k as $\varepsilon \to 0$. Using the asymptotic expansion of the Bessel function for large values of the argument for $D(x_k, \lambda, \varepsilon)$, in the first term of the asymptotic expansion we obtain

$$D(x_k, \lambda, \varepsilon) = \sin^2 x_k - x_k^2 + O(\varepsilon). \tag{1.2.14}$$

If $D(x_k, \lambda, \varepsilon) \to 0$ as $\varepsilon \to 0$, $x_k \to \infty$ then from (1.2.4) we get that $\sin^2 x_k \to x_k^2$. Such a limit passage is impossible when x_k continuously tends to infinity. Note that the zeros defined by (1.2.4), (1.2.10) and (1.2.12) at $\lambda = 0$ are completely transformed into zeros which are found in [1].

Let us consider the following cases: (a) $\mu = 0$, (b) $\lambda_0^2 = 1$ and (c) $\lambda_0^2 = \frac{1}{1-\nu^2}$.

The case $\mu = 0$ corresponding to the thickness resonance of a hollow cylinder is of independent interest and will be discussed later.

In case (b) Eq. (1.2.3) takes the form

$$\begin{aligned} D_0 = {} & 64(1+\nu)^2(1-\nu)^{-2}\varepsilon^2\Big\{\nu^2\left(1-\nu^2\right) + \frac{1}{3}\left[\mu^6 + 2(1+\nu)\mu^4\right. \\ & + 2(1+\nu)\left(2\nu^3 - 4\nu^2 + 2\nu + 1\right)\mu^2 + (1+\nu) \\ & \left.\times\left(8\nu^4 - 10\nu^3 + 2\nu^2 + 3\nu + 1\right)\right]\varepsilon^2 + \frac{1}{45}\left(-8\mu^8 + \cdots\right)\varepsilon^4 + \cdots\Big\} = 0 \end{aligned} \tag{1.2.15}$$

Hence, we get the following groups of roots:

$$\mu_p = \mu_{p0}\varepsilon^{-1/3} + \mu_{p1}\varepsilon^{1/3} + \mu_{p2}\varepsilon + \cdots \left(p = \overline{1,6}\right) \tag{1.2.16}$$

After substituting the last equation in (1.2.15) we obtain

$$\mu_{p0}^6 + 3\nu^2\left(1-\nu^2\right) = 0, \quad \mu_{p1} = -\frac{1+\nu}{3\mu_{p0}},$$

$$\mu_{p2} = -\frac{1+\nu}{135\mu_{p0}^3}\left(126\nu^3 - 216\nu^2 + 50\nu + 5\right). \tag{1.2.17}$$

As for the zeros defined by formula (1.2.12), they remain valid in this case too. Thus, in case $\lambda_0^2 = 1$ we have six zeros (two of which are purely imaginary) growing as $\varepsilon^{-1/3}$, $\varepsilon \to 0$ and a countable set of zeros defined by (1.2.13).

We will establish the connection of zeros defined by (1.2.16) with the zeros defined by the formulas (1.2.4) and (1.2.10). To this end we study the behaviour of the zeros of Eq. (1.2.3) in the neighbourhood of $\lambda_0^2 = 1$.

In Eq. (1.2.3) by setting $\lambda_0^2 - 1 = C_0\varepsilon^\alpha (\alpha > 0)$, $\mu_k = \mu_{k0}\varepsilon^{-\beta}$ and keeping only the leading order terms we obtain

$$\begin{aligned} D(\mu,\lambda,\varepsilon) &= 64(1+\nu)^2(1-\nu)^{-2}(1+C_0\varepsilon^\alpha)^2\varepsilon^2 \\ &\times \{[-(1-\nu^2)C_0\mu_{k0}^2\varepsilon^{\alpha-2\beta} + \nu^2(1-\nu^2) \\ &+ (1-\nu^2)(2\nu^2-1)C_0\varepsilon^\alpha + O(\varepsilon^{2\alpha})] \\ &+ \frac{1}{3}[\mu_{k0}^6\varepsilon^{2-6\beta} + 2(1+\nu)\varepsilon^{2-4\beta}\mu_{k0}^4 + O(\max(\varepsilon^{2-2\beta},\varepsilon^{4-8\beta}))]\} = 0 \end{aligned} \tag{1.2.18}$$

It is easy to establish that, here, the following cases are possible:

1. $a = 2\beta$, $0 < \alpha < \frac{2}{3}$.

In this case, we seek $\mu_k (k = 1, 2)$ in the following form:

$$\begin{aligned} \mu_k &= \mu_{k0}\varepsilon^{-\frac{\alpha}{2}} + \mu_{k1}\varepsilon^{+\frac{\alpha}{2}} + \cdots, \quad 0 < \alpha < \frac{1}{2}, \\ \mu_k &= \mu_{k0}\varepsilon^{-\frac{\alpha}{2}} + \mu_{k1}\varepsilon^{2-\frac{7\alpha}{2}} + \cdots, \quad \frac{1}{2} \le \alpha < \frac{2}{3}. \end{aligned} \tag{1.2.19}$$

Substituting (1.2.19) into (1.2.18), we obtain

$$\begin{aligned} \mu_{k0}^2 &= \frac{\nu^2}{C_0}, \mu_{k1} = \frac{2\nu^2-1}{2\mu_{k0}}, 0 < \alpha < \frac{1}{2} \\ \mu_{k1} &= \frac{1}{2\mu_{k0}}\left[2\nu^2 - 1 + \frac{\nu^6}{3(1-\nu^2)C_0^4}\right]\alpha = \frac{1}{2} \\ \mu_{k1} &= \frac{\nu^6}{6(1-\nu^2)C_0^4\mu_{k0}}, \frac{1}{2} < \alpha < \frac{2}{3} \end{aligned}$$

It is easy to see that these are the zeros defined by formulas (1.2.4) which grow as $\lambda_0^2 \to 1$. Thus, depending on the sign of C_0 they can be real or purely imaginary. Purely imaginary zeros correspond to a penetrating solution $\alpha = 2\beta$, $\alpha = \frac{2}{3}$.

2. $\alpha = 2\beta,\ \alpha = \frac{2}{3}$. In this case, from (1.2.18) we obtain

$$\mu_k = \mu_{k0}\varepsilon^{-\frac{1}{3}} + \mu_{k1}\varepsilon^{\frac{1}{3}} + \cdots \left(k = \overline{1,6}\right)$$

where

$$\begin{aligned} &\mu_{k0}^6 - 3\left(1 - v^2\right)C_0\mu_{k0}^2 + 3v^2\left(1 - v^2\right) = 0 \\ &\mu_{k1} = -2^{-1}\mu_{k0}^{-1}\left[\mu_{k0}^4 - \left(1 - v^2\right)C_0\right]^{-1} \\ &\quad \times \left[\left(1 - v^2\right)\left(2v^2 - 1\right)C_0 + \frac{2(1+v)}{3}\mu_{k0}^4\right] \end{aligned}$$

Here we get six increasing zeros, two of which, again, correspond to zeros determined by formula (1.2.4). The remaining four zeros correspond to the zeros given by (1.2.10). When $C_0 \to 0$ they completely coincide with the zeros defined by (1.2.16):

3. $\alpha = 2 - 4\beta,\ \frac{1}{2} < \alpha < \frac{2}{3}$

$$\begin{aligned} &\mu_k = \mu_{k0}\varepsilon^{\frac{\alpha}{4}-\frac{1}{2}} + \mu_{k1}\varepsilon^{\frac{1}{2}-\frac{5\alpha}{4}} + \cdots \\ &\mu_{k0}^4 \mp 3\left(1 - v^2\right)C_0 = 0,\ \mu_{k1} = -\frac{v^2}{4C_0\mu_{k0}}. \end{aligned} \tag{1.2.20}$$

If in formula (1.2.8) we assume that $\lambda_0^2 - 1 = C_0\varepsilon^\alpha$ then these zeros coincide with the zeros determined from formulas (1.2.10).

4. $\beta = \frac{1}{3},\ \alpha > \frac{2}{3}$

$$\begin{aligned} &\mu_k = \mu_{k0}\varepsilon^{-\frac{1}{3}} + \mu_{k1}\varepsilon^{\alpha-1} + \cdots \\ &\mu_{k0}^6 - 3v^2\left(1 - v^2\right) = 0,\ \mu_{k1} = -\frac{\left(1-v^2\right)C_0}{2\mu_{k0}^3} \end{aligned} \tag{1.2.21}$$

We will return later to the discussion of this unusual behaviour of the zeros of the dispersion equation when $\lambda_0^2 \to 1$.

In case (c) the expansion (1.2.3) takes the form:

$$\begin{aligned} D(\mu,\varepsilon) = 64\varepsilon^2\Bigg\{ &-v^2\mu^2 + \frac{1}{3}\left[\mu^6 + \frac{2\left(1 + 2v^2 - 2v^3\right)}{1 - v}\mu^4\right. \\ &\left. + \frac{11 - 16v - 3v^2 + 16v^3 - 9v^4}{(1 - v)^2}\mu^2 + \frac{29 - 54v + 21v^2}{(1 - v)^2}\right]\varepsilon^2 \\ &+ \frac{1}{45}\left(-8\mu^8 + \cdots\right)\varepsilon^4 + \cdots\Bigg\} = 0 \end{aligned}$$

Hence, we may obtain the following groups of zeros:

$$\mu_k = \mu_{k0}\varepsilon^{-\frac{1}{2}} + \mu_{k1} + \mu_{k2}\varepsilon^{+\frac{1}{2}} + \cdots$$
$$\mu_{k0}^4 - 3\nu^2 = 0,\, \mu_{k1} = 0,\, \mu_{k2} = \frac{6\nu^3 - 6\nu^2 - 5}{10(1-\nu)\mu_{k0}}$$

By simple verification it can be established that the zeros are defined by (1.2.10) at $\lambda_0^2 = \frac{1}{1-\nu^2}$. Thus, in this case we have four increasing zeros two of which are purely imaginary. As for the zeros defined by formula (1.2.12), they remain valid in the current case too.

We will show that Eq. (1.2.3) has two more bounded zeros with the asymptotic properties $\mu_k \to 0$ as $\varepsilon \to 0$. Assume that the principal terms of the asymptotic expressions for μ_k and λ have the form:

$$\mu_k = \mu_{k0}\varepsilon^{\beta},\ \lambda_0 = \lambda_0^*\varepsilon^q,\ \beta > 0, q > 0. \tag{1.2.22}$$

Substituting (1.2.22) into (1.2.3), on the condition of consistency of the built asymptotic process we see that only the case $q = \beta$ is possible.

We seek $\mu_k\ (k = 1, 2)$ in the form

$$\mu_k = \mu_{k0}\varepsilon^{\beta} + \mu_{k2}\varepsilon^{3\beta} + \cdots,\ \lambda_0 = \Lambda\varepsilon^{\beta} \tag{1.2.23}$$

After substituting (1.2.23) into (1.2.3) we obtain

$$\mu_{k0} = \pm\Lambda i,\ \mu_{k2} = \pm\frac{1}{2}\nu^2\Lambda^3 i,\ i = \sqrt{-1} \tag{1.2.24}$$

These roots correspond to ultra-low frequency vibrations of a hollow cylinder. The possible occurrence of such vibrations in the shells is discussed in detail in [2].

Consider the case when λ increases without bound as $\varepsilon \to 0$. Such vibrations will be called ultrahigh frequency [2].

We can show that all the zeros of (1.2.2) grow infinitely when $\lambda \to \infty$ as $\varepsilon \to 0$. Here we consider independently the following limiting cases: (a) $\lambda\varepsilon \to 0$ as $\varepsilon \to 0$ (b) $\lambda\varepsilon \to const$ as $\varepsilon \to 0$ (c) $\lambda\varepsilon \to \infty$ as $\varepsilon \to 0$.

We firstly define such μ_k at the limit $\lambda\varepsilon \to 0$ as $\varepsilon \to 0$. Assume that the principal terms of the asymptotic behaviour μ_k and λ have the form:

$$\begin{gathered}\mu_k = \mu_{k0}\varepsilon^{-\beta},\ \lambda_0 = \Lambda\varepsilon^{-q},\ \mu_{k0} = O(1),\ \Lambda = O(1)\\ 0<\beta<1, 0<q<1.\end{gathered} \tag{1.2.25}$$

It is easy to prove that $q \le \beta$. We consider the cases $q = \beta$ and $q < \beta$ separately. In the first case, we seek μ_k in the following form:

$$\begin{gathered}\mu_k = \mu_{k0}\varepsilon^{-\beta} + \mu_{k2}\varepsilon^{\beta} + \cdots,\ \lambda_0 = \Lambda\varepsilon^{-\beta},\ 0<\beta<\frac{1}{2}\\ \mu_k = \mu_{k0}\varepsilon^{-\beta} + \mu_{k2}\varepsilon^{2-3\beta} + \cdots,\ \frac{1}{2} \le \beta < 1.\end{gathered} \tag{1.2.26}$$

After substituting these expansions into (1.2.3), we obtain

$$\begin{aligned}
&\mu_{k0} = \pm\sqrt{1-\nu^2}\Lambda i,\ \mu_{k2} = \frac{\nu^2}{2\mu_{k0}},\ 0<\beta<\frac{1}{2}\\
&\mu_{k2} = \frac{\nu^2}{2\mu_{k0}} - \frac{\nu^2(1+\nu)^2\Lambda^4}{6\mu_{k0}},\ \beta=\frac{1}{2}\\
&\mu_{k2} = -6^{-1}\mu_{k0}^{-1}\nu^2\left(1+\nu^2\right)\Lambda^4,\ \frac{1}{2}<\beta<1.
\end{aligned} \tag{1.2.27}$$

In the case of $q<\beta$ substituting (1.2.25) into (1.2.3) and keeping only the leading order terms, we obtain, for μ_k, Λ, the following equation.

$$\begin{aligned}
D(\mu,\lambda,\varepsilon) = {}& 64(1+\nu)^2(1-\nu)^{-2}\Lambda_0^4\Big\langle\Big[-(1-\nu)^2\Lambda^2\mu_{k0}^2\\
&+O\left(\varepsilon^{2\beta-2q}\right)\Big]\varepsilon^{-2\beta-2q} + 3^{-1}\big\{\mu_{k0}^6\\
&+O\left[\max\left(\varepsilon^{2-2\beta},\varepsilon^{2\beta-2q}\right)\right]\big\}\varepsilon^{2-6\beta}\Big\rangle = 0
\end{aligned} \tag{1.2.28}$$

This implies that $q = 2\beta - 1$. From the condition $q>0$ it follows that $\beta > \frac{1}{2}$. Thus, $\frac{1}{2}<\beta<1$. Note that the case $q=0$, which corresponds to the value $\beta=\frac{1}{2}$, is investigated above.

Now seeking $\mu_k (k = 1,2,3,4)$ in the form

$$\begin{aligned}
&\mu_k = \mu_{k0}\varepsilon^{-\beta} + \mu_{k1}\varepsilon^{3\beta-2} + \cdots,\ \frac{1}{2}<\beta<\frac{2}{3}\\
&\mu_k = \mu_{k0}\varepsilon^{-\beta} + \mu_{k1}\varepsilon^{2-3\beta} + \cdots,\ \frac{2}{3}\le\beta<1
\end{aligned} \tag{1.2.29}$$

$$\lambda_0 = \Lambda\varepsilon^{1-2\beta}$$

then substituting into (1.2.3), we obtain

$$\begin{aligned}
&\mu_{k0}^4 - 3\left(1-\nu^2\right)\Lambda^2 = 0,\ \mu_{k1} = -\frac{1}{4}\frac{\mu_{k0}}{\Lambda^2},\ \frac{1}{2}<\beta<\frac{2}{3}\\
&\mu_{k1} = \frac{1}{4\cdot 5}\frac{1+\nu}{\mu_{k0}}(7\nu-17)\Lambda^2 - \frac{\mu_{k0}}{4\Lambda^2}\beta = \frac{2}{3}\\
&\mu_{k1} = \frac{1}{4\cdot 5}\frac{1+\nu}{\mu_{k0}}(7\nu-17)\Lambda^2,\ \frac{2}{3}<\beta<1.
\end{aligned} \tag{1.2.30}$$

So, in this case we have four zeros, increasing as $\varepsilon^{-\beta}$, two of which are real, but the remaining two are purely imaginary ones. The purely imaginary zeros correspond to the so-called irregular degeneration. Note that there is another possible case, namely

$$\mu_k = \frac{\delta_k}{\varepsilon} + O\left(\varepsilon^{1-2\beta}\right),\ \lambda = \Lambda\varepsilon^{-\beta},\ 0<\beta<1. \tag{1.2.31}$$

It is easy to see that, in this case, in the first term of asymptotic behaviour we obtain zeros defined by (1.2.13).

Consequently, in the case $\lambda = \Lambda\varepsilon^{-\beta}$ and $\lambda = \Lambda\varepsilon^{1-2\beta}$ we get, respectively, two and four zeros increasing as $\varepsilon^{-\beta}$ and a countable set of zeros increasing as ε^{-1}.

To construct the asymptotics of the zeros of the second group by assigning $\lambda = S\varepsilon^{-1}$ we seek μ_k in the following form:

$$\mu_k = \frac{\delta_k}{\varepsilon} + O(\varepsilon). \tag{1.2.32}$$

After substituting (1.2.32) in the dispersion Eq. (1.2.2) and converting it using the asymptotic expansions of the functions $J_v(x)$, $Y_v(x)$ for large values of the argument δ_k we obtain the equation

$$\begin{gathered}\left[\left(2\delta_k^2 + S^2\right)^2 \sin\alpha_{k0}\cos\gamma_{k0} - 4\alpha_{k0}\gamma_{k0}\delta_k^2\cos\alpha_{k0}\right.\\ \left.\times\sin\gamma_{k0}\right]\left[\left(2\delta_k^2 + S^2\right)^2\cos\alpha_{k0}\sin\gamma_{k0}\right.\\ \left.-4\alpha_{k0}\gamma_{k0}\delta_k^2\sin\alpha_{k0}\cos\gamma_{k0}\right] = 0\end{gathered} \tag{1.2.33}$$

where

$$\alpha_{k0} = \delta_k^2 + \frac{1-2v}{2(1-v)}S^2,\ \gamma_{k0} = \delta_k^2 + S^2.$$

For a given λ the transcendental Eq. (1.2.33) defines a countable set μ_k. It should be noted that Eq. (1.2.33) actually coincides with the Rayleigh-Lamb's dispersion equation for an elastic layer [3] and is rather well studied in the literature.

In case (c) we denote $\mu_k\varepsilon$ through x_k, $\lambda\varepsilon$ through Y. Again, using the asymptotic expansions of the Bessel functions in the first term of the asymptotics Eq. (1.2.2) can be written as:

$$\begin{gathered}\left[\left(2X_k + Y^2\right)^2\sin\alpha_k\cos\gamma_k - 4\alpha_k\gamma_k X_k^2\cos\alpha_k\sin\gamma_k\right]\\ \times\left[\left(2X_k + Y^2\right)^2\cos\alpha_k\sin\gamma_k - 4\alpha_k\gamma_k X_k^2\sin\alpha_k\cos\gamma_k\right] = 0\\ \alpha_k^2 = x_k^2 + \tfrac{1-2v}{2(1-v)}Y^2,\ \gamma_k^2 = X_k^2 + Y^2\end{gathered} \tag{1.2.34}$$

Thus, in case $\beta > 1$ Eq. (1.2.33) remains valid.

Now for comparison we present here the analysis of the dispersion equations obtained through Kirchhoff-Love and V. Z. Vlasov's theories and theories of the Timoshenko type.

In the first case [2], the dispersion equation has the form:

$$\begin{aligned} &-\left(1-\nu^2\right)\left(\lambda_0^2-1\right)\mu^2+\left(1-\nu^2\right)\lambda_0^2\left[1-\left(1-\nu^2\right)\lambda_0^2\right] \\ &+\frac{1}{3}\left[\mu^6+\left(1-\nu^2\right)\lambda_0^2\mu^4\right]\varepsilon^2=0. \end{aligned} \tag{1.2.35}$$

From (1.2.35) there may be obtained the following groups of zeros

1. $$\begin{aligned} &\mu_k=\mu_{k0}+\mu_{k2}\varepsilon^2+\cdots\ (k=1,2) \\ &\left(\lambda_0^2-1\right)\mu_{k0}^2+\lambda_0^2\left[\left(1-\nu^2\right)\lambda_0^2-1\right]=0 \\ &\mu_{k2}=-6^{-1}\left(1-\nu^2\right)^{-4}\left(\lambda_0^2-1\right)^{-4}\mu_{x0}^{-1}\nu^2\lambda_0^6 \\ &\times\left[\left(1-\nu^2\right)\lambda_0^4-2\left(1-\nu^2\right)\lambda_0^2+1\right] \end{aligned} \tag{1.2.36}$$

2. $$\begin{aligned} &\mu_k=\mu_{k0}\varepsilon^{-1/2}+\mu_{k1}\sqrt{\varepsilon}+\cdots\ (k=3,4,5,6) \\ &\mu_{k0}^4-3\left(1-\nu^2\right)\left(\lambda_0^2-1\right)=0 \\ &\mu_{k1}=-\frac{\nu^2}{4\mu_{k0}}\frac{\lambda_0^4}{\lambda_0^2-1} \end{aligned} \tag{1.2.37}$$

3. $$\begin{aligned} &\mu_p=\mu_{p0}\varepsilon^{-1/3}+\mu_{p1}\varepsilon^{1/3}+\varepsilon\mu_{p2}+\cdots\ \left(p=\overline{1,6}\right) \\ &\mu_{p0}^6+3\nu^2(1-\nu^2)=0,\ \mu_{p1}=-\tfrac{1-\nu^2}{6\mu_{p0}} \\ &\mu_{p2}=\tfrac{2\left(1-\nu^2\right)^2}{27\mu_{p0}^3} \end{aligned} \tag{1.2.38}$$

4. $$\begin{aligned} &\mu_k=\mu_{k0}\varepsilon^{\beta}+\mu_{k1}\varepsilon^{3\beta}+\cdots, \quad \lambda_0=\Lambda\varepsilon^{\beta} \\ &\mu_{k0}=\pm i\Lambda, \quad \mu_{k1}=\pm\frac{\nu^2}{2}\Lambda^3 i \end{aligned} \tag{1.2.39}$$

5. $$\begin{aligned} &\mu_k=\mu_{k0}\varepsilon^{-\beta}+\mu_{k1}\varepsilon^{\beta}+\cdots, \quad 0<\beta<1, \\ &\lambda_0=\Lambda\varepsilon^{-\beta} \\ &\mu_{k0}=\pm\sqrt{1-\nu^2}\Lambda i, \quad \mu_{k1}=\nu^2/2\mu_{k0} \end{aligned} \tag{1.2.40}$$

6. $$\begin{aligned} &\mu_k=\mu_{k0}\varepsilon^{-\beta}+\mu_{k1}\varepsilon^{2-3\beta}+\cdots,\ \tfrac{1}{2}<\beta\le\tfrac{2}{3}, \\ &\mu_k=\mu_{k0}\varepsilon^{-\beta}+\mu_{k1}\varepsilon^{3\beta-2}+\cdots,\ \tfrac{2}{3}<\beta<1 \\ &\lambda_0=\Lambda\varepsilon^{1-2\beta} \\ &\mu_{k0}^4-3(1-\nu^2)\Lambda^2=0,\ \mu_{k1}=-\tfrac{\mu_{k0}}{4\Lambda^2},\ \tfrac{1}{2}<\beta\le\tfrac{2}{3} \\ &\mu_{k1}=0,\ \tfrac{2}{3}<\beta<1 \end{aligned} \tag{1.2.41}$$

The dispersion equation obtained in accordance with V. Z. Vlasov's theory [4] has the form:

$$
\begin{aligned}
&-(1-\nu^2)(\lambda_0^2-1)\mu^2+(1-\nu^2)\lambda_0^2\left[1-(1-\nu^2)\lambda_0^2\right]\\
&\quad+\tfrac{1}{3}\left\{\mu^6+\left[2\nu+(1-\nu^2)\lambda_0^2\right]\mu^4+\mu^2\right.\\
&\quad\left.+(1-\nu^2)\lambda_0^2\right\}\varepsilon^2-\tfrac{1}{9}\mu^6\varepsilon^4=0.
\end{aligned}
\tag{1.2.42}
$$

Equation (1.2.42) has the following groups of zeros:

1.
$$
\begin{aligned}
&\mu_k=\mu_{k0}+\mu_{k2}\varepsilon^2+\cdots(k=1,2)\\
&\quad(\lambda_0^2-1)\mu_{k0}^2+\lambda_0^2\left[(1-\nu^2)\lambda_0^2-1\right]=0\\
&\mu_{k2}=-6(1-\nu^2)^{-1}(\lambda_0^2-1)^{-4}\mu_{k0}^{-1}\nu^2\lambda_0^2\\
&\quad\times\left[(\nu+2)(1-\nu^2)\lambda_0^8+2(1-\nu^2)(\nu^2-\nu-3)\lambda_0^6\right.\\
&\quad\left.-(4\nu^2-2\nu-6)\lambda_0^4-2(1+\nu)\lambda_0^2+\nu\right]
\end{aligned}
\tag{1.2.43}
$$

2.
$$
\begin{aligned}
&\mu_k=\mu_{k0}\varepsilon^{-1/2}+\mu_{k1}\sqrt{\varepsilon}+\cdots(k=3,4,5,6)\\
&\quad\mu_{k0}^4-3(1-\nu^2)(\lambda_0^2-1)=0\\
&\mu_{k1}=-\frac{\nu}{4(\lambda_0^2-1)\mu_{k0}}\left[2-(\nu+2)\lambda_0^2\right]
\end{aligned}
\tag{1.2.44}
$$

3.
$$
\begin{aligned}
&\mu_p=\mu_{p0}\varepsilon^{-1/3}+\mu_{p1}\varepsilon^{1/3}+\varepsilon\mu_{p2}+\cdots(p=\overline{1,6})\\
&\quad\mu_{p0}^6+3\nu^2(1-\nu^2)=0,\ \mu_{p1}=\frac{\nu^2-2\nu-1}{6\mu_{p0}}\\
&\mu_{k2}=\frac{1}{6\cdot 9\mu_{k0}^3}(4\nu^4-16\nu^3+8\nu^2+16\nu-5)
\end{aligned}
\tag{1.2.45}
$$

4.
$$
\begin{aligned}
&\mu_k=\mu_{k0}\varepsilon^{\beta}+\mu_{k1}\varepsilon^{3\beta}+\cdots,\quad\lambda_0=\Lambda\varepsilon^{\beta}\\
&\mu_{k0}=\pm\Lambda i,\quad\mu_{k1}=\pm\frac{\nu^2}{2}\Lambda^3 i
\end{aligned}
\tag{1.2.46}
$$

5.
$$
\begin{aligned}
&\mu_k=\mu_{k0}\varepsilon^{-\beta}+\mu_{k1}\varepsilon^{\beta}+\cdots,\quad 30<\beta<1,\\
&\lambda_0=\Lambda\varepsilon^{-\beta}\\
&\mu_{k0}=\pm\sqrt{1-\nu^2}\Lambda i,\quad\mu_{k1}=\frac{\nu^2}{2\mu_{k0}}
\end{aligned}
\tag{1.2.47}
$$

6.
$$
\begin{aligned}
&\mu_k=\mu_{k0}\varepsilon^{-\beta}+\mu_{k1}\varepsilon^{2-3\beta}+\cdots,\quad\tfrac{1}{2}<\beta\le\tfrac{2}{3},\\
&\lambda_0=\Lambda\varepsilon^{1-2\beta}\\
&\mu_k=\mu_{k0}\varepsilon^{-\beta}+\mu_{k1}\varepsilon^{2-3\beta}+\cdots,\tfrac{2}{3}<\beta<1\\
&\mu_{k0}^4-3(1-\nu^2)\Lambda^2=0,\ \mu_{k1}=-\tfrac{\mu_{k0}}{4\Lambda^2},\ \tfrac{1}{2}<\beta\le\tfrac{2}{3}\\
&\mu_{k1}=0,\ \tfrac{2}{3}<\beta<1
\end{aligned}
\tag{1.2.48}
$$

Comparing (1.2.36), (1.2.37), (1.2.38), (1.2.39), (1.2.40), (1.2.41) and (2.1.43), (1.2.44), (1.2.45), (1.2.46), (1.2.47), (1.2.48) with the exact expansions (1.2.4), (1.2.10), (1.2.16), (1.2.23), (1.2.26), (1.2.29), we see that the first terms of the asymptotic expansions are the same, but subsequent terms differ significantly. However, in the case of ultra-low vibrations two terms of the expansions coincide which is to be expected. But in the case of super-high frequency vibrations (in the sense of the applied theory of shells), applied theory gives an even distortion of the second order term.

We produce an asymptotic analysis of the dispersion equation based on Timoshenko's model.

P. M. Naghdi and R. M. Kuper derived equations describing the dynamics of a cylindrical shell subject to rotational inertia and transverse dislocation. These equations in the axisymmetric case have the form [5].

$$\begin{gathered} v\frac{\partial U}{\partial \xi} + \frac{\partial^2 W}{\partial \xi^2} - \gamma^2\frac{\partial^2 W}{\partial t^2} = 0 \\ k_1^2\frac{\partial^2 U}{\partial \xi^2} - \gamma^2\frac{\partial^2 U}{\partial t^2} - U - v\frac{\partial W}{\partial \xi} + k_1^2 R_0\frac{\partial \psi}{\partial \xi} = 0 \\ -k_1^2\frac{\partial^2 U}{\partial \xi} + \frac{1}{3}\varepsilon^2 R_0\left(\frac{\partial^2 \psi}{\partial \xi^2} - \gamma^2\frac{\partial^2 \psi}{\partial t^2} - k_1^2 R_0\psi\right) = 0 \\ \gamma^2 = \frac{1-v^2}{E} g R_0^2,\ k_1^2 = \frac{1-v}{2}k^2 \end{gathered} \tag{1.2.49}$$

where k^2 is the shear coefficient.

The solution of (1.2.49) will be sought in the form:

$$U = Ae^{\mu\xi + i\omega t},\ W = Be^{\mu\xi + i\omega t},\ \psi = Ce^{\mu\xi + i\omega t}$$

From the condition of existence of nontrivial solutions, we obtain the following dispersion equation

$$\begin{aligned} D^T(\mu, \lambda_0, \varepsilon) = {} & (1 - v^2)k_1^2\left[\lambda_0^2\mu^2 - \mu^2 + (1 - v^2)\lambda_0^4 - \lambda_0^2\right] \\ & - \frac{1}{3}\left\{k_1^2\mu^6 + (1 - v^2)\left[(2k_1^2 + 1)\lambda_0^2 - 1\right]\mu^4 + (1 - v^2)\lambda_0^2\right. \\ & \times\left[(k_1^2 + 2)(1 - v^2)\lambda_0^2 + v^2 - 2\right]\mu^2 + (1 - v^2)^2\lambda_0^4 \\ & \left.\times\left[(1 - v^2)\lambda_0^2 - 1\right]\right\}\varepsilon^2 = 0. \end{aligned} \tag{1.2.50}$$

Hence, we get the following sets of roots:

1.
$$\begin{aligned} \mu_k &= \mu_{k0} + \mu_{k2}\varepsilon^2 + \cdots,\ \lambda_0 = O(1) \\ & (\lambda_0^2 - 1)\mu_{k0}^2 + \lambda_0^2\left[(1 - v^2)\lambda_0^2 - 1\right] = 0 \\ \mu_{k2} &= 6^{-1}(\lambda_0^2 - 1)^{-1}(1 - v^2)^{-1}k_1^{-2}\mu_{k0}^{-1} \\ & \times\left\{k_1^2\mu_{k0}^6 + (1 - v^2)\left[(2k_1^2 + 1)\lambda_0^2 - 1\right]\mu_{k0}^4 + (1 - v^2)\right. \\ & \times\left[(k_1^2 + 2)(1 - v^2)\lambda_0^4 + (v^2 - 2)\lambda_0^2\right]\mu_{k0}^2 \\ & \left.+ (1 - v^2)^2\lambda_0^4\left[(1 - v^2)\lambda_0^2 - 1\right]\right\} \end{aligned} \tag{1.2.51}$$

2. $$\mu_k = \frac{\mu_{k0}}{\sqrt{\varepsilon}} + \mu_{k2}\sqrt{\varepsilon} + \cdots \tag{1.2.52}$$

where

$$\begin{aligned}&\mu_{k0}^4 - 3(1-\nu^2)\left(\lambda_0^2 - 1\right) = 0,\\&\mu_{k2} = -\frac{1}{4k_1^2\left(\lambda_0^2-1\right)\mu_{k0}}\\&\times\left\{\left[(1-\nu^2)\lambda_0^4 + (2\nu^2-1)\lambda_0^2\right]k_1^2 + (1-\nu^2)\left(\lambda_0^2-1\right)^2\right\}\end{aligned}$$

Comparing (1.2.51) and (2.1.52) with the exact expansion (1.2.4) and (1.2.10) we see that the first terms are the same, the following terms essentially depend on the factor k that cannot be adjusted so that the next terms coincide.

3. $$\begin{gathered}\mu_p = \mu_{p0}\varepsilon^{-1/3} + \mu_{p1}\varepsilon^{1/3} + \varepsilon\mu_{p2} + \cdots, \lambda_0^2 = 1\\ \mu_{p0}^6 + 3\nu^2\left(1-\nu^2\right) = 0, \mu_{p1} = \frac{(1-\nu^2)}{3\mu_{p0}}\\ \mu_{p2} = \frac{1}{6\mu_{p0}^3}\left[\frac{7(1-\nu^2)^2}{9} + \frac{\nu^2(1-\nu^2)}{k_1^2}\right]\end{gathered} \tag{1.2.53}$$

4. $$\begin{aligned}&\mu_k = \mu_{k0}\varepsilon^{\beta} + \mu_{k1}\varepsilon^{3\beta} + \cdots, \quad \lambda_0 = \Lambda\varepsilon^{\beta}\\&\mu_{k0} = \pm\Lambda i, \quad \mu_{k1} = \pm\frac{\nu^2}{2}\Lambda^3 i\end{aligned} \tag{1.2.54}$$

Comparing the expansion (1.2.54) with the exact expansion (1.2.23) we see that in the case of ultra-low vibrations in the theory of Timoshenko type as in Kirchhoff-Love and V. Z. Vlasov's theory the first two terms of the expansions coincide

5. $$\begin{aligned}&\mu_k = \mu_{k0}\varepsilon^{-\beta} + \mu_{k1}\varepsilon^{\beta} + \cdots, \lambda_0 = \Lambda\varepsilon^{-\beta},\\&0<\beta<1,\\&\mu_{k0} = \pm\sqrt{1-\nu^2}\Lambda i, \quad \mu_{k1} = \frac{\nu^2}{2\mu_{k0}}\end{aligned} \tag{1.2.55}$$

This shows that in the case of $0<\beta<\frac{1}{2}$ two members of the expansion (1.2.55) coincide with two members of the exact expansion (1.2.26). But in the domain $\frac{1}{2}\leq\beta<1$ the Timoshenko-type theory gives some distortion at the order of the second term of the expansion of the roots of the dispersion equation. A similar result follows from the Kirchhoff-Love and V. Z. Vlasov's theory:

6.
$$
\begin{aligned}
&\mu_k = \mu_{k0}\varepsilon^{-\beta} + \mu_{k1}\varepsilon^{3\beta-2} + \cdots, \lambda_0 = \Lambda\varepsilon^{1-2\beta},\\
&\tfrac{1}{2} < \beta \le \tfrac{2}{3},\\
&\mu_k = \mu_{k0}\varepsilon^{-\beta} + \mu_{k1}\varepsilon^{2-3\beta} + \cdots, \tfrac{2}{3} \le \beta < 1
\end{aligned}
\tag{1.2.56}
$$

where

$$
\begin{aligned}
&\mu_{k0}^4 - 3(1-\nu^2)\Lambda^2 = 0, \mu_{k1} = -\tfrac{\mu_{k0}}{4\Lambda^2}, \tfrac{1}{2} < \beta \le \tfrac{2}{3}\\
&\mu_{k1} = -\tfrac{\mu_{k0}}{4\Lambda^2} - \tfrac{1-\nu^2}{4\mu_{k0}}\Lambda^2\left(1 + \tfrac{1}{k_1^2}\right), \beta = \tfrac{2}{3}\\
&\mu_{k1} = -\tfrac{1-\nu^2}{4\mu_{k0}}\Lambda^2\left(1 + \tfrac{1}{k_1^2}\right), \tfrac{2}{3} < \beta < 1
\end{aligned}
$$

In the case of $\lambda_0 = \Lambda\varepsilon^{1-2\beta}$, $\frac{1}{2} < \beta \le \frac{2}{3}$ the Timoshenko type theory gives the same result. As in the Kirchhoff-Love and V. Z. Vlasov's theory, here two terms of the expansion coincide with two members of the exact expansion (1.2.29). In the case $\lambda_0 = \Lambda\varepsilon^{1-2\beta}$, $\frac{2}{3} < \beta < 1$, unlike Kirchhoff-Love and V. Z. Vlasov's theory the Timoshenko-type theory correctly determines the order of the second term of the asymptotic expansion and two members of the expansion in Timoshenko's theory at $k^2 = 5/(6-\nu)$ coincide with two members of the exact expansions (1.2.29). The value of the shear coefficient $k^2 = 5/(6-\nu)$ is given, for the first time, in the work of P. A. Zhilin [6].

The above-mentioned analysis shows that for thin single-layer shells the Timoshenko-type theory actually has no advantage over the Kirchoff-Love and V. Z. Vlasov's theory. Furthermore, in an asymptotic sense, it is less consistent, due to the uncertainty of the shear coefficient selection.

Thus, the analysis of existing applied theories shows that in the first approximation all the theories accurately approximate the solution corresponding to the zeros defined by Eqs. (1.2.4), (1.2.10), (1.2.16), (1.2.23), (1.2.26), (1.2.29), and at the same time they never assume a more accurate definition in the consequent approximations.

Note that the analysis of various theories and their place in the theory of shells were discussed thoroughly in the above-mentioned work of P. A. Zhilin.

1.3 Construction of Asymptotic Formulas for Displacements and Stresses

In this section, assuming that ε is a small parameter, we present the asymptotic construction of homogeneous solutions corresponding to different groups of roots of the dispersion equation. We firstly consider the case $\lambda = O(1)$. As shown above, in this case, the dispersion equation has three groups of zeros. Assuming that $\rho = 1 + \varepsilon\eta$, $-1 \le \eta \le 1$ and expanding in the small parameter ε the solution of the first group, we find, for the roots in question, the following asymptotic expressions (hereinafter the peak values of the displacements and stresses are given; summation sign is omitted throughout and it is assumed that the summation is carried out on the specified index):

$$\begin{aligned}
U_{\rho k} &= C_k \nu\left[-1 + \left\{\left[(1+\nu)\lambda_0^2 - 1\right]\eta + 2 - (1+\nu)\lambda_0^2\right\} \times * \right.\\
&\quad \left. \times\, \varepsilon + \cdots\right] m_k'(\xi)\\
U_{\xi k} &= C_k\left[1 - \left(1 - \nu^2\right)\lambda_0^2\right.\\
&\quad \left. \times \left\{1 + \varepsilon\left[\frac{\nu\lambda_0^2}{\lambda_0^2 - 1}\eta + (1+\nu)\lambda_0^2 - 2\right]\right\}\right] m_k(\xi)\\
\sigma_z^{(k)} &= 2GC_k\left[(1+\nu)\left(1 - \lambda_0^2\right)\right.\\
&\quad + \varepsilon\left\{-\frac{\lambda_0^2}{\lambda_0^2 - 1}\frac{\nu}{1-\nu}\left[(1+\nu)(1-2\nu)\lambda_0^2 + \nu^2 + \nu - 1\right]\eta\right.\\
&\quad \left.\left. + 2(1+\nu)\left(\lambda_0^2 - 1\right) - (1+\nu)^2\lambda_0^2\left(\lambda_0^2 - 1\right)\right\} + \cdots\right] m_k'(\xi)\\
\sigma_\varphi^{(k)} &= 2GC_k\left[-\nu(1+\nu)\lambda_0^2 + \varepsilon\left\{(1+\nu)\lambda_0^2 - (1+\nu)^2\lambda_0^4\right.\right.\\
&\quad \left.\left. + \frac{\lambda_0^2}{\lambda_0^2 - 1}\frac{\nu}{1-\nu}\left[(1+\nu)^2\lambda_0^4 - (1+\nu)(2+\nu)\lambda_0^2 + \nu(2+\nu)\right]\eta\right\} + \cdots\right] m_k'(\xi)\\
\sigma_r &= O\left(\varepsilon^2\right), \tau_{rz} = O\left(\varepsilon^2\right)
\end{aligned} \tag{1.3.1}$$

where

$$m_k(\xi) = E_k ch\mu_k\xi + N_k sh\mu_k\xi$$

μ_k is given by (1.2.4), and C_k are arbitrary constants. Thus, the zeros given by (1.2.4) correspond to some penetrating solutions.

We now turn to the study of homogeneous solutions corresponding to the second group of zeros, which are defined by (1.2.10). As follows from (1.2.11), this group of four solutions correspond to the roots. Substituting the expressions into (1.1.15) instead of ρ_1, ρ_2, ρ and then expanding the obtained formulas in powers of ε, we obtain

$$
\begin{aligned}
U_{\rho k} &= \sqrt{\varepsilon} C_k \{-\mu_{k0}^2 + \varepsilon\left[\nu\mu_{k0}^2\eta + (1-\nu)\mu_{k0}^2\right.\\
&\quad \left. - (1-\nu^2)\lambda_0^2 - 2\mu_{k0}\mu_{k2}\right] + \cdots\} m_k'(\xi)\\
U_{\xi k} &= \sqrt{\varepsilon} C_k \{3(1-\nu^2)(\lambda_0^2 - 1)\eta + \nu\mu_{k0}^2 + \varepsilon\\
&\quad \times \left[(1-\nu^2)\lambda_0^2\mu_{k0}\eta + 4\mu_{k0}^3\mu_{k2}\eta + 2\nu\mu_{k0}\mu_{k2}\right.\\
&\quad \left. + \nu(1+\nu)\lambda_0^2\mu_{k0}^2 - 2\nu\mu_{k0}^2\right] + \cdots\} m_k(\xi)\\
\sigma_z^{(k)} &= 2GC_k\sqrt{\varepsilon}\{3(1+\nu)(\lambda_0^2 - 1)\eta + \varepsilon\\
&\quad \times \left[(1-\nu)^{-1}(1+\nu)\lambda_0^2\mu_{k0}\eta + 4(1-\nu)^{-1}\mu_{k0}^3\mu_{k2}\eta - \nu(1+\nu)\lambda_0^2 + \cdots\right\} m_k'(\xi)\\
\sigma_\varphi^{(k)} &= 2GC_k\sqrt{\varepsilon}\{-(1+\nu)\mu_{k0}^2 + 3\nu(1+\nu)(\lambda_0^2 - 1)\eta\\
&\quad + \varepsilon\left[-2(1+\nu)\mu_{k0}\mu_{k2} - (1+\nu)\lambda_0^2\right.\\
&\quad + 4\nu(1-\nu)^{-1}\mu_{k0}^3\mu_{k2}\eta + \nu(1+\nu)(2-\nu)\lambda_0^2\mu_{k0}^2\eta\\
&\quad \left. + (1-\nu)^2\mu_{k0}^2\eta + (1+\nu)\mu_{k0}^2 - \nu(1+\nu)\lambda_0^2\mu_{k0}^2\right] + \cdots\} m_k'(\xi)\\
\tau_{rz}^{(k)} &= \frac{1-2\nu}{2} GC_k\sqrt{\varepsilon}\left[3(1+\nu)(\lambda_0^2 - 1)(\eta^2 - 1) + O(\varepsilon)\right] m_k(\xi)
\end{aligned}
\tag{1.3.2}
$$

where $\sigma_r = O(\varepsilon)$,

$$
m_k(\xi) = E_k \exp(\mu_k\xi) + N_k \exp(-\mu_k\xi)(k = 3,4,5,6)
$$

and μ_k is given by (1.2.10).

From (1.3.2) it can be seen that $U_{\rho k}$, $\sigma_z^{(k)}$, $\sigma_\varphi^{(k)}$ have the order of unity for small ε; $U_{\xi k}$, $\tau_{rz}^{(k)}$ have the order of $\sqrt{\varepsilon}$, $\sigma_r^{(k)}$ has the order of ε.

Hence, we see that in the case $\lambda_0^2 < 1$ we get four damped solutions similar to the simple edge effect in the applied theory of shells. In the case $\lambda_0^2 > 1$ we have two damped and two penetrating solutions.

The expression for the displacements and stresses in the case $\lambda_0^2 = 1$ are given by

$$
\begin{aligned}
U_{\rho k} &= C_k\left\{-\mu_{k0}^2 - \sqrt[3]{\varepsilon^2}\left[1 - \nu^2 + 2\mu_{k0}\mu_{k1}\right]\right.\\
&\quad \left. + \varepsilon\left[\nu\mu_{k0}^2\eta + (1-\nu)\mu_{k0}^2 - 2\mu_{k0}\mu_{k2}\right] + \cdots\right\} m_k'(\xi)\\
U_{\xi k} &= C_k\left\{\nu\mu_{k0}^2 + \sqrt[3]{\varepsilon}(1-\nu)^{-1}\mu_{k0}^4\eta\right.\\
&\quad \left. + 2\nu\mu_{k0}\mu_{k1}\sqrt[3]{\varepsilon^2} + \cdots\right\} m_k(\xi)
\end{aligned}
\tag{1.3.3}
$$

$$\sigma_z^{(k)} = 2GC_k\sqrt[3]{\varepsilon}\Big\{(1-\nu)^{-1}\mu_{k0}^4\eta - \nu(1+\nu)\sqrt[3]{\varepsilon} + \sqrt[3]{\varepsilon^2}$$
$$\times\left[4\mu_{k0}^3\mu_{k1}(1-\nu)^{-1}\eta + (1+\nu)(1-\nu)^{-1}\mu_{k0}^2\eta + \cdots\right] + \cdots\Big\}m_k'(\xi)$$
$$\sigma_\varphi^{(k)} = 2GC_k\{-(1+\nu)\mu_{k0}^2 + \nu(1-\nu)^{-1}\mu_{k0}^4\eta\sqrt[3]{\varepsilon}$$
$$+[-2(1+\nu)\mu_{k0}\mu_{k1} + \ldots]\sqrt[3]{\varepsilon^2} + \cdots\Big\}m_k'(\xi)$$
$$\tau_{zk}^{(k)} = \frac{1-2\nu}{2}\left[-\nu(1+\nu)\left(\eta^2-1\right) + O\left(\sqrt[3]{\varepsilon}\right)\right]m_k(\xi)$$
$$\sigma_r^{(k)} = O(\varepsilon)$$

where

$$m_k(\xi) = E_k\exp(\mu_k\xi) + N_k\exp(-\mu_k\xi), k = 1,2,3,4$$
$$m_k(\xi) = E_k\cos\mu_k\xi + N_k\sin\mu_k\xi\; k = 5,6.$$

Here μ_k is given by (1.2.16).

Once again, we have four damped and two penetrating solutions.

The third group of zeros using the first term of the asymptotic expansion of Bessel function for the displacements and stresses in the first approximation we obtain two classes of solutions, the first of which corresponds to the zeros of the function $\sin 2\delta + 2\delta$, and the second one corresponds to the zeros of $\sin 2\delta - 2\delta$. They have the same structure and can be represented by the following expressions:

$$U_{\rho n} = -2\varepsilon^2C_n\left[(2-\nu)F_n'(\eta) + (1-\nu)\delta_n^{-2}F_n''(\eta) + O(\varepsilon)\right]m_n'(\xi)$$
$$U_{\xi n} = 2\varepsilon C_n\left[(1-\nu)\delta_n^{-1}F''(\eta) - \nu\delta_nF_n(\eta) + O(\varepsilon)\right]m_n'(\xi) \tag{1.3.4}$$

$$\sigma_z^{(n)} = 2GC_n\varepsilon\left[F_n''(\eta) + O(\varepsilon)\right]m_n'(\xi)$$
$$\sigma_\varphi^{(n)} = 2G\varepsilon C_n\left[F_n''(\eta) + \delta_n^2F_n(\eta) + O(\varepsilon)\right]m_n'(\xi)$$
$$\sigma_r^{(n)} = -2G\varepsilon C_n\left[\delta_n^2F_n(\eta) + O(\varepsilon)\right]m_n'(\xi)$$
$$\tau_{rz}^{(n)} = -2GC_n\left[F_n'(\eta) + O(\varepsilon)\right]m_n(\xi)$$

where $F_n(\eta)$ are P. F. Papkovich's functions that are solutions of the following spectral problem:

$$F^{IV} + 2\delta^2 F^{II} + \delta^4 F = 0,\ F(\pm 1) = F^I(\pm 1) = 0$$

$$F_n(\eta) = \left(\delta_n^{-1} \sin\delta_n + \cos\delta_n\right) \cos\delta_n\eta + \eta \sin\delta_n \sin\delta_n\eta\ n = 2, 4, \ldots$$

$$F_n(\eta) = \left(\sin\delta_n - \delta_n^{-1} \cos\delta_n\right) \sin\delta_n\eta + \eta \cos\delta_n \cos\delta_n\eta\ n = 1, 3, \ldots$$

Note that for $\varepsilon \to 0$ this group of solutions corresponds to edge effects localized in the vicinity of the surfaces $z = \pm l$ and almost coinciding with the edge effects in the Saint-Venant's theory of plates.

As noted above, the dispersion Eq. (1.2.2) has two zeros characterized by the fact that they tend to zero at the limit $\varepsilon \to 0$. They correspond to the solutions describing the ultra-low frequency vibrations of a hollow cylinder having the form:

$$\begin{aligned}
U_{\rho k} &= C_k\left[v^2 - 1 + (\eta - 2)\Lambda^2\varepsilon + \cdots\right]m_k'(\xi)\\
U_{\xi k} &= C_k\left\{-v\Lambda^2 + \Lambda^2\left[v\Lambda^2\eta - v(1+v)\Lambda^2 + 2v\right]\varepsilon^{1+2\beta}\right\}m_k(\xi)\\
\sigma_z^{(k)} &= 2GC_k\Big\{-v(1+v)\Lambda^2 + \Big[-v^2(1-v)^{-1}\left(v^2+v-1\right)\Lambda^4\eta\\
&\qquad + 2v(1+v)\Lambda^2 - v(1+v)^2\Lambda^4\Big]\varepsilon^{1+2\beta} + \cdots\Big\}m_k'(\xi)\\
\sigma_\varphi^{(k)} &= 2GC_k\varepsilon^{2\beta}[1 - (\eta+1)\varepsilon + \cdots]v^2(1+v)\Lambda^4 m_k'(\xi)\\
\sigma_r^{(k)} &= O\left(\varepsilon^{2+2\beta}\right),\ \tau_{rz}^{(k)} = O\left(\varepsilon^{3+4\beta}\right)
\end{aligned} \tag{1.3.5}$$

where μ_k is given by (1.2.23).

We now give expressions for displacements and stresses in case of microwave vibrations of the cylinder. Here we will separately consider the cases when

(a) $\lambda\varepsilon \to 0$ as $\varepsilon \to 0$ (b) $\lambda\varepsilon \to const$ as $\varepsilon \to 0$,

and (c) $\lambda\varepsilon \to \infty$ as $\varepsilon \to 0$.

In case (a) we have

$$\begin{aligned}
U_{\rho k} &= C_k\left[-\mu_{k0}^2 + O\left(\varepsilon^{2-2\beta}\right)\right]m_k'(\xi)\\
U_{\xi k} &= C_k\left[3\left(1-v^2\right)\Lambda^2\eta + O\left(\varepsilon^{1-2\beta}\right)\right]\varepsilon^{1-2\beta}m_k(\xi)\\
\sigma_z^{(k)} &= 2GC_k\varepsilon^{1-2\beta}\left[3(1+v)\Lambda^2\eta + O\left(\varepsilon^{1-2\beta}\right)\right]m_k'(\xi)\\
\sigma_\varphi^{(k)} &= 2GC_k\varepsilon^{1-2\beta}\left[3v(1+v)\Lambda^2\eta + O\left(\varepsilon^{1-2\beta}\right)\right]m_k'(\xi)
\end{aligned} \tag{1.3.6}$$

$$
\begin{aligned}
&\tau_{rz}^{(k)} = G\frac{1-2v}{2(1-v)}C_k\left(\eta^2-1\right)\varepsilon^{3-4\beta}\left[3\left(1-v^2\right)\Lambda^2\mu_{k0}^2+O\left(\varepsilon^{2-2\beta}\right)\right]m_k(\xi)\\
&\sigma_r^{(k)} = O\left(\varepsilon^{4-5\beta}\right)\lambda = O\left(\varepsilon^{1-2\beta}\right)\mu = O\left(\varepsilon^{-\beta}\right)\\
&\quad\frac{1}{2}<\beta<1, k=1,2,3,4\\
&U_{\rho k} = C_k\left[v^2\left(1+v^2\right)\Lambda^2(\eta-1)+O\left(\varepsilon^{2\beta}\right)\right]\varepsilon m_k'(\xi)\\
&U_{\xi k} = C_k\left[-v\left(1-v^2\right)\Lambda^2+O\left(\varepsilon^{1-2\beta}\right)\right]m_k(\xi)\\
&\sigma_z^{(k)} = 2GC_k\left[-v(1+v)\Lambda^2+O\left(\varepsilon^{1-2\beta}\right)\right]m_k'(\xi)\\
&\sigma_\varphi^{(k)} = 2GC_k\varepsilon^{1-2\beta}\Big[-(1-v)^{-1}v^2(1+v)^2\\
&\quad\times\Lambda^4(\eta+1)+O\left(\varepsilon^{2\beta}\right)\Big]m_k'(\xi)\\
&\sigma_r^{(k)} = O\left(\varepsilon^{2-3\beta}\right), \tau_{rz}^{(k)} = O\left(\varepsilon^{3-4\beta}\right), \lambda = O\left(\varepsilon^{-\beta}\right)\\
&\quad\mu = O\left(\varepsilon^{-\beta}\right), 0<\beta\le\frac{1}{2}k=1,2
\end{aligned}
\tag{1.3.7}
$$

$$
\begin{aligned}
U_{\rho k} &= C_k\left[v^2\left(1+v^2\right)\Lambda^2(\eta-1)+O\left(\varepsilon^{1-2\beta}\right)\right]m_k'(\xi)\\
U_{\xi k} &= C_k\varepsilon^{1-2\beta}\Bigg[-v(1+v)(1-v)\Lambda^4\\
&\quad+\begin{cases} O\left(\varepsilon^{2\beta-1}\right), & \frac{1}{2}<\beta\le\frac{3}{4}\\ O\left(\varepsilon^{2\beta-2}\right), & \frac{3}{4}<\beta<1\end{cases}\Bigg]\\
\sigma_z^{(k)} &= 2GC_k\varepsilon^{1-2\beta}\Bigg[-v(1+v)^2\Lambda^4\\
&\quad+\begin{cases} O\left(\varepsilon^{2\beta-1}\right), & \frac{1}{2}<\beta\le\frac{3}{4}\\ O\left(\varepsilon^{2\beta-2}\right), & \frac{3}{4}<\beta<1\end{cases}\Bigg]m_k'(\xi)\\
\sigma_\varphi^{(k)} &= 2GC_k\varepsilon^{1-2\beta}\Bigg[-(1-v)^{-1}(1+v)^2(\eta+1)\Lambda^4\\
&\quad+\begin{cases} O\left(\varepsilon^{2\beta-1}\right), & \frac{1}{2}<\beta\le\frac{3}{4}\\ O\left(\varepsilon^{2\beta-2}\right), & \frac{3}{4}<\beta\le1\end{cases}\Bigg]m_k'(\xi)\\
\sigma_r^{(k)} &= O\left(\varepsilon^{2-3\beta}\right), \tau_{rz}^{(k)} = O\left(\varepsilon^{3-4\beta}\right), \frac{1}{2}<\beta<1, k=1,2
\end{aligned}
\tag{1.3.8}
$$

In case $\lambda = \Lambda\varepsilon^{-\beta}$, $0<\beta<1$ and $\mu_n = \frac{\delta_k}{\varepsilon}+O(\varepsilon)$, the behaviour of the solution (1.3.4) remains valid in the first term of the asymptotic expansion.

In case (b), using the first term of the asymptotic of the Bessel function for displacements and stresses in the first approximation, we find two classes of solutions, the first of which corresponds to the zeros of the function

$$\left(2\delta_n^2+s^2\right)^2 \sin\alpha_{n0}\cos\gamma_{n0} - 4\alpha_{n0}\gamma_{n0}\delta_n^2\cos\alpha_{n0}\sin\gamma_{n0}$$

and the second one corresponds to the zeros of the function

$$\left(2\delta_n^2+s^2\right)^2 \sin\gamma_{n0}\cos\alpha_{n0} - 4\alpha_{n0}\gamma_{n0}\delta_n^2\sin\alpha_{n0}\cos\gamma_{n0}$$

They have the same structure and can be represented by the following expressions

$$\begin{aligned}
U_{\rho n} &= \varepsilon B_{12}\Big[\left(2\delta_n^2+s^2\right)^2\cos\gamma_{n0}\cos\alpha_{n0}\eta \\
&\quad - \delta_n^2\cos\alpha_{n0}\cos\gamma_{n0}\eta + O(\varepsilon)\Big]m_n'(\xi) \\
U_{\xi n} &= B_n\delta_n^2\left[\frac{2\delta_n^2+s^2}{\alpha_{n0}}\cos\gamma_{n0}\sin\alpha_{n0}\eta\right. \\
&\quad \left. - 2\gamma_{n0}\cos\alpha_{n0}\sin\gamma_{n0}\eta + O(\varepsilon)\right]m_n(\xi)
\end{aligned} \tag{1.3.9}$$

$$\begin{aligned}
\sigma_z^{(n)} &= G\frac{B}{\alpha_{n0}}\left[\left(2\delta_n^2+\frac{\nu}{\nu-1}s^2\right)\left(2\delta_n^2+s^2\right)\cos\gamma_{n0}\right. \\
&\quad \times \left.\sin\alpha_{n0}\eta - 4\alpha_{n0}\gamma_{n0}\delta_n^2\cos\alpha_{n0}\sin\gamma_{n0}\eta + O(\varepsilon)\right]m_n'(\xi) \\
\sigma_\varphi^{(n)} &= -G\frac{\nu}{\nu-1}\frac{B_n}{\alpha_{n0}}\left[s^2\left(2\delta_n^2+s^2\right)\cos\gamma_{n0}\sin\alpha_{n0}\eta + O(\varepsilon)\right]m_n'(\xi) \\
\sigma_r^{(n)} &= -GB_n\left[\left(2\delta_n^2+s^2\right)\sin\gamma_{n0}\cos\alpha_{n0}\eta\right. \\
&\quad - \left.4\alpha_{n0}\gamma_{n0}\delta_n^2\sin\alpha_{n0}\cos\gamma_{n0}\eta + O(\varepsilon)\right]m_n'(\xi) \\
\tau_{rz}^{(n)} &= 2G\varepsilon^{-1}\delta_n^2\left(2\delta_n^2+s^2\right)B_n \\
&\quad \times \left[\cos\gamma_{n0}cos\alpha_{n0}\eta - \cos\alpha_{n0}\cos\gamma_{n0}\eta + O(\varepsilon)\right]m_n(\xi) \\
n &= 1,3,5,\ldots
\end{aligned}$$

where B_n are arbitrary constants.

The expressions for $n = 2,4,6$ are obtained from (1.3.9) by replacing $\cos x$ with $\sin x$ and $\sin x$ with $-\cos x$ respectively. The propagation constant δ_n in the expression (1.3.9) can be real, purely imaginary or a complex one.

In the case of a purely imaginary δ_n the solution (1.3.9) is a traveling wave that transfers energy to the cylinder, and the time-averaged flow of energy in such a wave is independent of the coordinate ξ, what is natural for a lossless medium. Real zeros correspond to nonhomogeneous waves not transferring energy along the cylinder. Different signs of δ_n correspond to waves, exponentially decaying to the left or right of the section $\xi = const$. The dispersion Eq. (1.2.33) also has a countable set of complex roots appearing in fours in each quadrant of the complex plane $\delta = \pm x \pm iy$. Each value of δ chosen from a quadrant, when inserted in (1.3.9), corresponds to a travelling wave decaying or increasing in its amplitude. For a standing wave to emerge (two waves traveling towards each other) it is necessary to take those zeros on one edge for which $\operatorname{Re}\delta_k > 0$ and on the other edge those zeros for which $\operatorname{Re}\delta_k < 0$.

Similarly, in case (c) we obtain

$$
\begin{aligned}
U_{\rho n} &= \varepsilon B_n\left[\left(2x_n^2+y^2\right)\cos\gamma_n\cos\alpha_n\eta\right.\\
&\quad\left. - 2x_n^2\alpha_n\gamma_n\cos\alpha_n\cos\gamma_n\eta + O(\varepsilon)\right]m_n'(\xi)\\
U_{\xi n} &= B_n x_n^2\left[\left(2x_n^2+y^2\right)\cos\gamma_n\sin\alpha_n\eta\right.\\
&\quad\left. - 2\gamma_n\cos\alpha_n\cos\gamma_n\eta + O(\varepsilon)\right]m_n(\xi)\\
\sigma_z^{(n)} &= G\frac{B}{\alpha_n}\left[\left(2x_n^2+\frac{\nu}{\nu-1}y^2\right)\left(2x_n^2+y^2\right)^2\cos\gamma_n\right.\\
&\quad\times\left.\sin\alpha_n\eta - 4\alpha_n\gamma_n x_n^2\cos\alpha_n\sin\gamma_n\eta + O(\varepsilon)\right]m_n'(\xi)\\
\sigma_\varphi^{(n)} &= G\frac{\nu}{\nu-1}\frac{B_n}{\alpha_n}\left[y^2\left(2x_n^2+y^2\right)\cos\gamma_n\sin\alpha_n\eta + O(\varepsilon)\right]m_n'(\xi)\\
\sigma_r^{(n)} &= -GB_n\left[\left(2x_n^2+y^2\right)^2\sin\gamma_n\cos\alpha_n\eta\right.\\
&\quad\left. - 4\alpha_n\gamma_n x_n^2\sin\alpha_n\cos\gamma_n\eta + O(\varepsilon)\right]m_n'(\xi)\\
\tau_{rz}^{(n)} &= 2G\varepsilon^{-1}x_n^2\left(2x_n^2+y^2\right)^2 B_n\\
&\quad\times\left[\cos\gamma_n\cos\alpha_n\eta - \cos\alpha_n\cos\gamma_n\eta + O(\varepsilon)\right]m_n(\xi)\\
n &= 1,3,5,\ldots
\end{aligned}
\tag{1.3.10}
$$

The expressions for $n = 2,4,6,\ldots$ are obtained from (1.3.10) as before by replacing of $\cos x$ with $\sin x$ and $\sin x$ with $-\cos x$ respectively. We emphasize that the above mentioned remarks about the nature of solutions remain valid in this case as well.

It is important to note that the solutions corresponding to the cases (b) and (c) in the applied theory of shells are not available.

As noted above, the solutions corresponding to the case (b) describe the stress-deformed state of a boundary type layer. Their role in the theory of shells is as follows: they are firstly localized near the edges, defects and other concentrates and behave as an independent stress-strain state along with internal stresses. Moreover, for some of the components of the stress tensor the stresses are basic ones. Therefore, in the calculation of the stress-strain state near the concentrates the accounting of the boundary layer impact is significant [7]. Secondly, they allow more accurate and proper satisfaction of arbitrary boundary conditions at the ends of the cylinder by means of a class of homogeneous solutions. Thus, their impact on the internal stress-strain state already has a global character [7].

As for the solutions corresponding to the case (c), as is well-known, the Rayleigh-Lamb equation has a finite number of purely imaginary roots which correspond to penetrating solutions. Therefore, we cannot speak here about the local character of the stress-strain state; the range of applicability of the applied theories of shells is exhausted and Saint-Venant's principle becomes invalid.

In the study of the roots of the dispersion equation and the construction of the waveform it was shown that the case $\lambda_0^2 = 1$ is special. Naturally, the question arises: what caused this pecularity? The fact is that for $\lambda_0^2 < 1$ the nature of the integrals of the dynamical theory of elasticity is the same as in the static elasticity. When $\lambda_0^2 > 1$ the nature of the integrals of dynamic elasticity theory differs sharply from the static nature of the integrals of elasticity theory. Therefore, it is natural to consider $\lambda_0^2 = 1$ as a "transition" point at which the change in the nature of the dynamic theory of elasticity integrals takes place.

1.4 Generalized Orthogonality Condition of Homogeneous Solutions: Satisfaction of Boundary Conditions at the Cylinder Ends

It is well-known that Schiff-Papkovich orthogonality relations played an important role in the development of methods for solving the basic boundary value problems of elasticity theory. In [8], these relationships are generalized to the dynamic case for an elastic half-strip. On employing the mentioned relations, the mixed boundary value problem for the forced harmonic vibrations of an elastic band was exactly solved. The validity of these relations for a semi-infinite cylinder is also affirmed, although the validity of these relations for a hollow cylinder is not obvious. Below we will derive the generalized orthogonality condition of homogeneous solutions at the end of a hollow cylinder allowing an accurate solution of the problem of forced vibrations of a hollow cylinder with mixed boundary conditions at cylinder ends.

Let us consider the following homogeneous boundary conditions on the side surfaces

$$\sigma_r = 0, \tau_{rz} = 0 \quad \text{at} \quad \rho = \rho_1 \rho_2 \tag{1.4.1}$$

$$U_r = 0, U_z = 0 \quad \text{at} \quad \rho = \rho_1, \quad \rho_2 \tag{1.4.2}$$

$$U_r = 0, \tau_{rz} = 0 \quad \text{at} \quad \rho = \rho_1, \quad \rho_2 \tag{1.4.3}$$

$$\sigma_r = 0, U_z = 0 \quad \text{at} \quad \rho = \rho_1, \quad \rho_2 \tag{1.4.4}$$

We prove that for any of the boundary conditions (1.4.1)–(1.4.4), the following orthogonality relation holds true

$$\int_{\rho_1}^{\rho_2} \left(U_{rp} \tau_{rz}^k - \sigma_{zp} U_{zk} \right) \rho d\rho = 0 \quad (\rho \neq k). \tag{1.4.5}$$

Note that the ratio of (1.4.5) is the direct consequence of Betty theorem and does not depend on the type of boundary conditions on the side surface of the cylinder. In fact, suppose that U_r^i, U_θ^i, σ_z^i, τ_{rz}^i (i = 1,2) are displacements and strains of the first and second states. Then, according to Betty theorem, on the surface $\xi = const$ the following equality is valid:

$$\int\limits_{\rho_1}^{\rho_2} (U_z^1\sigma_z^2 + \tau_{rz}^2 U_z^1)\rho d\rho = \int\limits_{\rho_1}^{\rho_2} (U_z^2\sigma_z^1 + U_z^2\tau_{rz}^1)\rho d\rho \tag{1.4.6}$$

As the first state we take the k-th elementary solution and as the second one p-th elementary solution. Then, substituting (1.1.15) into (1.4.6), we obtain

$$\begin{aligned} &m_k(\xi)\tfrac{dm_p}{d\xi}\int\limits_{\rho_1}^{\rho_2}\left[W_k(\rho)Q_{zp}(\rho) - \tau_k(\rho)U_\rho(\rho)\right]\rho d\rho \\ &+ m_p(\xi)\tfrac{dm_k}{d\xi}\int\limits_{\rho_1}^{\rho_2}\left[\tau_p(\rho)U_k(\rho) - Q_{zk}(\rho)W_\rho(\rho)\right]\rho d\rho = 0. \end{aligned}$$

Since this equality holds true for any ξ, we obtain

$$\int\limits_{\rho_1}^{\rho_2}\left[\tau_p(\rho)U_k(\rho) - Q_{zk}(\rho)W_\rho(\rho)\right]\rho d\rho = 0.$$

The relationship (1.4.5) allows us to obtain an exact solution to the problem of forced vibrations of a cylinder with mixed boundary conditions at cylinder ends.

For example, let us consider the first mixed end conditions, and for simplicity assume them to be symmetrical relative to the plane $\xi = 0$. The skew-symmetric case is treated in a similar manner (in the symmetric case one can take $m_k = ch\mu_k\xi$, whereas in the skew-symmetric case one should take $m_k = sh\mu_k\xi$).

Let there be given the following conditions:

$$\sigma_z = Q(\rho)e^{i\omega t}, U_r = a(\rho)e^{i\omega t} \quad for \quad \xi = \pm l_0. \tag{1.4.7}$$

To satisfy the boundary conditions (1.4.7) it is necessary to perform the expansion

$$\begin{aligned} Q(\rho) &= \sum_{k=1}^{\infty} C_k\sigma_{zk}(\rho)ch\mu_k l_0 \\ a(\rho) &= \sum_{k=1}^{\infty} C_k U_k(\rho)ch\mu_k l_0 \end{aligned} \tag{1.4.8}$$

where C_k are arbitrary constants to be determined from the end conditions, i.e., the constants C_k may be found from Eq. (1.4.8) if the relation (1.4.5) is employed.

Let us multiply the first Eq. (1.4.8) by $\rho W_p(\rho) sh\mu_p l_0$, and the second by $\rho\tau_p(\rho) sh\mu_p l_0$, add the resulting product and integrate over ρ from ρ_1 to ρ_2. In view of the relations of the generalized orthogonality, the required constants take the form

$$C_k = 2^{-1}\Delta_k^{-1} ch^{-1}\mu_k l_0 \int\limits_{\rho_1}^{\rho_2} [a(\rho)\tau_k(\rho) - Q(\rho)W_k(\rho)]\rho d\rho \tag{1.4.9}$$

where Δ_k is the value of the integral (1.4.5) at $p = k$.

$$\begin{aligned}\Delta_k = \lambda^2\mu_k^2\Big\{ & 2\left[z_1(\gamma_k\rho)z_0(\alpha_k\rho) - \frac{\alpha_k}{\gamma_k}z_0(\gamma_k\rho)z_1(\alpha_k\rho)\right]\rho \\ & + \frac{1}{2}\mu_k^2\rho^2\left[z_0^2(\alpha_k\rho) + z_1^2(\alpha_k\rho)\right] \\ & - \frac{1}{2}\rho^2\left[z_0^2(\gamma_k\rho) + z_1^2(\gamma_k\rho)\right]\Big\}\rho = \rho_1, \rho_2.\end{aligned}$$

Note that it is easy to obtain the asymptotic formula for Δ_k by using the asymptotic behaviour of the solutions corresponding to different groups of roots. Here we do not consider all the cases discussed above and give the expressions for Δ_k corresponding to the solutions determined by the formula (1.3.9).

$$\begin{aligned}\Delta_k^0 = s^2\delta_k^4\Big[& -\frac{2}{\gamma_k}(3\delta_k^2 + 2s^2)\cos^2\alpha_k \sin 2\gamma_k \\ & + \frac{(2\delta_k^2 + s^2)^2}{\alpha_k^2}\cos^2\gamma_k - 4\delta_k^2\cos^2\alpha_k \quad k = 1, 3, 5, \ldots \\ & + \frac{2\delta_k^2 + s^2}{2\alpha_k^3}\left(6\delta_k^2 + \frac{3 - 7\nu}{1 - \nu}s^2\right)\cos^2\gamma_k \sin 2\alpha_k\Big]\end{aligned}$$

$$\begin{aligned}\Delta_k^0 = s^2\delta_k^4\Big[& \frac{2}{\gamma_k}(3\delta_k^2 + 2s^2)\sin^2\alpha_k \sin 2\gamma_k \\ & + \frac{(2\delta_k^2 + s^2)^2}{\alpha_k^2}\sin^2\gamma_k - 4\delta_k^2\sin^2\alpha_k + \quad k = 2, 4, 6, \ldots \\ & + \frac{2\delta_k^2 + s^2}{2\alpha_k^3}\left(6\delta_k^2 + \frac{3 - 7\nu}{1 - \nu}\right)\sin^2\gamma_k \sin 2\alpha_k\Big].\end{aligned}$$

As noted above, the generalized orthogonality condition of homogeneous solutions allows to accurately solve the problem of forced vibrations of the cylinder only in the case of mixed end conditions. In all other cases, to satisfy the boundary conditions at the cylinder ends we have to resort to various approximate approaches. Therefore, we consider the question of satisfying the boundary conditions at the ends of cylinder by means of a class of homogeneous solutions. Let the sets of strains σ_z^i, τ_{rz}^i (i = 1,2) be given for $\xi = \pm l_0$ $\left(l_0 = R_0^{-1} l\right)$. Thus, as noted above, it is sufficient to consider the case when the load is symmetrical with respect to the plane $\xi = 0$:

$$\sigma_z = Q(\rho)e^{i\omega t},\ \tau_{rz} = \tau(\rho)e^{i\omega t} \quad for \quad \xi = \pm l_0. \tag{1.4.10}$$

We seek a solution in the form (1.1.15). To determine the arbitrary constants C_k (k = 1,2,…), the variation of which will be regarded as independent, we use Lagrange variational principle. According to this principle, we have

$$\int\limits_{v} \left(g\omega^2 U_p^* \delta U_p^* - \sigma_{pj}^* \delta\varepsilon_{pj}^*\right) dv + \int\limits_{s} T_p^* \delta U_p^* ds = 0 \tag{1.4.11}$$

U_p^*,…, σ_{pj} are amplitude values of the corresponding quantities.

Since homogeneous solutions satisfy the equation of motion and the boundary conditions on the cylindrical surface then the variational principle takes the following form:

$$\int\limits_{\rho_1}^{\rho_2} \left[\left(G_z^* - Q\right)\delta W^* + \left(\tau_{rz}^* - \tau\right)\delta U^*\right]\rho d\rho = 0. \tag{1.4.12}$$

Here σ_z^*, τ_{rz}^*, U^*, W^* are the amplitudes of stresses and displacements, respectively. From (1.4.12) we obtain an infinite system of linear algebraic equations:

$$\sum_{k=1}^{\infty} M_{kp} C_k = N_p \quad (p = 1, 2, \ldots) \tag{1.4.13}$$

where

$$
\begin{aligned}
M_{kp} = & \Big[\rho\Big\{2\alpha_k\alpha_p\left(\alpha_p^2-\alpha_k^2\right)^{-1}\left[\alpha_k z_1(\alpha_p\rho)z_1'(\alpha_k\rho)\right.\\
& \left.-\alpha_p z_1(\alpha_k\rho)z_1'(\alpha_p\rho)\right]\left(\mu_k^2 ch\mu_p l_0 sh\mu_k l_0\right.\\
& \left.+\mu_p^2 ch\mu_k l_0 sh\mu_p l_0\right)+\left(\gamma_p^2-\gamma_k^2\right)^{-1}\left[\gamma_k z_1(\gamma_p\rho)z_1'(\gamma_k\rho)\right.\\
& \left.-\gamma_p z_1(\gamma_k\rho)z_1'(\gamma_p\rho)\right]\ \left[(2\mu_k^2+\lambda^2)\times\right. \qquad (k\neq p)\\
& \left.\times\, ch\mu_p l_0 sh\mu_k l_0+\left(2\mu_p^2+\lambda^2\right)ch\mu_k l_0 sh\mu_p l_0\right]\\
& +\alpha_k\left(\alpha_k^2-\gamma_p^2\right)^{-1}\left[\gamma_p z_1(\alpha_k\rho)z_1'(\gamma_p\rho)-\alpha_k z_1(\gamma_p\rho)z_1'(\alpha_k\rho)\right]\\
& \times\left[2\mu_k^2 ch\mu_p l_0 sh\mu_k l_0+(2\mu_n^2+\lambda^2)ch\mu_k l_0 sh\mu_p l_0\right]\\
& +\alpha_p\left(\alpha_p^2-\gamma_k^2\right)^{-1}\left[\gamma_k z_1(\alpha_p\rho)z_1'(\gamma_k\rho)-\alpha_p z_1(\gamma_k\rho)z_1'(\alpha_p\rho)\right]\\
& \times\left[(2\mu_k^2+\lambda^2)ch\mu_p l_0 sh\mu_k l_0+2\mu_n^2 ch\mu_k l_0 sh\mu_p l_0\right]\Big\}_{\rho_1}^{\rho_2}\Big]\\
M_{kp} = & \Big\{\mu_k^2\alpha_k^2\rho^2\left[z_1'^2(\alpha_k\rho)+\frac{\alpha_k^2\rho^2-1}{\alpha_k^2\rho^2}z_1^2(\alpha_k\rho)\right]\\
& +\frac{2\mu_k^2+\lambda^2}{2}\rho^2\left[z_1'^2(\gamma_k\rho)+\frac{\gamma_k^2\rho^2-1}{\gamma_k^2\rho^2}z_1^2(\gamma_k\rho)\right]\\
& +\frac{\alpha_k(4\mu_k^2+\lambda^2)}{\alpha_k^2-\gamma_k^2}\rho\left[\gamma_k z_1(\alpha_k\rho)z_1'(\gamma_k\rho)\right.\\
& \left.-\alpha_k z_1(\gamma_k\rho)z_1'(\alpha_k\rho)\right]\Big\}_{\rho_1}^{\rho_2}\sin 2\mu_k l_0\\
N_p = & \int\limits_{\rho_1}^{\rho_2}\left[Q(\rho)W(\alpha_p\rho,\gamma_p\rho,l_0)+\tau(\rho)U(\alpha_p\rho,\gamma_p\rho,l_0)\right]\rho d\rho.
\end{aligned}
$$

Using the smallness of the parameter ε of thin shells we can construct an asymptotic solution of (1.4.13). Firstly, we consider the case $\lambda = O(1)$. But first of all, let us determine more precisely the suggestions for the external load. Assume that $Q = O(1)$. Taking into account that the magnitude σ_z and τ_{rz} corresponding to the roots of the first and second groups have different orders ($\sigma_z^{(1)} \sim 1$, $\tau_{rz}^{(1)} \sim \varepsilon^3$, $\sigma_z^{(2)} \sim 1$, $\tau_{rz}^{(2)} \sim \sqrt{\varepsilon}$). On choosing the order of T one should be guided by the following considerations. Using formulas (1.1.15), (1.3.1), (1.3.2), (1.3.4) and also the relation $F_k(\pm 1) = 0$ we obtain

$$
\int\limits_{-1}^{1}\tau_{rz}d\eta = -2(1-2\nu)G(1+\nu)\left(\lambda_0^2-1\right)sh\frac{\mu_k}{\sqrt{\varepsilon}}\xi C_k\sqrt{\varepsilon}. \qquad (1.4.14)
$$

If now we represent the tangential strains given on the boundary in the form

$$T = T_1 + T_2, \quad T_1 = \frac{1}{2}\int_{-1}^{1} T d\eta, \quad T_2 = T - T_1 \tag{1.4.15}$$

then, on the basis of the asymptotic formula (1.4.14), it is necessary to assume that T_1 has order $\sqrt{\varepsilon}$, the magnitude T_2 can be of the same order as Q, i.e. $T \sim 1$. Thus, in this case, using formulas (1.3.1), (1.3.2) and (1.3.4) the solution may be represented in the form

$$\begin{aligned}
U_r &= R_0\left[U_1 C_1 + \sum_{k=1}^{2} U_{2k} A_k + \sum_{n=1}^{\infty} U_{3n} B_n\right] \\
U_z &= R_0\left[W_1 C_1 + \sum_{k=1}^{2} W_{2k} A_k + \sum_{n=1}^{\infty} W_{3n} B_n\right] \\
\sigma_z &= 2G\left[\sigma_z^{(1)} C_1 + \sum_{k=1}^{2} \sigma_z^{(k)} A_k + \sum_{n=1}^{\infty} \sigma_z^{(n)} B_n\right] \\
\sigma_\varphi &= 2G\left[\sigma_\varphi^{(1)} + \sum_{k=1}^{2} \sigma_\varphi^{(k)} A_k + \sum_{n=1}^{\infty} \sigma_\varphi^{(n)} B_n\right] \\
\tau_{rz} &= G\left[\tau_1 C_1 + \sum_{k=1}^{2} \tau_{2k} A_k + \sum_{n=1}^{\infty} \tau_{3n} B_n\right].
\end{aligned} \tag{1.4.16}$$

Here C_1, A_k, B_n are arbitrary constants which will be sought in the form

$$\begin{aligned}
C_1 &= C_{10} + \sqrt{\varepsilon} C_{11} + \varepsilon C_{12} + \cdots \\
A_k &= A_{k0} + \sqrt{\varepsilon} A_{k1} + \varepsilon A_{k2} + \cdots \\
B_n &= B_{n0} + \sqrt{\varepsilon} B_{n1} + \varepsilon B_{n2} + \cdots .
\end{aligned} \tag{1.4.17}$$

Taking into account the accepted order of the strains given on the boundary, we obtain, on the basis of the variational principle, the following system of equations

$$\begin{aligned}
&\prod_0 C_{10} = E_0 \\
&\sum_{k=1}^{2} \prod_{pk} A_{k0} = E_p \quad (p = 1, 2)
\end{aligned} \tag{1.4.18}$$

$$\sum_{n=1,3}^{\infty} g_{tn}B_{n0} = H_t(t = 1,3,\ldots)$$
$$\sum_{n=2,4}^{\infty} g_{tn}B_{n0} = H_t(t = 2,4,\ldots). \tag{1.4.19}$$

Here

$$\prod_0 = 2G(1+v)\left(1-\lambda_0^2\right)\left[1-\left(1-v^2\right)\lambda_0^2\right] \times sh\left(\lambda_0\frac{\sqrt{1-(1-v^2)\lambda_0^2}}{\lambda_0^2-1}\right)l_0 ch\left(\lambda_0\frac{\sqrt{1-(1-v^2)\lambda_0^2}}{\lambda_0^2-1}\right)l_0$$

$$E_0 = \int_{-1}^{1}\{\left[(1-v^2)\lambda_0^2\right]Q(\eta)ch\left(\lambda_0\frac{\sqrt{1-(1-v^2)\lambda_0^2}}{\lambda_0^2-1}\right)l_0 - vsh\left(\lambda_0\frac{\sqrt{1-(1-v^2)\lambda_0^2}}{\lambda_0^2-1}l_0\right) \tag{1.4.20}$$

$$\prod_{pk} = 2G(1+v)\left(\lambda_0^2-1\right)\left[6\left(1-v^2\right)\left(\lambda_0^2-1\right)\mu_{k0} \times ch\frac{\mu_{k0}}{\sqrt{\varepsilon}}l_0 sh\frac{\mu_{p0}}{\sqrt{\varepsilon}}l_0 + (1-2v)\mu_{p0}^3 sh\frac{\mu_{k0}}{\sqrt{\varepsilon}}l_0 ch\frac{\mu_{p0}}{\sqrt{\varepsilon}}l_0\right]$$

$$E_p = \int_{-1}^{1}\left\{\left[3\left(1-v^2\right)\left(\lambda_0^2-1\right)\eta + v\mu_{p0}^2\right]Q(\eta)sh\frac{\mu_{p0}}{\sqrt{\varepsilon}}l_0 - \mu_{p0}^3 ch\frac{\mu_{p0}}{\sqrt{\varepsilon}}l_0 T(\eta)\right\}d\eta.$$

$$g_{tn} = 8G\delta_t^2\delta_n^2\left(\delta_t^2-\delta_n^2\right)^{-2}(\delta_t-\delta_n)^{-1}\left(\sin^2\delta_t-\sin^2\delta_n\right) \times\left[v\left(\delta_t^2-\delta_n^2\right)^2+2\delta_t\delta_n\right]\left(\delta_t ch\frac{\delta_t}{\varepsilon}l_0 sh\frac{\delta_n}{\varepsilon}l_0 + \delta_n ch\frac{\delta_n}{\varepsilon}l_0 sh\frac{\delta_t}{\varepsilon}l_0\right),\quad t = 1,3,\ldots \tag{1.4.21}$$

$$g_{tt} = 4G\delta_t^3\left(1-2/3\sin^2\delta_t\right)sh\frac{2\delta_t}{\varepsilon}l_0$$

$$H_t = \int_{-1}^{1}\{Q(\eta)\left[(1-v)\delta_t^{-1}F_t''(\eta)-v\delta_t F_t(\eta)\right] - T(\eta)\left[(1-v)\delta_t^{-2}F_t'''(\eta)+(2-v)F_t'(\eta)\right]d\eta.$$

For $t, n = 2, 4, \ldots$ the corresponding expressions g_{tn} are obtained by replacing in the above formula $\cos\delta_n$ with $\sin\delta_n$ and $\sin\delta_n$ with $\cos\delta_n$, respectively.

From the structure of the resulting system it can be concluded that the unknowns C_{10}, A_{k0} and B_{n0} are found independently.

The definition of C_{ii}, A_{ki}, B_{ni} $(i = 1, 2, \ldots)$ is consistently reduced to the inversion of the same matrix which coincide with the matrices (1.4.20), (1.4.21). In turn, the elements of these matrices are not dependent on the type of loading on the ends of the cylinder and, therefore, the inversion can be performed once and for all. It is appropriate to note that the systems (1.4.19) have already been met in the theory of thick plates and thin shells in statics [9–11, 97], and based on these results the numerical analysis of various tasks has already been performed. For a semi-infinite cylinder $m_k = \exp(-\mu_k\xi)$ the systems are similar to the systems (1.4.18), (1.4.19) so we will not consider them here.

In the case $\lambda_0^2 = 1$ the system (1.4.19) remains valid and the system (1.4.18) takes the form

$$\sum_{k=1}^{3} \prod\nolimits_{pk} A_{k0} = E_p (p = 1, 2, 3) \tag{1.4.22}$$

where

$$\prod\nolimits_{pk} = 2/3G(1-2\nu)\nu^2(1+\nu)^2\mu_{p0}^2 ch\frac{\mu_{p0}}{\sqrt[3]{\varepsilon}} l_0 sh\frac{\mu_{k0}}{\sqrt[3]{\varepsilon}} l_0$$

$$E_p = -\int_{-1}^{1}\left[\mu_{p0}T(\eta)sh\frac{\mu_{p0}}{\sqrt{\varepsilon}}l_0 - \nu Q(\eta)\sqrt[3]{\varepsilon}ch\frac{\mu_{p0}}{\sqrt[3]{\varepsilon}}l_0\right]d\eta.$$

Similarly, in the case of ultra-low vibrations of a cylinder we have:

$$\begin{gathered}\left(\nu^2(1+\nu)\Lambda^4\mu_{10}\varepsilon^{\beta} ch\mu_{10}l_0 sh\mu_{10}l_0\right)A_{10} \\ = \int_{-1}^{1}\left[\nu\Lambda^2 Q(\eta)ch\mu_{10}l_0 + \mu_{10}(1-\nu^2)T(\eta)sh\mu_{10}l_0\right]d\eta.\end{gathered} \tag{1.4.23}$$

Consider the case $\mu = O(\varepsilon^{-\beta})$, $\lambda = O(\varepsilon^{1-2\beta})$, $\frac{1}{2} < \beta < 1$.

On the basis of formulas (1.3.9), from the variational principle we obtain

$$\sum_{k=1}^{2} \prod\nolimits_{pk} A_{k0} = E_p (p = 1, 2). \tag{1.4.24}$$

Here

$$\prod_{pk} = 12G(1+v)\left(1-v^2\right)\Lambda^4\mu_{k0}\varepsilon^{1-3\beta}ch\frac{\mu_{k0}}{\varepsilon^\beta}l_0 sh\frac{\mu_{p0}}{\varepsilon^\beta}l_0$$

$$E_p = \int_{-1}^{1}\left[3\left(1-v^2\right)\Lambda^2\eta Q(\eta)sh\frac{\mu_{p0}}{\varepsilon^\beta}l_0 - \mu_{p0}^2\varepsilon^{1-\beta}T(\eta)ch\frac{\mu_{p0}}{\varepsilon^\beta}l_0\right]d\eta.$$

In other cases, the system has a similar form and hence it is not considered here. We write down the infinite system when $\mu_n = \varepsilon^{-1}\delta_n$, $\lambda = \varepsilon^{-1}s$.
Taking into account (1.3.9), from (1.4.13) we obtain

$$\begin{aligned} &\sum_{k=1,3,\ldots}^{\infty} g_{pk}B_{k0} = H_p(p=1,3,\ldots) \\ &\sum_{k=2,4,\ldots}^{\infty} g_{pk}B_{k0} = H_p(p=2,4,\ldots) \end{aligned} \tag{1.4.25}$$

where

$$\begin{aligned} g_{pk} = 4G\Big\{ &\left(2\delta_k^2+s^2\right)\left(2\delta_p^2+s^2\right)\left(\delta_k^2-\delta_p^2\right)^{-1}\cos\gamma_{k0}\cos\gamma_{p0} \\ &\times\left(\alpha_{k0}\sin\alpha_{k0}\cos\alpha_{p0}-\alpha_{p0}\sin\alpha_{p0}\cos\alpha_{k0}\right) \\ &\times\left(\delta_k^2 ch\frac{\delta_k}{\varepsilon}sh\frac{\delta_p}{\varepsilon}l_0+\delta_p^2 ch\frac{\delta_p}{\varepsilon}l_0 sh\frac{\delta_k}{\varepsilon}l_0\right) \\ &+2\delta_k^2\delta_p^2\left(\delta_k^2-\delta_p^2\right)^{-1}\cos\alpha_{k0}\cos\alpha_{p0} \\ &\times\left(\gamma_{k0}\sin\gamma_{k0}\cos\gamma_{p0}-\gamma_{p0}\sin\gamma_{p0}\cos\gamma_{k0}\right) \\ &\times\left[\left(2\delta_k^2+s^2\right)ch\frac{\delta_k}{\varepsilon}l_0 sh\frac{\delta_p}{\varepsilon}l_0+\left(2\delta_p^2+s^2\right)ch\frac{\delta_p}{\varepsilon}l_0 sh\frac{\delta_k}{\varepsilon}l_0\right] \\ &+\delta_k^2\left(2\delta_p^2+s^2\right)\left(\delta_k^2-\delta_p^2+\chi^2\right)^{-1}\left(\alpha_{p0}\sin\alpha_{p0}\cos\gamma_{k0}\right. \\ &\left.-\gamma_{k0}\sin\gamma_{k0}\cos\alpha_{p0}\right)\left[\left(2\delta_k^2+s^2\right)ch\frac{\delta_k}{\varepsilon}l_0 sh\frac{\delta_p}{\varepsilon}l_0+\right. \quad k\neq p \\ &\left.+2\delta_k^2 ch\frac{\delta_p}{\varepsilon}l_0 sh\frac{\delta_k}{\varepsilon}l_0\right]-2\delta_p^2\left(2\delta_k^2+s^2\right)\left(\delta_k^2-\delta_p^2+\chi^2\right)^{-1} \\ &\times\left(\alpha_{k0}\sin\alpha_{k0}\cos\gamma_{p0}-\gamma_{p0}\sin\gamma_{p0}\cos\alpha_{k0}\right) \\ &\times\left[2\delta_k^2 ch\frac{\delta_k}{\varepsilon}l_0 sh\frac{\delta_p}{\varepsilon}l_0+\left(2\delta_p^2+s^2\right)ch\frac{\delta_p}{\varepsilon}l_0 sh\frac{\delta_k}{\varepsilon}l_0\right] \end{aligned} \tag{1.4.26}$$

$$\chi^2 = \frac{1}{2(1-v)}\lambda^2$$

$$
\begin{aligned}
g_{kk} &= 4G\delta_k^2\left(2\delta_k^2+s^2\right)\left\{2\delta_k^2\cos^2\alpha_{k0}+\left(2\delta_k^2+s^2\right)\cos^2\gamma_{k0}\right.\\
&\quad+\left[\frac{2\delta_k^2+s^2}{2\alpha_{k0}}+\frac{2(1-\nu)}{s^2}\alpha_{k0}\right]\cos^2\gamma_{k0}\sin 2\alpha_{k0}\\
&\quad\left.+\left[\frac{\delta_k^2}{\gamma_{k0}}-\frac{2(1-\nu)}{s^2}\gamma_{k0}\right]\cos^2\alpha_{k0}\sin 2\gamma_{k0}\right\}sh2\frac{\delta_k}{\varepsilon}l_0 \qquad (1.4.27)\\
H_t &= \int\left\{\delta_t\left[\left(2\delta_t^2+s^2\right)\cos\gamma_{t0}\cos\alpha_{t0}\eta-2\delta_t^2\cos\alpha_{t0}\cos\gamma_{t0}\eta\right]\tau(\eta)\right.\\
&\quad+\delta_t^2\left[\left(2\delta_t^2+s^2\right)\cos\gamma_{t0}\sin\alpha_{t0}\eta-2\gamma_{t0}\cos\alpha_{t0}\sin\gamma_{t0}\eta\right]\\
&\quad\left.\times Q(\eta)\right\}d\eta,\, t=1,3,\ldots
\end{aligned}
$$

Note that, as shown above, in the cases when $\lambda = O(\varepsilon^{-\beta})$ and $\lambda = O(\varepsilon^{1-2\beta})$ $(0 \le \beta < 1)$, the system (1.4.13) splits into two subsystems, one of which contains a finite number of unknowns related to the dynamic problem of the theory of elasticity. The other subsystem coincides with the system obtained in N. A. Bazarenko, I. I. Vorovich's work in a static case. Matrices of these systems do not depend on either the type of external loads or the frequency of the exciting force. In the case $\lambda = \Lambda O(\varepsilon^{-\beta})$ $(\beta \ge 1)$ we obtain an infinite system associated with the dynamic problem of elasticity theory for an elastic strip.

Expressions for $n = 2, 4, 6, \ldots$ are obtained from (1.4.26) and (1.4.27) by replacing $\cos x$ with $\sin x$ and $\sin x$ with $\cos x$ respectively.

Thus, the solution of boundary value problems of dynamic elasticity theory in general, as in statics, is reduced to solving a system of linear algebraic equations. However, we should pay attention to the following circumstance which drastically complicates the solution of infinite systems (1.4.13) by the cut-off method. The fact is that, as can be seen from the expressions for the components of the displacement vector and of the stress tensor, the amplitude characteristics depend on the frequency parameter in a complicated way. The physical essence of the phenomenon under consideration here is that for some values of frequency the amplitudes of stresses and displacements and, of course, unknowns, determined from the infinite systems, must become infinitely large. This circumstance leads to the fact that the conventional methods of solving infinite systems based on the preliminary proof of their regularity are not applicable in the analysis of infinite systems (1.4.13) as, from the start, the impossibility of obtaining estimates of regularity, uniform with respect to frequency, is obvious. However, as mentioned above, the system (1.4.19) coincides with the system which is obtained in shell statics and the solvability and convergence of the method of reduction was proved for the latter in [11]. As for the system (1.4.25), the solvability and convergence of the reduction method follows from the results of [11] for the frequency range that does not contain its natural frequency.

1.5 Construction of Dynamic Refined Applied Theories of a Hollow Cylinder

A number of currently existing applied theories of shells are based on certain hypotheses. Lack of a unified theory necessitates the analysis of the stress-strain state of shells based on the three-dimensional equations of elasticity, as well as developing applied theories relying on asymptotic methods with a predetermined or desired accuracy. A number of applied theories of shells currently exist based on certain hypotheses.

The asymptotic behaviour of solutions of the dynamic problem of elasticity theory for a hollow cylinder under the action of axisymmetric loads applied to the ends of cylinder was investigated by means of homogeneous solutions in 1.1–1.4.

In this section, we present the asymptotic method to remove the load from the cylindrical part of the boundary. The method is based on the reduction of the three-dimensional elasticity problem to a two dimensional one where a two-dimensional differential equation is obtained for each characteristic of stress strain state in the axisymmetric case. The higher is the desired accuracy the higher is the order of the equation.

Such an approach allows, together with the homogeneous solutions previously constructed in 1.1–1.4, to solve the inhomogeneous problem for a hollow cylinder relatively easily.

1. Consider the construction of the applied theories intended to relieve the stress from the side surface of the cylinder. Here we illustrate the construction of such theories for displacements U, W and concerning the stresses, they can be found using the generalized Hooke's law.

For this purpose, we write the equations of motion in the form

$$\begin{aligned} &\tfrac{2(1-v)}{1-2v}\left(U'' + \tfrac{1}{\rho}U' - \tfrac{U}{\rho^2} + \gamma^2 U + \tfrac{1}{1-2v}PW'\right) = 0 \\ &\tfrac{1}{1-2v}P\left(U' + \tfrac{U}{\rho}\right) + W'' + \tfrac{1}{\rho}W' + \tfrac{2(1-v)}{1-2v}\alpha^2 W = 0 \end{aligned} \tag{1.5.1}$$

where

$$P = \frac{\partial}{\partial \xi}, \sqrt{\frac{m_0}{G}}R_0\frac{\partial}{\partial t} = i\lambda, \rho = \frac{r}{R_0}, \xi = \frac{z}{R_0},$$
$$R_0 = \frac{1}{2}(R_1 + R_2), \gamma^2 = p^2 + \lambda^2, \alpha^2 = p^2 + \frac{1-2v}{2(1-v)}\lambda^2,$$

and m_0 is the density of the shell material.

Considering the variables of differentiation p, λ as numbers, the general solutions of the system (1.5.1) can be represented as:

$$
\begin{aligned}
U &= -\alpha J_1(\alpha\rho)C_1 - \alpha Y_1(\alpha\rho)C_2 - pJ_1(\gamma p)C_3 - pY(\gamma\rho)C_4 \\
W &= pJ_0(\alpha\rho)C_1 + pY_0(\alpha\rho)C_2 + \gamma J_0(\gamma p)C_3 + \gamma Y_0(\gamma\rho)C_4.
\end{aligned}
\tag{1.5.2}
$$

Strains are given by the formulas

$$
\begin{aligned}
\sigma_r &= 2G\left\{\left[\frac{\alpha}{\rho}J_1(\alpha\rho) - \delta^2 J_0(\alpha\rho)\right]C_1 + \left[\frac{\alpha}{\rho}Y_1(\alpha\rho) - \delta^2 Y_0(\alpha\rho)\right]C_2\right. \\
&\quad \left. + P\left[\frac{1}{\rho}J_1(\gamma\rho) - \gamma J_0(\gamma\rho)\right]C_3 + P\left[\frac{1}{\rho}Y_1(\gamma\rho) - \gamma Y_0(\gamma\rho)\right]C_4\right\} \\
\tau_{rz} &= -G[2\alpha pJ_1(\alpha\rho)C_1 + 2\alpha pY_1(\alpha\rho)C_2 \\
&\quad + (p^2+\gamma^2)J_1(\gamma p)C_3 + (p^2+\gamma^2)Y_1(\gamma p)C_4] \\
\sigma_z &= 2G\left\{\left[p^2 - \frac{v}{2(1-v)}\lambda^2\right]J_0(\alpha\rho)C_1\right. \\
&\quad \left. + \left[p^2 - \frac{v}{2(1-v)}\lambda^2\right]Y_0(\alpha\rho)C_2 + \gamma J_0(\gamma\rho)C_3 + \gamma Y_0(\gamma\rho)C_4\right\} \\
\sigma_\varphi &= 2G\left\{-\left[\frac{\alpha}{\rho}J_1(\alpha\rho) + \frac{v}{2(1-v)}\lambda^2 J_0(\alpha\rho)\right]C_1\right. \\
&\quad \left. - \left[\frac{\alpha}{\rho}Y_1(\alpha\rho) + \frac{v}{2(1-v)}\lambda^2 Y_0(\alpha\rho)\right]C_2 - \frac{1}{\rho}J_1(\gamma\rho)C_3 - \frac{1}{\rho}Y_1(\gamma\rho)C_4\right\}.
\end{aligned}
\tag{1.5.3}
$$

Here $J_k(x)$, $Y_k(x)$ are Bessel functions of the first and second kinds, respectively.

2. The functions C_1, C_2, C_3, C_4 are determined by the boundary conditions on the cylindrical surfaces $r = R_1$, R_2 and on the faces $z = \pm l$ and from the initial conditions.

Discarding strains from the ends can be produced by using homogeneous solutions constructed in 1.1–1.3. Therefore, ignoring the conditions at the ends, we satisfy the boundary conditions on the side surfaces

$$
\sigma_r = \sigma_k(\xi, t), \tau_{rz} = \tau(\xi, t) \quad \text{for} \quad r = r_k (k = 1, 2). \tag{1.5.4}
$$

Based on the fact that the differentiation operators p, λ are subject to the same formal rules of addition and multiplication as numbers, the system (1.5.4) can be solved with respect to C_i and may be reduced to the form

$$
DC_i = a_i(\xi, t)(i = 1, 2, 3, 4) \tag{1.5.5}
$$

where $a_i(\xi, t)$ are functions defined by the external load, and D is the determinant operator of system (1.5.4). If, now, the operator D is applied to both sides of (1.5.2) and DC_i is replaced with their values from (1.5.5) then we obtain the differential dependence of the displacements U, W as:

$$\begin{aligned} D_u &= -\alpha J_1(\alpha\rho)a_1(\xi,t) - \alpha Y_1(\alpha\rho)a_2(\xi,t) \\ &\quad - pJ_1(\gamma\rho)a_3(\xi,t) - pY_1(\gamma\rho)a_4(\xi,t) \\ D_w &= pJ_0(\alpha\rho)a_1(\xi,t) + Y_0(\alpha\rho)a_2(\xi,t) \\ &\quad + \gamma J_0(\gamma\rho)a_3(\xi,t) + \gamma Y_0(\gamma\rho)a_4(\xi,t) \end{aligned} \tag{1.5.6}$$

where

$$\begin{aligned} D &= \frac{8}{\pi^2\rho_1\rho_2}p^2\left(2p^2+\lambda^2\right)^2 - \frac{\alpha^2\lambda^2}{\rho_1\rho_2}L_{11}(\alpha)L_{11}(\gamma) \\ &\quad + 2^{-1}\frac{\alpha\lambda^2\left(2p^2+\lambda^2\right)^2}{\rho_2}L_{01}(\alpha)L_{11}(\gamma) + 2^{-1}\frac{\alpha\lambda^2\left(2p^2+\lambda^2\right)^2}{\rho_1} \\ &\quad \times L_{10}(\alpha)L_{11}(\gamma) - \frac{2\gamma\lambda^2p^2\alpha^2}{\rho_1}L_{10}(\gamma)L_{11}(\alpha) - \frac{2\gamma\lambda^2p^2\alpha^2}{\rho_2} \\ &\quad \times L_{01}(\gamma)L_{11}(\alpha) - 4^{-1}\left(2p^2+\lambda^2\right)^4 L_{00}(\alpha)L_{11}(\gamma) \\ &\quad - 4p^4\alpha^2\left(p^2+\lambda^2\right)L_{00}(\gamma)L_{11}(\alpha) + \alpha\gamma p^2\left(2p^2+\lambda^2\right)^2 \\ &\quad \times \left[L_{01}(\gamma)L_{10}(\alpha) + L_{01}(\alpha)L_{10}(\gamma)\right] \end{aligned}$$

$$\begin{aligned} a_1(\xi,t) &= \Delta_{11}^{(1)}\sigma_1(\xi,t) - \Delta_{12}^{(1)}\tau_1(\xi,t) + \Delta_{13}^{(1)}\sigma_2(\xi,t) - \Delta_{14}^{(1)}\tau_2(\xi,t) \\ a_2(\xi,t) &= -\Delta_{11}^{(2)}\sigma_1(\xi,t) - \Delta_{12}^{(2)}\tau_1(\xi,t) + \Delta_{13}^{(2)}\sigma_2(\xi,t) - \Delta_{14}^{(2)}\tau_2(\xi,t) \\ a_3(\xi,t) &= \Delta_{11}^{(3)}\sigma_1(\xi,t) - \Delta_{12}^{(3)}\tau_1(\xi,t) + \Delta_{13}^{(3)}\sigma_2(\xi,t) - \Delta_{14}^{(3)}\tau_2(\xi,t) \\ a_4(\xi,t) &= -\Delta_{11}^{(4)}\sigma_1(\xi,t) + \Delta_{12}^{(4)}\tau_1(\xi,t) - \Delta_{13}^{(4)}\sigma_2(\xi,t) + \Delta_{14}^{(4)}\tau_2(\xi,t) \end{aligned}$$

$$\begin{aligned} \Delta_{11}^{(1)} &= \frac{4\alpha p^2\left(2p^2+\lambda^2\right)}{\pi\rho_2}Y_1(\alpha\rho_1) - \left(2p^2+\lambda^2\right)l_{11} \\ &\quad \times [y(\alpha\rho_2)]L_{11}(\gamma) + 2\alpha p^2\left(2p^2+\lambda^2\right)l_{12}(\gamma\rho_2)Y_1(\alpha\rho_2) \\ \Delta_{12}^{(1)} &= \frac{2p\left(2p^2+\lambda^2\right)}{\pi\rho_2}l_{11}[y(\alpha\rho_1)] - p\left(2p^2+\lambda^2\right)l_{11} \\ &\quad \times [y(\alpha\rho_2)]l_{12}^*(\gamma\rho_2) + 2\alpha p^3 y_1(\alpha\rho_2)l_{13}(\gamma) \\ \Delta_{13}^{(1)} &= \frac{4\alpha p^2\left(2p^2+\lambda^2\right)}{\pi\rho_1}Y_1(\alpha\rho_2) - \left(2p^2+\lambda^2\right)^2 l_{11} \\ &\quad \times [y(\alpha\rho_2)]L_{11}(\gamma) - 2\alpha p^2\left(2p^2+\lambda^2\right)y_1(\alpha\rho_1)l_{12}^*(\gamma\rho_1) \\ \Delta_{14}^{(1)} &= \frac{-2p\left(2p^2+\lambda^2\right)}{\pi\rho_1}l_{11}[y(\alpha\rho_2)] + p\left(2p^2+\lambda^2\right)l_{11} \\ &\quad \times [y(\alpha\rho_1)]l_{12}(\gamma\rho_2) - 2\alpha p^3 y_1(\alpha\rho_2)l_{13}(\gamma). \end{aligned} \tag{1.5.7}$$

The expressions $\Delta_{11}^{(2)},\ldots,\Delta_{14}^{(2)}$ are obtained from the expression $\Delta_{11}^{(1)},\ldots,\Delta_{14}^{(1)}$ by replacing the functions $y_k(x)$ with $J_k(x)$.

$$\begin{aligned}\Delta_{11}^{(3)} &= \frac{2p\left(2p^2+\lambda^2\right)^2}{\pi\rho_2} Y_1(\gamma\rho_1) - 4\alpha^2 p^3 l_{11}[y(\gamma\rho_2)]L_{11}(\alpha)\\ &\quad + 2p\alpha\left(2p^2+\lambda^2\right)Y_1(\gamma\rho_2)l_{15}(\alpha\rho_2)\end{aligned} \tag{1.5.8}$$

$$\begin{aligned}\Delta_{12}^{(3)} &= \frac{2p^2\left(2p^2+\lambda^2\right)}{\pi\rho_2} l_{11}[y(\gamma\rho_1)] + \left(2p^2+\lambda^2\right)y_1(\gamma\rho_2)l_{14}(\alpha)\\ &\quad - 2\alpha p^2 l_{11}[y(\gamma\rho_2)]l_{15}^*(\alpha\rho_1)\end{aligned}$$

$$\begin{aligned}\Delta_{13}^{(3)} &= \frac{2p\left(2p^2+\lambda^2\right)^2}{\pi\rho_1} Y_1(\gamma\rho_2) + 4\alpha^2 p^3 l_{11}[y(\gamma\rho_1)]L_{11}(\alpha)\\ &\quad - 2\alpha p\left(2p^2+\lambda^2\right)y_1(\gamma\rho_1)l_{15}^*(\alpha\rho_1)\end{aligned}$$

$$\begin{aligned}\Delta_{14}^{(3)} &= \frac{2p^2\left(2p^2+\lambda^2\right)}{\pi\rho_1} l_{11}[y(\gamma\rho)] + 2\alpha p^2 l_{11}[y(\gamma\rho_1)]l_{15}(\alpha\rho_2)\\ &\quad - \left(2p^2+\lambda^2\right)Y_1(\gamma\rho_1)l_{14}(\alpha).\end{aligned}$$

As before, the expressions $\Delta_{11}^{(4)},\ldots,\Delta_{14}^{(4)}$ are obtained from the expression $\Delta_{11}^{(3)}$, $\ldots,\Delta_{14}^{(3)}$ on replacing the functions $y_k(x)$ with $J_k(x)$.

$$\begin{aligned}& l_{11}[x_k(\alpha\rho)] = \tfrac{\alpha}{\rho}x_k(\alpha\rho) - \delta^2 x_{k-1}(\alpha\rho)\\ & l_{12}(\alpha\rho) = \tfrac{1}{\rho}L_{11}(\alpha) - \alpha L_{10}(\alpha)\\ & l_{12}^*(\alpha\rho) = \tfrac{1}{\rho}L_{11}(\alpha) - \alpha L_{01}(\alpha)\\ & l_{13}(x) = \tfrac{L_{11}(x)}{\rho_1\rho_2} - \tfrac{x}{\rho_1}L_{10}(x) - \tfrac{x}{\rho_2}L_{01}(x) + x^2 L_{00}(x)\\ & l_{14}(x) = \tfrac{x^2}{\rho_1\rho_2}L_{11}(x) - \tfrac{x\delta^2}{\rho_1}L_{10}(x) - \tfrac{x\delta^2}{\rho_2}L_{01}(x) + \delta^4 L_{00}(x)\\ & l_{15}(x) = \tfrac{x}{\rho}L_{11}(x) - \delta^2 L_{10}(x)\\ & l_{15}^*(x) = \tfrac{x}{\rho_2}L_{11}(x) - \delta^2 L_{01}(x)\\ & \delta^2 = p^2 - \tfrac{\lambda^2}{2}\end{aligned}$$

$$\begin{aligned}& L_{ii}(x) = J_i(x\rho_1)Y_i(x\rho_2) - J_i(x\rho_2)Y_i(x\rho_1)\\ & L_{ij}(x) = J_i(x\rho_1)Y_j(x\rho_2) - J_j(x\rho_2)Y_i(x\rho_1)(i,j = 0,1).\end{aligned}$$

Considering $\rho = 1 + \varepsilon y,\ -1 \le y \le 1$, where $\varepsilon = h/R_0$ is the parameter of thin-walledness, and expanding (1.5.6) with respect to the small parameter ε, taking into account (1.5.7), we obtain

$$
\begin{aligned}
\sum_{k=1}^{N} D_k\left(\frac{\partial}{\partial\xi},\frac{\partial}{\partial t}\right)U\varepsilon^k = \sum_{k=1}^{N}\Bigg[&A_{1k}\left(\frac{\partial}{\partial\xi},\frac{\partial}{\partial t}\right)\sigma(\xi,t)\\
&+A_{2k}\left(\frac{\partial}{\partial\xi},\frac{\partial}{\partial t}\right)\tau_1(\xi,t)+A_{3k}\left(\frac{\partial}{\partial\xi},\frac{\partial}{\partial t}\right)\sigma_2(\xi,t)\\
&+A_{4k}\left(\frac{\partial}{\partial\xi},\frac{\partial}{\partial t}\right)\tau_2(\xi,t)\Bigg]\varepsilon^k\\
\sum_{k=1}^{N} D_k\left(\frac{\partial}{\partial\xi},\frac{\partial}{\partial t}\right)W\varepsilon^k = \sum_{k=1}^{N}\Bigg[&B_{1k}\left(\frac{\partial}{\partial\xi},\frac{\partial}{\partial t}\right)\sigma_1(\xi,t)\\
&+B_{2k}\left(\frac{\partial}{\partial\xi},\frac{\partial}{\partial t}\right)\tau_1(\xi,t)+B_{3k}\left(\frac{\partial}{\partial\xi},\frac{\partial}{\partial t}\right)\sigma_2(\xi,t)\\
&+B_{4k}\left(\frac{\partial}{\partial\xi},\frac{\partial}{\partial t}\right)\tau_2(\xi,t)\Bigg]\varepsilon^k
\end{aligned}
$$

where

$D_k\left(\frac{\partial}{\partial\xi},\frac{\partial}{\partial t}\right)A_{ik}\left(\frac{\partial}{\partial\xi},\frac{\partial}{\partial t}\right), B_{ik}\left(\frac{\partial}{\partial\xi},\frac{\partial}{\partial t}\right)$ are polynomials of variables $\frac{\partial}{\partial\xi},\frac{\partial}{\partial t}$.

Discarding the terms of order higher than ε^n on both sides, we arrive at the applied theory for a circular cylindrical shell with the above accuracy in dynamic problems of the theory of shells. Thus, we obtain a series of applied theories with any prescribed accuracy depending on ε. For simplicity, as a specific example, based on Eq. (1.5.8), we propose a theory in which $N = 2$ and $\tau_1 = \sigma_2 = 0$:

$$
\begin{aligned}
&\Big\langle(1-\nu)\Big\{2\Big(2+2\nu-\tfrac{\partial^2}{\partial t^2}\Big)\tfrac{\partial^2}{\partial\xi^2}-\tfrac{\partial^2}{\partial t^2}\Big[(1-\nu)\tfrac{\partial^2}{\partial t^2}-2\Big]\Big\}\\
&+\tfrac{1}{3}\Big\{4\tfrac{\partial^6}{\partial\xi^6}+4\Big[(3-2\nu)\tfrac{\partial^2}{\partial t^2}-4(1-\nu^2)\Big]\tfrac{\partial^4}{\partial\xi^4}\\
&+\Big[(4\nu^2-16\nu+11)\tfrac{\partial^4}{\partial t^4}+2(4\nu^2+18\nu-18)\tfrac{\partial^2}{\partial t^2}+36(1-\nu^2)\Big]\tfrac{\partial^2}{\partial\xi^2}\\
&+(3-4\nu)(1-\nu)\tfrac{\partial^6}{\partial t^6}-2(6\nu^2-14\nu+7)\tfrac{\partial^4}{\partial t^4}\\
&+18(1-\nu)\tfrac{\partial^2}{\partial t^2}\Big\}\varepsilon^2+\cdots\Big\rangle U=\tfrac{(1-\nu)}{\varepsilon}\Big\{-2\tfrac{\partial^2}{\partial\xi^2}\\
&-(1-\nu)\tfrac{\partial^2}{\partial t^2}+\varepsilon\Big[\nu\Big(2\tfrac{\partial^2}{\partial\xi^2}+\tfrac{\partial^2}{\partial t^2}\Big)y-16(1-\nu)\tfrac{\partial^2}{\partial\xi^2}\\
&+4(1-2\nu)\tfrac{\partial^2}{\partial t^2}\Big]+\cdots\Big\}\sigma_1(\xi,t)
\end{aligned}
$$

$$\begin{aligned}&\Big\langle (1-v)\Big\{2\Big(2+2v-\tfrac{\partial^2}{\partial t^2}\Big)\tfrac{\partial^2}{\partial \xi^2}-\tfrac{\partial^2}{\partial t^2}\Big[(1-v)\tfrac{\partial^2}{\partial t^2}-2\Big]\Big\}\\&+\tfrac{1}{3}\Big\{4\tfrac{\partial^6}{\partial \xi^6}+4\Big[(3-2v)\tfrac{\partial^2}{\partial t^2}-4(1-v^2)\Big]\tfrac{\partial^4}{\partial \xi^4}\\&+\Big[(4v^2-16v+11)\tfrac{\partial^4}{\partial t^4}+2(4v^2+18v-18)\tfrac{\partial^2}{\partial t^2}+36(1-v^2)\Big]\tfrac{\partial^2}{\partial \xi^2}\\&+(3-4v)(1-v)\tfrac{\partial^6}{\partial t^6}-2(6v^2-14v+7)\tfrac{\partial^4}{\partial t^4}\\&+18(1-v)\tfrac{\partial^2}{\partial t^2}\Big\}\varepsilon^2+\cdots\Big\rangle W=\tfrac{(1-v)}{\varepsilon}\Big\{2v\tfrac{\partial^2}{\partial \xi^2}\\&+\varepsilon\Big[\Big(2\tfrac{\partial^4}{\partial \xi^4}+\tfrac{1-v}{2}\tfrac{\partial^4}{\partial \xi^2\partial t^2}\Big)y+v\Big(\tfrac{1}{2}\tfrac{\partial^2}{\partial t^2}-2\Big)\tfrac{\partial^2}{\partial \xi^2}\Big]+\cdots\Big\}\sigma_1(\xi,t).\end{aligned}\tag{1.5.9}$$

Note that this theory is, in fact, an extension of the method developed in [1, 12] to dynamic problems.

Keeping more number of terms in (1.5.8) we may get a more accurate theory. Note that these applied theories are generally intended to remove the stresses from the cylindrical portion of the boundary, as removing the stresses from the ends is produced by homogeneous solutions. Nevertheless, one can raise the question of the relationships between the solutions given by applied theories of (1.5.8)-type and the exact solutions defined by the dispersion Eq. (1.1.13). Therefore, if we seek the homogeneous solutions of Eq. (1.5.8) in the form $U, W \sim e^{p\xi + i\omega t}$, we obtain

$$D_n(p,\lambda,\varepsilon) = D_1(p,\lambda)\varepsilon + D_2(p,\lambda)\varepsilon^q + \cdots D_n(p,\lambda)\varepsilon^n = 0. \tag{1.5.10}$$

As it may be easily seen from Eq. (1.5.10) we can determine all zeros of the dispersion equation, defined by the formulas (1.2.4), (1.2.10), (1.2.16), (1.2.23), (1.2.26) (1.2.29). As for the zeros of the dispersion equation defined by (1.2.12), (1.2.32), (1.2.34), their definition from Eq. (1.5.10) is not possible.

1.6 Torsional Vibrations of an Isotropic Hollow Cylinder

As is known, for axisymmetric wave motions two different types of motion are possible. The first of these is associated with the propagation of longitudinal waves and the other one—with torsional waves in the hollow cylinder. The propagation of a longitudinal wave in a hollow cylinder was studied above. Therefore, now we take up the study of torsional vibrations of a hollow cylinder. A similar problem for the solid cylinder is investigated in [3].

The equations of motion in case of torsional vibrations have the following form:

$$\frac{\partial^2 U_\varphi}{\partial \rho^2} + \frac{1}{\rho}\frac{\partial U_\varphi}{\partial \rho} - \frac{1}{\rho^2}U_\varphi + \frac{\partial^2 U_\varphi}{\partial \xi^2} - gG^{-1}\frac{\partial^2 U_\varphi}{\partial t^2} = 0 \tag{1.6.1}$$

$$\tau_{r\varphi} = G\left(\frac{\partial U_\varphi}{\partial \rho} - \frac{U_\varphi}{\rho}\right), \tau_{z\varphi} = G\frac{\partial U_\varphi}{\partial \xi} \tag{1.6.2}$$

It is assumed that the lateral side of the cylinder is free from loads,

$$\tau_{r\varphi} = 0 \quad \text{at} \quad \rho = \rho_s \quad (s = 1, 2) \tag{1.6.3}$$

At the ends of the cylinder the following boundary conditions hold true:

$$\tau_{z\varphi} = \tau^{\pm}(\rho)e^{i\omega t} \quad at \quad \xi = \pm l_0 \tag{1.6.4}$$

The solution to Eq. (1.6.1) will be sought in the form

$$U_{\varphi} = U(\rho)m(\xi)e^{i\omega t} \tag{1.6.5}$$

Substituting (1.6.5) into (1.6.1) and (1.6.3), after separating the variables we obtain

$$\frac{d^2m}{d\xi^2} - \mu^2 m(\xi) = 0 \tag{1.6.6}$$

$$U'' + \frac{1}{\rho}U' + \left(\gamma^2 - \frac{1}{\rho^2}\right)U = 0 \tag{1.6.7}$$

$$\left(U' - \frac{1}{\rho}U\right)_{\rho=\rho_s} = 0 \tag{1.6.8}$$

Let us define an operator T by the formula

$$TU = \left[-U'' - \frac{1}{\rho}U' + \frac{1}{\rho^2}U, \left(U' - \frac{1}{\rho}U\right)_{\rho=\rho_s}\right] = 0$$

In the space $L_2(\rho_1, \rho_2)$ the problems (1.6.7) and (1.6.8) may be represented in the operator form as:

$$TU = \gamma^2 U \tag{1.6.9}$$

where $L_2(\rho_1, \rho_2)$ is a Hilbert space with weight ρ and scalar product

$$(U_1, U_2) = \int_{\rho_1}^{\rho_2} U_1 \bar{U}_2 \rho d\rho \tag{1.6.10}$$

It is easy to prove that the operator T is self-adjoint. Thus, the spectrum of the operator (1.6.9) is real and the eigenvectors are orthogonal, complete and form a basis in $L2$. On the other hand, from the real-value of γ^2 it follows that magnitude μ^2 is real too, i.e. μ is either a real or a purely imaginary quantity. In [3], proceeding from physical considerations for the solid cylinder, it was correctly noted the invalidity of statement of the author of the work [13] was in error in that the parameter μ can be a complex one.

Now let us consider the problem (1.6.6), (1.6.4):

$$\frac{d^2m}{d\xi^2} - \mu^2 m(\xi) = 0$$
$$\tau_{z\varphi} = \tau^{\pm}(\rho).$$

Let us expand $\tau^{\pm}(\rho)$ by the eigenfunctions of the operator (1.6.9):

$$\tau^{\pm}(\rho) = \sum_{k=1}^{\infty} a_k^{\pm} U_k(\rho)$$

where

$$\|U_k\|^2 = 1 = \int_{\rho_1}^{\rho_2} U_k^2 \rho d\rho, (U_k, U_n) = \delta_{kn},$$

Then the general solution of (1.6.6), (1.6.4) can be represented as follows

$$m_k(\xi) = C_{1k} e^{\mu_k \xi} + C_{2k} e^{-\mu_k \xi} \quad (1.6.11)$$

Finally, $U_\varphi(\rho, \xi, t)$ can be represented in the form

$$U_\varphi = \sum_{k=1}^{\infty} C_k m_k(\xi) U_k(\rho) e^{i\omega t} \quad (1.6.12)$$

where the constants C_k are determined from the conditions

$$m_k'\big|_{\xi = \pm l_0} = a_k^{\pm}$$

Note that the eigenvalues and eigenvectors of this problem can be found explicitly by means of the asymptotic method. Therefore, now, we may construct the asymptotic formulas for the eigenvalues and eigenvectors. The general solution of Eq. (1.6.7) may be given in the form

$$U(\rho) = C_1 J_1(\gamma\rho) + C_2 Y_1(\gamma\rho) \quad (1.6.13)$$

where $J_1(\gamma\rho)$, $Y_1(\gamma\rho)$ are Bessel functions of the first and second kind respectively; C_1, C_2 are arbitrary constants determined from the homogeneous boundary conditions (1.6.8). From the condition of existence of nontrivial solutions, we obtain the following dispersion equation

$$D(\mu, \lambda, \varepsilon) = \gamma^2 D_0(\mu, \lambda, \varepsilon) \tag{1.6.14}$$

where

$$\begin{aligned} D_0 = {} & L_{00}(\gamma\rho_1, \gamma\rho_2) - \frac{2}{\gamma\rho_1} L_{10}(\gamma\rho_1, \gamma\rho_2) \\ & - \frac{2}{\gamma\rho_2} L_{01}(\gamma\rho_1, \gamma\rho_2) + \frac{1}{\gamma^2\rho_1\rho_2} L_{11}(\gamma\rho_1, \gamma\rho_2) \end{aligned}$$

Equation (1.6.14) has a countable set of zeros with an accumulation point at infinity. Summing up all the roots we get homogeneous solutions in the following form:

$$\begin{aligned} U_\varphi &= R_0 \sum_{k=1}^{\infty} C_k \left[\gamma_k L_{10}(\gamma_k\rho, \gamma_k\rho_2) - \frac{2}{\rho_2} L_{11}(\gamma_k\rho, \gamma_k\rho_2) \right] m_k(\xi) e^{i\omega t} \\ \tau_{z\varphi} &= G \sum_{k=1}^{\infty} C_k \left[\gamma_k L_{10}(\gamma_k\rho, \gamma_k\rho_2) - \frac{2}{\rho_2} L_{11}(\gamma_k\rho, \gamma_k\rho_2) \right] \frac{dm_k}{d\xi} e^{i\omega t} \\ \tau_{r\varphi} &= G \sum_{k=1}^{\infty} C_k \left[\gamma_k^2 L_{00}(\gamma_k\rho, \gamma_k\rho_2) - \frac{2\gamma_k}{\rho} L_{10}(\gamma_k\rho, \gamma_k\rho_2) \right. \\ & \left. - \frac{2\gamma_k}{\rho} L_{01}(\gamma_k\rho, \gamma_k\rho_2) + \frac{4}{\rho\rho_2} L_{11}(\gamma_k\rho, \gamma_k\rho_2) \right] m_k(\xi) e^{i\omega t} \end{aligned} \tag{1.6.15}$$

Now we turn to the study of the roots of Eq. (1.6.14). It can be shown that

$$\lim_{\gamma \to 0} D_0(\mu, \lambda, \varepsilon) = C \quad 0 < C < \infty \tag{1.6.16}$$

In fact, expanding ε in a series $D_0(\mu, \lambda, \varepsilon)$, we obtain:

$$D_0(\mu, \lambda, \varepsilon) = 4\pi^{-1}\varepsilon \left[1 + \frac{1}{3}(9 - 2\gamma^2)\varepsilon^2 + \cdots \right] \tag{1.6.17}$$

Thus, we obtain that $\gamma^2 = 0$ is a double root of Eq. (1.6.14). Here we get two purely imaginary zeros which correspond to some penetrating solutions. The remaining zeros of the function $D(\mu, \lambda, \varepsilon)$ increase indefinitely when $\varepsilon \to 0$. To construct this group of zeros by means of the asymptotic expansions of functions, $J_\nu(x)$, $Y_\nu(x)$ for large values of the argument, Eq. (1.6.14) can be written in the following form in the first term of the asymptotics:

$$D(\mu, \lambda, \varepsilon) = \frac{2}{\pi} \sin 2\gamma\varepsilon + O(\varepsilon) = 0. \tag{1.6.18}$$

Hence, setting $\lambda = \lambda_0 \varepsilon^{-1}$, we seek

$$\mu_n = \delta_n \varepsilon^{-1} + O(\varepsilon)(n = k - 2, k = 3, 4, \ldots) \tag{1.6.19}$$

Employing (1.6.18) we get

$$\delta_n^2 = \frac{n^2 \pi^2}{4} - \lambda_0^2 (n = 1, 2, \ldots) \tag{1.6.20}$$

From (1.6.20) it follows that from the inequality $\lambda_0^2 < \frac{n^2\pi^2}{4}$ we obtain real zeros, which correspond to evanescent solutions, and for $\lambda_0^2 > \frac{n^2\pi^2}{4}$ we obtain a series of purely imaginary ones, here the case $\lambda = \lambda_0 \varepsilon^{-\beta}$, $\mu_n = \mu_{n0} \varepsilon^{-\beta}$, $\beta > 1$ is also possible. It is easy to see that in this case we again obtain (1.6.20). The only difference is that, in this case, n is a sufficiently large number.

Now we give an asymptotic formula for U_φ, $\tau_{z\varphi}$ and $\tau_{r\varphi}$. The zero $\gamma = 0$ corresponds to the following stress-strain state:

$$\begin{aligned} U_\varphi &= C_0 \rho m_0(\xi) e^{i\omega t} \\ \tau_{z\varphi} &= G C_0 \rho m_0(\xi) e^{i\omega t} \\ \tau_{r\varphi} &= 0 \end{aligned} \tag{1.6.21}$$

Here m_0 is given by $m_0(\xi) = a_1 \cos \lambda \xi + a_2 \sin \lambda \varepsilon$

Finally, for the second group, similarly, we obtain the asymptotic behaviour in the first term given as

$$\begin{aligned} U_\varphi &= R_0 \varepsilon B_n \cos \varepsilon \gamma_n (\eta - 1) m_n(\xi) e^{i\omega t} \\ \tau_{z\varphi} &= G \varepsilon B_n \cos \gamma_n (\eta - 1) \frac{d m_n}{d\xi} e^{i\omega t} \\ \tau_{r\varphi} &= -G B_n \sin \varepsilon \gamma_n (\eta - 1) m_n(\xi) e^{i\omega t} \end{aligned} \tag{1.6.22}$$

where B_n are new arbitrary constants and $\eta = \frac{\rho - 1}{\varepsilon}$.

Thus, we arrive at simple asymptotic formulas for the eigenvalues and eigenfunctions.

Note that the case $\mu = 0$ corresponds to torsional vibrations of a hollow cylinder and will be discussed in Chap. 3.

1.7 Elastic Vibrations of a Hollow Cylinder with a Fixed Side Surface

Here, the asymptotic behaviour of the dynamic problem of elasticity theory is studied for a hollow cylinder with a rigid sealing of the side surface on the basis of homogeneous solutions. A similar problem for the solid cylinder is considered in another way in [3].

Let us consider axisymmetric dynamic problem of elasticity theory for a hollow cylinder. It is assumed that the side surface is rigidly sealed, i.e.

$$U_r = 0, U_z = 0 \quad \text{for} \quad r = R_1, R_2. \tag{1.7.1}$$

The nature of the boundary conditions at the end of the boundary are to be specified, but we consider them to be sufficiently smooth.

From (1.7.1) we obtain the following dispersion equation:

$$\begin{aligned}\Delta = 8\pi^{-2}\rho_1^{-1}\rho_2^{-1}\mu^2 - \alpha^2\gamma^2 L_{00}(\gamma)L_{11}(\alpha) \\ - \mu^4 L_{00}(\alpha)L_{11}(\gamma) + \alpha\gamma\mu^2[L_{01}(\gamma)L_{10}(\alpha) + L_{01}(\alpha)L_{10}(\gamma)] = 0\end{aligned} \tag{1.7.2}$$

Equation (1.7.2) has a countable set of zeros whose asymptotic behaviour is examined below. The zeros of the dispersion Eq. (1.7.2) correspond to the following homogeneous solutions:

$$U_r = -R_0 \sum_{k=1}^{\infty} C_k[\alpha_k z_1(\alpha_k\rho) + z_1(\gamma_k\rho)]\frac{dm_k}{d\xi}e^{i\omega t} \tag{1.7.3}$$

$$U_z = R_0 \sum_{k=1}^{\infty} C_k\left[\mu_k^2 z_0(\alpha_k\rho)\gamma_k z_0(\gamma_k\rho)\right]m_k(\xi)e^{i\omega t}$$

where

$$z_v(\alpha_k\rho) = C_{1k}J_v(\alpha_k\rho) + C_{2k}Y_v(\alpha_k\rho)$$
$$z_v(\gamma_k\rho) = C_{3k}J_v(\gamma_k\rho) + C_{4k}Y_v(\gamma_k\rho)$$

$$\begin{aligned}C_{1k} &= -2^{-1}\pi^{-1}\rho_2^{-1}\mu_k^2 Y_0(\alpha_k\rho_1) + \alpha_k\gamma_k^2 Y_1(\alpha_k\rho_2)L_{00}(\gamma_k) \\ &\quad - \mu_k^2\gamma_k Y_0(\alpha_k\rho_2)L_{01}(\gamma_k) \\ C_{2k} &= -2\pi^{-1}\rho_2^{-1}\mu_k^2 J_0(\alpha_k\rho_1) + \alpha_k\gamma_k^2 J_1(\alpha_k\rho_2)L_{00}(\gamma_k) \\ &\quad - \mu_k^2\gamma_k J_0(\alpha_k\rho_2)L_{01}(\gamma_k) \\ C_{3k} &= -2\pi^{-1}\rho_2^{-1}\mu_k^2\gamma_k Y_0(\gamma_k\rho_1) + \mu_k^4 Y_1(\gamma_k\rho_2)L_{00}(\alpha_k) \\ &\quad - \alpha_k\gamma_k\mu_k^2 Y_0(\gamma_k\rho_2)L_{01}(\alpha_k) \\ C_{4k} &= -2\pi^{-1}\rho_2^{-1}\mu_k^2\gamma_k J_0(\gamma_k\rho_1) + \mu_k^4 J_1(\gamma_k\rho_2)L_{00}(\alpha_k) \\ &\quad - \alpha_k\gamma_k\mu_k^2 J_0(\gamma_k\rho_2)L_{01}(\alpha_k)\end{aligned}$$

It can be proved that all zeros of (1.7.2) increase without bound when $\varepsilon \to 0$. Depending on the given λ they can be divided into two groups:

(a) $\lambda = O(\varepsilon^{-\beta})$, $0 \le \beta < 1$, $\varepsilon\mu_k \to const$ for $\varepsilon \to 0$

(b) $\lambda = O(\varepsilon^{-\beta})$, $\beta \ge 1$, $\mu_k = O(\varepsilon^{-\beta})$ for $\varepsilon \to 0$

In case (a) choosing $\lambda = \lambda_0 \varepsilon^{-\beta}$ we seek

$$\mu_n = \frac{\delta_n}{\varepsilon} + O\left(\varepsilon^{1-2\beta}\right)(n = 1, 2, \ldots) \tag{1.7.4}$$

On substituting (1.7.4) into the dispersion relation (1.7.2), and employing the asymptotic expansions of functions $J_\nu(x)$, $Y_\nu(x)$, for δ_n we obtain the following equation:

$$(3 - 4\nu)^2 \sin^2 2\delta_n - 4\delta_n^2 = 0. \tag{1.7.5}$$

This equation has a countable set of zeros and it actually coincides with the characteristic equation of a similar problem for an elastic strip in the static problem of elasticity theory [14] where the values of several roots are given for different values of ν.

Similarly, to construct a second group of zeros by setting $\lambda = s\varepsilon^{-\beta}$ $(\beta \geq 1)$ we find them in the form

$$\mu_n = \delta_n \varepsilon^{-\beta} + O\left(\varepsilon^{\beta}\right). \tag{1.7.6}$$

Substituting (1.7.6) into (1.7.2) we end up with

$$\begin{aligned} &\left(\alpha_{n0}\gamma_{n0} \sin \gamma_{n0} \cos \alpha_{n0} - \delta_n^{*2} \cos \gamma_{n0} \sin \alpha_{n0}\right) \\ &\times \left(\alpha_{n0}\gamma_{n0} \sin \alpha_{n0} \cos \gamma_{n0} - \delta_n^{*2} \cos \alpha_{n0} \sin \gamma_{n0}\right) = 0 \end{aligned} \tag{1.7.7}$$

where

$$\alpha_{n0}^2 = \left[\delta_n^2 + \frac{1 - 2\nu}{2(1 - \nu)} s^2\right] \varepsilon^{2-2\beta}, \quad \gamma_{n0}^2 = \left(\delta_n^2 + s^2\right)\varepsilon^{2-2\beta},$$
$$\delta_n^{*2} = \delta_n \varepsilon^{2-2\beta}$$

Equation (1.7.7) is the Rayleigh-Lamb equation for an elastic strip [3].

We now give the asymptotic construction of homogeneous solutions corresponding to the two groups of zeros the asymptotic properties of which have been studied in the foregoing discussions.

Expanding the solution of the first group with respect to the small parameter ε, we find for them the following asymptotic expressions (here for simplicity only peak values of displacements are given; concerning the stresses, they can be identified using the generalized Hooke's law). Here we have two classes of solutions the first of which corresponds to the zeros $(3 - 4\nu)\sin 2\delta_n - 2\delta_n$,, and the second one—to the zeros of the function $(3 - 4\nu)\sin 2\delta_n + 2\delta_n$.

$$
\begin{aligned}
U_r^{(1)} &= R_0\varepsilon^2 \sum_{n=1}^{\infty} C_{n1}\{[\delta_n \cos\delta_n - (3-4\nu)\sin\delta_n]\cos\delta_n\eta \\
&\quad + \eta\delta_n \sin\delta_n \sin\delta_n\eta + O(\varepsilon)\}\frac{dm_n}{d\xi} \\
U_z^{(1)} &= R_0\varepsilon \sum_{n=1}^{\infty} C_{n1}\delta_n^2[\cos\delta_n \sin\delta_n\eta \\
&\quad - \eta\,\sin\delta_n \cos\delta_n\eta + O(\varepsilon)]m_n(\xi)
\end{aligned} \tag{1.7.8}
$$

$$
\begin{aligned}
U_r^{(2)} &= R_0\varepsilon^2 \sum_{n=1}^{\infty} C_{n2}\{[\delta_n \sin\delta_n + (3-4\nu)\cos\delta_n]\sin\delta_n\eta \\
&\quad + \eta\delta_n \cos\delta_n \cos\delta_n\eta + O(\varepsilon)\}\frac{dm_n}{d\xi} \\
U_z^{(2)} &= R_0\varepsilon \sum_{n=1}^{\infty} C_{n2}\delta_n^2[\eta\,\cos\delta_n \sin\delta_n\eta \\
&\quad - \sin\delta_n \cos\delta_n\eta + O(\varepsilon)]m_n(\xi)
\end{aligned} \tag{1.7.9}
$$

The solutions corresponding to the second group of zeros of the dispersion equation are given here. As in the first case here in the first approximation we obtain two classes of solutions, the first corresponds to the zeros of

$$
\begin{aligned}
&\alpha_{n0}\gamma_{n0}\sin\gamma_{n0}\cos\alpha_{n0} - \delta_n^{*2}\cos\gamma_{n0}\sin\alpha_{n0}, \\
&\alpha_{n0}\gamma_{n0}\cos\gamma_{n0}\sin\alpha_{n0} - \delta_n^{*2}\sin\gamma_{n0}\cos\alpha_{n0},
\end{aligned}
$$

and the second one to the zeros of the functions

$$
\begin{aligned}
U_r^{(1)} &= R_0\varepsilon^2 \sum_{n=1}^{\infty} B_{n1}\frac{1}{\alpha_{n0}}[\alpha_{n0}\gamma_{n0}\sin\gamma_{n0}\cos\alpha_{n0}\eta \\
&\quad - \delta_n^{*2}\sin\alpha_{n0}\cos\gamma_{n0}\eta + O(\varepsilon)]\frac{dm_n}{d\xi} \\
U_z^{(1)} &= R_0\varepsilon \sum_{n=1}^{\infty} B_{n1}\frac{\delta_n^{*2}\gamma_{n0}}{\alpha_{n0}}[\sin\gamma_{n0}\sin\alpha_{n0}\eta \\
&\quad - \sin\alpha_{n0}\sin\gamma_{n0}\eta + O(\varepsilon)]m_n(\xi)
\end{aligned} \tag{1.7.10}
$$

$$
\begin{aligned}
U_r^{(2)} &= R_0\varepsilon^2 \sum_{n=1}^{\infty} B_{n2}\frac{1}{\alpha_{n0}}[\alpha_{n0}\gamma_{n0}\cos\gamma_{n0}\sin\alpha_{n0}\eta \\
&\quad - \delta_n^{*2}\cos\alpha_{n0}\sin\gamma_{n0}\eta + O(\varepsilon)]\frac{dm_n}{d\xi} \\
U_z^{(2)} &= R_0\varepsilon \sum_{n=1}^{\infty} B_{n2}\frac{\delta_n^{2}\gamma_{n0}}{\alpha_{n0}}[\cos\alpha_{n0}\cos\gamma_{n0}\eta \\
&\quad - C_{n1}, C_{n2}, B_{n1}, B_{n2}
\end{aligned} \tag{1.7.11}
$$

are arbitrary constants, which are determined by the boundary conditions at cylinder ends.

As can be seen from formulas (1.7.8), (1.7.9) and (1.7.10), (1.7.11), the solutions defined by formulas (1.7.8), (1.7.10) correspond to bending vibrations of a shell and solutions defined by the formulas (1.7.9), (1.7.11) correspond to vibrations of a shell tension-compression. Note that none of the applied theories of shells is able to describe such kinds of solutions.

1.8 Forced Vibrations of a Hollow Cylinder with Mixed Boundary Conditions on the Side Surface

Allocation of a class of boundary conditions allowing us to solve exactly the dynamic problem for solids of finite size is extremely important. In 1.4 a class of boundary conditions is specified where the problem of cylinder vibrations can be solved exactly. Another class of boundary conditions assuming an exact solution is pointed out below.

Consider the forced vibrations of a hollow cylinder with homogeneous mixed boundary conditions on the side surface

$$U_r = 0, \tau_{rz} = 0 \quad \text{for} \quad \rho = \rho_1, \rho_2 \tag{1.8.1}$$

and on the rest of the boundary one of the boundary conditions of the following types holds true

$$\sigma_z = Q^{\pm}(r)e^{i\omega t}, \tau_{rz} = \tau(r)e^{i\omega t} \quad \text{for} \quad \xi = \pm l_0 \tag{1.8.2}$$

$$\sigma_z = Q^{\pm}(r)e^{i\omega t}, U_r = a^{\pm}(r)e^{i\omega t} \quad \text{for} \quad \xi = \pm l_0 \tag{1.8.3}$$

$$U_z = b_0^{\pm}(r)e^{i\omega t}, \tau_{rz} = \tau^{\pm}(r)e^{i\omega t} \quad \text{for} \quad \xi = \pm l_0 \tag{1.8.4}$$

General analysis of this problem shows that it is convenient to use the Helmholtz theorem of vector analysis:

$$\begin{aligned} \bar{U} &= grad\phi + rot\bar{F}, \quad div\bar{F} = 0 \\ \bar{F} &= \bar{F}(0, F, 0) \end{aligned} \tag{1.8.5}$$

Equation (1.8.5) in the axisymmetric case in cylindrical coordinates has the form:

$$U_r = R_0\left(\frac{\partial\phi}{\partial\rho} - \frac{\partial F}{\partial\xi}\right), U_z = R_0\left(\frac{\partial\phi}{\partial\xi} + \frac{\partial F}{\partial\rho} - \frac{F}{\rho}\right) \tag{1.8.6}$$

Substituting (1.8.6) into (1.1.4), we obtain

$$\begin{gathered}\frac{\partial^2\phi}{\partial\rho^2} + \frac{1}{\rho}\frac{\partial\phi}{\partial\rho} + \frac{\partial^2\phi}{\partial\xi^2} - \frac{1-2v}{2(1-v)}G^{-1}gR_0^2\frac{\partial^2\phi}{\partial t^2} = 0 \\ \frac{\partial^2\psi}{\partial\rho^2} + \frac{1}{\rho}\frac{\partial\psi}{\partial\rho} + \frac{\partial^2\psi}{\partial\xi^2} - G^{-1}gR_0^2\frac{\partial^2\psi}{\partial t^2} = 0\end{gathered} \tag{1.8.7}$$

$$F = \frac{\partial\psi}{\partial\rho}$$

The general solution of (1.8.7) is sought in the form

$$\phi = a(\rho)\frac{dm}{d\xi}e^{i\omega t}, \psi = b(\rho)m(\xi)e^{i\omega t} \tag{1.8.8}$$

where the function $m(\xi)$ is subject to the condition

$$\frac{d^2m}{d\xi^2} - \mu^2 m(\xi) = 0 \tag{1.8.9}$$

Substituting (1.8.8) into (1.8.7), (1.8.1) respectively and utilizing (1.8.9), we obtain

$$a'' + \frac{1}{\rho}a' + \alpha^2 a = 0, \alpha^2 = \mu^2 + \frac{1-20}{2(1-v)}\lambda^2 \tag{1.8.10}$$

$a' = 0$ for $\rho = \rho_1, \rho_2$

$$b'' + \frac{1}{\rho}b' + \gamma^2 b = 0, \gamma^2 = \mu^2 + \lambda^2 \tag{1.8.11}$$

$b' = 0$ at $\rho = \rho_1, \rho_2$. According to formulas (1.8.6) and (1.1.5), the components of the displacement vector and the stress tensor may be written in the form:

$$
\begin{aligned}
&U_r = R_0(a' - b')\frac{dm}{d\xi}e^{i\omega t},\\
&U_z = R_0\left(\mu^2 a - \gamma^2 b\right)m(\xi)e^{i\omega t}\\
&\sigma_z = 2G\left[\left(\mu^2 - \frac{\nu}{2-2\nu}\lambda^2\right)a - \gamma^2 b\right]\frac{dm}{d\xi}e^{i\omega t}\\
&\sigma_\varphi = 2G\left[\frac{1}{\rho}(a' - b') - \frac{\nu}{2-2\nu}\lambda^2 a\right]\frac{dm}{d\xi}e^{i\omega t}\\
&\sigma_r = 2G\left[-\frac{1}{\rho}(a' - b') + \gamma^2 b - \delta^2 a\right]\frac{dm}{d\xi}e^{i\omega t}\\
&\tau_{rz} = G\left[2\mu^2 a' - \left(2\mu^2 + \lambda^2\right)b'\right]m(\xi)e^{i\omega t},\ \delta^2 = \mu^2 + \frac{\lambda^2}{2}
\end{aligned}
\tag{1.8.12}
$$

Thus, the general boundary value problem in part is divided into two problems: (1.8.10) and (1.8.11). We define an operator T by the formula

$$
Ty = \left\{-\frac{1}{\rho}(\rho y')', y'\big|_{\rho=\rho_1,\rho_2} = 0\right. \tag{1.8.13}
$$

Then, in the space $L_2(\rho_1, \rho_2)$ the problems (1.8.10) and (1.8.11) can be given in the operator form:

$$
Ta = \alpha^2 a \tag{1.8.14}
$$

$$
Tb = \gamma^2 b \tag{1.8.15}
$$

Operator Eqs. (1.8.14) and (1.8.15) are mathematically identical.

It is easy to prove that the operator T is a self-adjoint. Consequently, the spectrum of the operator T is real, and the eigenvectors are orthogonal, complete and form a basis in the space $L_2(\rho_1, \rho_2)$. On the other hand, from the real value of α^2, γ^2 it follows that the value of μ^2 is also real, that is, the value of μ is either real or purely imaginary.

Now let us consider the problem

$$
\begin{aligned}
&\frac{d^2 m_k^{(i)}}{d\xi^2} - \mu_{ki}^2 m_k^{(i)}(\xi) = 0\\
&\sigma_z(\rho, \pm l_0) = Q^{\pm}(\rho), \quad \tau_{rz}(\rho, \pm l_0) = \tau^{\pm}(\rho)
\end{aligned}
\tag{1.8.16}
$$

where the eigenvalues of operators (1.8.14) and (1.8.15) are designated by means of $\mu_{ki}(i = 1, 2)$ respectively. The functions $Q^{\pm}(\rho)$, $\tau^{\pm}(\rho)$ are expanded by the eigenfunctions of Eqs. (1.8.14) or (1.8.15). Choosing eigenfunctions of the operator (1.8.14) we obtain

$$Q^{\pm}(\rho) = \sum_{k=1}^{\infty} a_k^{\pm}(\rho), \tau^{\pm}(\rho) = \sum_{k=1}^{\infty} b_k^{\pm} a_k(\rho) \tag{1.8.17}$$

where $a_k^{\pm} = \int\limits_{\rho_1}^{\rho_2} Q^{\pm}(\rho) a_k(\rho)\rho d\rho, b_k^{\pm} = \int\limits_{\rho_1}^{\rho_2} r^{\pm}(\rho) a_k(\rho)\rho d\rho$

$$\|a_k\|^2 = 1 = \int\limits_{\rho_1}^{\rho_2} a_k^2(\rho)\rho d\rho, (a_k, a_n) = \delta_{kn}$$

$$\|b_k\|^2 = 1 = \int\limits_{\rho_1}^{\rho_2} b_k^2(\rho)\rho d\rho, (b_k, b_n) = \delta_{kn}$$

The general solution of (1.8.16) can be represented as follows:

$$m_k^{(i)} = c_{k1}^{(i)} e^{\mu_{ki}\xi} + c_{k2}^{(i)} e^{-\mu_{ki}\xi} \tag{1.8.18}$$

Therefore, the general solution of the given problem can be represented in the following form:

$$U_r = R_0 \sum_{k=1}^{\infty} \left[C_k a_k'(\rho) \frac{dm_k^{(1)}}{d\xi} - d_k b_k'(\rho) \frac{dm_k^{(2)}}{d\xi} \right] e^{i\omega t} \tag{1.8.19}$$

$$U_z = R_0 \sum_{k=1}^{\infty} \left[C_k \mu_{k1}^2 a_k(\rho) m_k^{(1)} - \left(\mu_{k2}^2 + \lambda^2\right) d_k b_k(\rho) m_k^{(2)} \right] e^{i\omega t}$$

where the constants C_k, d_k are determined from one of the conditions (1.8.2)–(1.8.4). Satisfying the conditions (1.8.2) with respect to C_k, d_k we obtain the following system of equations:

$$\begin{aligned} &2GC_k\left(\mu_{k1}^2 - \tfrac{\nu}{2-2\nu}\lambda^2\right)\frac{dm_k^{(1)}}{d\xi}\bigg|_{\xi=\pm l_0} = a_k^{\pm} \\ &G\Bigg[2\mu_{k1}^2 m_k^{(1)}(\xi) \int\limits_{\rho_1}^{\rho_2} a_k(\rho) a_k'(\rho)\rho d\rho C_k \\ &-\left(2\mu_{k2}^2 + \lambda^2\right) m_k^{(2)} \int\limits_{\rho_1}^{\rho_2} a_k(\rho) b_k'(\rho)\rho d\rho d_k\Bigg]_{\xi=\pm l_0} = b_k^{\pm}. \end{aligned} \tag{1.8.20}$$

The remaining boundary conditions at the ends of the cylinder can be satisfied similarly.

We proceed to the construction of the asymptotic formulas for the eigenvalues and eigenvectors. General solutions of Eqs. (1.8.10) and (1.8.11), respectively, have the form:

$$a = C_1 J_0(\alpha\rho) + C_2 Y_0(\alpha\rho) \tag{1.8.21}$$

$$b = D_1 J_0(\gamma\rho) + D_2 Y_0(\gamma\rho) \tag{1.8.22}$$

Satisfying homogeneous boundary conditions on the side surface, we obtain a system of linear algebraic equations with respect to some arbitrary constants. From the condition of existence of nontrivial solutions of the system we get the dispersion equation.

Equations (1.8.21) and (1.8.22) have a countable set of roots with a limit point at infinity. They correspond to the following homogeneous solutions:

$$\begin{aligned}
\sigma_z^{(1)} &= 2G\sum_{k=1}^{\infty} C_k\left[\mu_{k1}^2 - \frac{\nu}{2-2\nu}\lambda^2\right] \\
&\quad \times L_{01}(\alpha_k\rho, \alpha_k\rho_2)\frac{dm_k^{(1)}}{d\xi}e^{i\omega t} \\
\sigma_\varphi^{(1)} &= -2G\sum_{k=1}^{\infty} C_k\left[\frac{\nu}{2-2\nu}\lambda^2 L_{01}(\alpha_k\rho, \alpha_k\rho_2)\right. \\
&\quad \left. + \frac{\alpha_k}{\rho}L_{11}(\alpha_k\rho, \alpha_k\rho_2)\right]\frac{dm_k^{(1)}}{d\xi}e^{i\omega t} \\
\sigma_r^{(1)} &= -2G\sum_{k=1}^{\infty} C_k\left[\delta_k^{*2} L_{01}(\alpha_k\rho, \alpha_k\rho_2)\right. \\
&\quad \left. - \frac{\alpha_k}{\rho}L_{11}(\alpha_k\rho, \alpha_k\rho_2)\right]\frac{dm_k^{(1)}}{d\xi}e^{i\omega t} \\
\tau_{rz}^{(1)} &= -2G\sum_{k=1}^{\infty} C_k\mu_{k1}^2\alpha_k L_{11}(\alpha_k\rho, \alpha_k\rho_2)m_k^{(1)}(\xi)e^{i\omega t} \\
U_r^{(2)} &= R_0\sum_{k=1}^{\infty} D_k\gamma_k L_{11}(\gamma_k\rho, \gamma_k\rho_2)\frac{dm_k^{(2)}}{d\xi}e^{i\omega t} \\
U_z^{(2)} &= -R_0\sum_{k=1}^{\infty} D_k\gamma_k L_{01}(\gamma_k\rho, \gamma_k\rho_2)m_k^{(2)}(\xi)e^{i\omega t}
\end{aligned} \tag{1.8.23}$$

$$\sigma_z^{(2)} = -2G\sum_{k=1}^{\infty} D_k\gamma_k L_{01}(\gamma_k\rho, \gamma_k\rho_2)\frac{dm_k^{(2)}}{d\xi}e^{i\omega t}$$

$$\sigma_\varphi^{(2)} = 2G\sum_{k=1}^{\infty}\frac{D_k\gamma_k}{\rho}L_{11}(\gamma_k\rho, \gamma_k\rho_2)\frac{dm_k^{(2)}}{d\xi}e^{i\omega t}$$

$$\sigma_r^{(2)} = -2G\sum_{k=1}^{\infty} D_k\gamma_k\left[\frac{1}{\rho}L_{11}(\gamma_k\rho, \gamma_k\rho_2)\right. \tag{1.8.24}$$

$$\left.-\gamma_k L_{01}(\gamma_k\rho, \gamma_k\rho_2)\right]\frac{dm_k^{(2)}}{d\xi}e^{i\omega t}$$

$$\tau_{rz}^{(2)} = G\sum_{k=1}^{\infty} D_k\gamma_k(2\mu_{k2}^2 + \lambda^2)L_{11}(\gamma_k\rho, \gamma_k\rho_2)m_k^{(2)}(\xi)e^{i\omega t}$$

Here C_k and D_k are arbitrary constants.

Equation (1.8.21) has a bounded root $\alpha = 0$. From (1.8.23) we observe that this root corresponds to the following choice of displacements and stress components:

$$U_z = R_0(1-2\nu)C_0 m_0(\xi)e^{i\omega t}$$

$$\sigma_z = 2G(1-\nu)C_0\frac{dm_0}{d\xi}e^{i\omega t}$$

$$\sigma_\varphi = 2G\nu C_0\frac{dm_0}{d\xi}e^{i\omega t}$$

$$\sigma_r = -2G\nu C_0\frac{dm_0}{d\xi}e^{i\omega t} \tag{1.8.25}$$

$$U_r = 0, \tau_{rz} = 0$$

$$m_0(\xi) = C_1\cos\sqrt{\frac{1-2\nu}{2(1-\nu)}}\lambda\xi + C_2\sin\sqrt{\frac{1-2\nu}{2(1-\nu)}}\lambda\xi$$

It can be shown that all other zeros of the functions $\Delta_1(\mu, \lambda, \varepsilon)$,$\Delta_2(\mu, \lambda, \varepsilon)$ grow infinitely as $\varepsilon \to 0$. Using the asymptotic behaviour of the Bessel functions for large values of the argument, Eqs. (1.8.21) and (1.8.22) can be written in the form:

$$
\begin{aligned}
L_{11}(\alpha\rho_1, \alpha\rho_2) &= \frac{2}{\pi\alpha}\frac{1}{\sqrt{\rho_1\rho_2}}\left[\left(1+\frac{3}{8\alpha^2}\right)\sin 2\alpha\varepsilon \right. \\
&\quad \left. -\frac{3}{4}\varepsilon\cos 2\alpha\varepsilon + O\left(\frac{1}{\alpha^3}\right)\right] = 0 \\
L_{11}(\gamma\rho_1, \gamma\rho_2) &= \frac{2}{\pi\gamma}\frac{1}{\sqrt{\rho_1\rho_2}}\left[\left(1+\frac{3}{8\gamma^2}\right)\sin 2\gamma\varepsilon \right. \\
&\quad \left. -\frac{3}{4}\varepsilon\cos 2\gamma\varepsilon + O\left(\frac{1}{\gamma^3}\right)\right] = 0
\end{aligned} \tag{1.8.26}
$$

Giving $\lambda = S\varepsilon^{-1}$ we find μ_{ni} as follows:

$$
\mu_{ni} = \delta_{ni}\varepsilon^{-1} + \chi_{ni}\varepsilon + \cdots \quad (i = 1, 2) \tag{1.8.27}
$$

Substituting (1.8.27) to (1.8.26), respectively, we obtain

$$
\sin 2\sqrt{\delta_{n1}^2 + \frac{1-2v}{2(1-v)}S^2} = 0 \tag{1.8.28}
$$

$$
\chi_{n1} = \frac{3}{8}\delta_{n1}^{-1}
$$

$$
\sin 2\sqrt{\delta_{n2}^2 + S^2} = 0 \tag{1.8.29}
$$

$\chi_{n2} = \frac{3}{8}\delta_{n2}^{-1}$ As is seen from (1.8.28) and (1.8.29) for $S^2 > \frac{1-v}{1-2v}\frac{k^2\pi^2}{2}$ and $S^2 > \frac{k^2\pi^2}{2}$ accordingly, we have two series of purely imaginary roots. Using the first term in the asymptotic expansion of Bessel function, we get, from (1.8.23) and (1.8.24), that these zeros correspond to following groups of solutions:

$$
\begin{aligned}
U_r^{(1)} &= -R_0\varepsilon^2\sum_{k=1}^{\infty} C_k[\alpha_{k0}\sin\alpha_{k0}(\eta-1) + O(\varepsilon)]\frac{dm_k^{(1)}}{d\xi}e^{i\omega t} \\
U_z^{(1)} &= R_0\varepsilon\sum_{k=1}^{\infty} C_k\left[\delta_{k1}^2\cos\alpha_{k0}(\eta-1) + O(\varepsilon)\right]m_k^{(1)}(\xi)e^{i\omega t} \\
\sigma_z^{(1)} &= 2G\varepsilon\sum_{k=1}^{\infty} C_k\left[\left(\delta_{k1}^2 - \frac{v}{2-2v}\right)\cos\alpha_{k0}(\eta-1) + O(\varepsilon)\right]\frac{dm_k^{(1)}}{d\xi}e^{i\omega t}
\end{aligned}
$$

$$\sigma_{\varphi}^{(1)} = -G\frac{\nu}{1-\nu}S^2\sum_{k=1}^{\infty}C_k[\cos\alpha_{k0}(\eta-1)+O(\varepsilon)]\frac{dm_k^{(1)}}{d\xi}e^{i\omega t}$$

$$\sigma_{r}^{(1)} = -2G\varepsilon\sum_{k=1}^{\infty}C_k\left[\left(\delta_{k1}^2+\frac{S^2}{2}\right)\cos\alpha_{k0}(\eta-1)+O(\varepsilon)\right]\frac{dm_k^{(1)}}{d\xi}e^{i\omega t}$$

$$\tau_{rz}^{(1)} = -2C\sum_{k=1}^{\infty}C_k\left[\delta_{k1}^2\alpha_{k0}\sin\alpha_{k0}(\eta-1)+O(\varepsilon)\right]m_k^{(1)}(\xi)e^{i\omega t} \quad (1.8.30)$$

$$U_r^{(2)} = R_0\varepsilon^2\sum_{k=1}^{\infty}D_k[\sin\gamma_{k0}(\eta-1)+O(\varepsilon)]\frac{dm_k^{(2)}}{d\xi}e^{i\omega t}$$

$$U_z^{(2)} = -R_0\varepsilon\sum_{k=1}^{\infty}D_k[\cos\gamma_{k0}(\eta-1)+O(\varepsilon)]m_k^{(2)}(\xi)e^{i\omega t}$$

$$\sigma_{z}^{(2)} = -2G\varepsilon\sum_{k=1}^{\infty}D_k[\gamma_{k0}\cos\gamma_{k0}(\eta-1)+O(\varepsilon)]\frac{dm_k^{(2)}}{d\xi}e^{i\omega t}$$

$$\sigma_{\varphi}^{(2)} = 2G\varepsilon^2\sum_{k=1}^{\infty}D_k[\sin\gamma_{k0}(\eta-1)+O(\varepsilon)]\frac{dm_k^{(2)}}{d\xi}e^{i\omega t}$$

$$\sigma_{r}^{(2)} = 2G\varepsilon\sum_{k=1}^{\infty}D_k[\cos\gamma_{k0}(\eta-1)+O(\varepsilon)]\frac{dm_k^{(2)}}{d\xi}e^{i\omega t} \quad (1.8.31)$$

$$\tau_{rz}^{(2)} = G\sum_{k=1}^{\infty}D_k\left[\left(2\delta_{k2}^2+S^2\right)\sin\gamma_{k0}(\eta-1)+O(\varepsilon)\right]\frac{dm_k^{(2)}}{d\xi}e^{i\omega t}$$

$$\alpha_{k0}^2 = \delta_{k1}^2+\frac{1-20}{2(1-\nu)}S^2,\ \gamma_{k0} = \delta_{k2}^2+S^2$$

Equation (1.8.22) has one finite root $\gamma = 0$. Direct verification shows that this root corresponds to the trivial solution. Thus, we obtain simple asymptotic formulas allowing us to calculate the stress-strain state of the shell.

References

1. Bazarenko, N.A., Vorovich, I.I.: Asymptotic behaviour of the solution of the problem of elasticity theory for a hollow cylinder of finite length with a small thickness. J. Appl. Math. Mech. **29**(6), 1035–1052 (1965)
2. Goldenveizer A.L., Lidskiy V.B., Tovstik P.E.: Free vibrations of thin elastic shells p 383. Nauka, Moscow (1979)
3. Hrinchenko V.T., Myaleshka V.V.: Harmonic vibrations and waves in elastic bodies p 283. Naukova Dumka, Kiev (1981)

4. Vlasov V.Z.: The general theory of shells and its applications in engineering p 783 (1949)
5. Grigolyuk E.N., Selezov I.T.: Non-classical theory of vibrations of rods, plates and shells p 272. VINITI (1973)
6. Zhilin P.A.: The theory of simple shells and its applications: Diss. Doctor Phys.-Mat. Sci. – Leningrad p 348 (1983)
7. Goldenveizer, A.L.: Methods of study and refinement of the theory of shells (overview of recent works). J. Appl. Math. Mech. **32**(4), 684–695 (1968)
8. Zilbergleit A.P., Nuller B.M.: Generalized orthogonality of homogeneous solutions in dynamic problems of elasticity theory. Dokl. AN USSR **234**(2), pp 325–333 (1977)
9. Aksentyan, O.C., Vorovich, I.I.: Stress state a of plate with small thinness. PMM **27**(6), 1057–1074 (1963)
10. Bazarenko N.A., Vorovich I.I.: Analysis of three-dimensional stress and strain states of circular cylindrical shells. Construction of refined applied theories. J. Appl.Math. Mech. **33**(3) 495–510 (1969)
11. Ustinov, Y.A., Yudovich, V.I.: On the completeness of the system of elementary solutions of the biharmonic equation in a half-band. J. Appl. Math. Mech. **37**(4), 706–714 (1973)
12. Wilensky, T.V., Vorovich, I.I.: Asymptotic behavior of the solution of the problem of elasticity theory for a spherical shell of small thickness. J. Appl. Math. Mech. **30**(2), 278–295 (1966)
13. McMahon, G.W.: Experimental study of the vibrations of solid, isotropic, elastic cylinders. J. Acoust. Soc. Amer. **36**(1), 87–94 (1964)
14. Uflyand Y.S.: Integral transforms in the theory of elasticity (2nd ed., Rev. and add.) p 402. Nauka, Leningrad (1967)

Chapter 2
Asymptotic Analysis of Dynamic Elasticity Problem for a Hollow Sphere

Abstract The second chapter deals with a three-dimensional dynamic problem of elasticity theory for a spherical layer. In the case of axisymmetric vibrations homogeneous solutions are constructed. One way for the construction of heterogeneous solutions is pointed out. An asymptotic analysis of homogeneous solutions for a spherical shell corresponding to different groups of roots of the dispersion equation is carried out. Non-axisymmetric dynamic problem of elasticity theory for a spherical layer is considered. Due to spherical symmetry, the general boundary value problem is divided into two problems: one coincides with the boundary value problem for axisymmetric vibrations of a hollow sphere, and the second describes the vortex motion of a hollow sphere and coincides with the boundary value problem for purely torsional vibrations of a hollow sphere.

2.1 The General Representation of the Solution to the Equations of Axisymmetric Dynamic Elasticity Theory in Spherical Coordinates

Let $\Gamma = [R_1, R_2] \times [\theta_1, \theta_2] \times [0, 2\pi]$ be the volume occupied by a spherical layer. The layer is related to the spherical coordinate system r, θ, φ (Fig. 2.1). Spherical parts of the layer boundary $r = R_s$ $(s = 1, 2)$ will be called the front surfaces and the rest of the boundary will be called the side surface. We assume that the displacement vector components do not depend on the variable φ:

$$U_r = U_r(r, \theta, t),\ U_\theta = U_\theta(r, \theta, t),\ U_\varphi = U_\varphi(r, \theta, t)$$

We present here a complete system of equations describing the dynamic behaviour of a spherical layer.

M. F. Mekhtiev, *Vibrations of Hollow Elastic Bodies*,
Advanced Structured Materials 88,
https://doi.org/10.1007/978-3-319-74354-7_2

Fig. 2.1 Hollow sphere

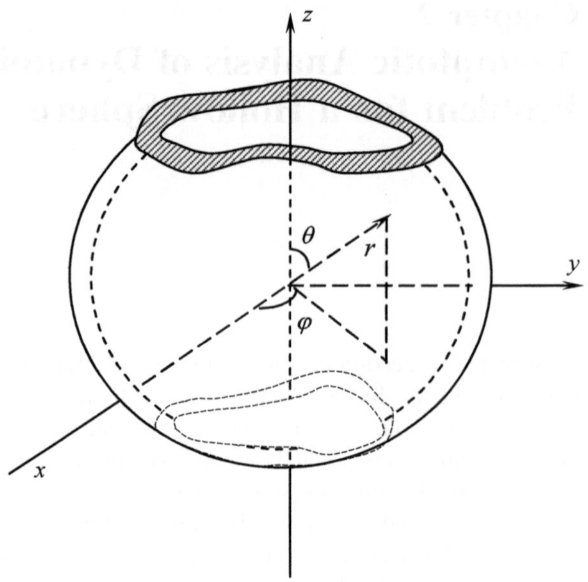

The equations of motion in the axisymmetric case in a spherical coordinate system have the following form:

$$\frac{1}{1-2\nu}\frac{\partial X}{\partial r}+\Delta U_r-\frac{2}{r^2}U_r-\frac{2}{r^2}\frac{\partial U_\theta}{\partial \theta} -\frac{2ctg\theta}{r^2}U_\theta=G^{-1}g\frac{\partial^2 U_r}{\partial t^2} \tag{2.1.1}$$

$$\frac{1}{1-2\nu}\frac{1}{r}\frac{\partial X}{\partial \theta}+\Delta U_\theta+\frac{2}{r^2}\frac{\partial U_r}{\partial \theta} -\frac{1}{r^2\sin^2\theta}U_\theta=G^{-1}g\frac{\partial^2 U_\theta}{\partial t^2} \tag{2.1.2}$$

$$\Delta U_\varphi-\frac{1}{r^2\sin^2\theta}U_\varphi=G^{-1}g\frac{\partial^2 U_\varphi}{\partial t^2}.$$

Here X is a volumetric expansion, Δ is the Laplace operator in spherical coordinates, i.e.

$$X=\frac{\partial U_r}{\partial r}+\frac{2}{r}U_r+\frac{ctg\theta U_\theta}{r}+\frac{1}{r}\frac{\partial U_\theta}{\partial \theta}$$

$$\Delta=\frac{\partial^2}{\partial r^2}+\frac{2}{r}\frac{\partial}{\partial r}+\frac{1}{r^2}ctg\theta\frac{\partial}{\partial \theta}+\frac{1}{r^2}\frac{\partial^2}{\partial \theta^2}.$$

The relationships of the generalized Hooke's law are

$$\begin{aligned}\sigma_r &= 2G\left(\frac{\partial U_r}{\partial r}+\frac{v}{1-2v}X\right),\\ \sigma_\theta &= 2G\left(\frac{1}{r}\frac{\partial U_\theta}{\partial\theta}+\frac{U_r}{r}+\frac{v}{1-2v}X\right)\\ \sigma_\varphi &= 2G\left(\frac{U_r}{r}+\frac{ctg\theta}{r}U_\theta+\frac{v}{1-2v}X\right)\end{aligned}\tag{2.1.3}$$

$$\begin{aligned}\tau_{r\theta} &= G\left(\frac{\partial U_\theta}{\partial r}+\frac{1}{r}\frac{\partial U_r}{\partial\theta}-\frac{U_\theta}{r}\right),\\ \tau_{r\varphi} &= G\left(\frac{\partial U_\varphi}{\partial r}-\frac{U_\varphi}{r}\right),\\ \tau_{\theta\varphi} &= G\left(\frac{1}{r}\frac{\partial U_\varphi}{\partial\theta}-\frac{ctg\theta}{r}U_\varphi\right)\end{aligned}\tag{2.1.4}$$

Assume that the following load from the surfaces is acting on the layer

$$\sigma_r = Q_s(\theta)e^{i\omega t}, \tau_{r\theta} = \tau_s(\theta)e^{i\omega t} \text{ at } r = R_s \tag{2.1.5}$$

$$\tau_{r\varphi} = T_s(\theta)e^{i\omega t} \text{ at } r = R_s. \tag{2.1.6}$$

We will not specify the nature of the boundary conditions on the side surfaces for now. It is easy to see that the solution of the problem (2.1.2), (2.1.6) describes the torsional vibrations of a spherical layer which will be discussed later. Now we turn to the study of the problem (2.1.1) together with the boundary condition (2.1.5).

2.2 Inhomogeneous Solutions

The partial solutions of equations of motion (2.1.1) which satisfy the inhomogeneous boundary conditions (2.1.5) on the front surfaces of the layer will be called inhomogeneous solutions.

The methods suggested for the cylinder introduced in the first chapter may be employed to construct the inhomogeneous solutions. However, this is not the only technique to derive stresses on the faces. One of the known methods is the following: the domain Γ is extended up to a closed spherical shell $F_0 = [R_1, R_2]\,[0,\pi]\,[0,2\pi]$ and the load $[Q_s(\theta), \tau_s(\theta)]$ specified on the front surfaces $S_{1,2}$ is extending it to the closed spherical surfaces $S_{1,2}^0(r = R_1, R_2)$ rather than arbitrarily. External forces prescribed on $S_{1,2}$ are denoted by $Q_s^*(\theta)$, $\tau_s^*(\theta)$, respectively. Thus, $(\theta,\varphi) \in S_{1,2}$ $Q_s^* = Q_s$, $\tau_s^* = \tau_s$.

Imagine an external load in the form of a series of Legendre functions

$$Q_s^*(\theta,t)=\sum_{n=0}^{\infty}\sigma_{ns}P_n(\cos\theta)e^{i\omega t},\ \tau_s^*=\sum_{n=1}^{\infty}\tau_{ns}\frac{dP_n}{d\theta}e^{i\omega t}. \quad (2.2.1)$$

The coefficients of these series are determined through the well-known formulas for Legendre functions generally presented in analysis

$$\begin{aligned}\sigma_{ns}&=\frac{2n+1}{2}\int_0^{\pi}Q_s^*(\theta)P_n(\cos\theta)\sin\theta d\theta\\ \tau_{ns}&=\frac{2n+1}{2n(n+1)}\int_0^{\pi}\tau_s^*(\theta)\frac{dP_n}{d\theta}\sin\theta d\theta.\end{aligned} \quad (2.2.2)$$

Then the components of the displacement vector may be sought in the form of the series:

$$\begin{aligned}U_r&=\left(\sum_{n=0}^{\infty}U_{rn}(r)P_n(\cos\theta)\right)e^{i\omega t}\\ U_\theta&=\left[\sum_{n=0}^{\infty}U_{\theta n}(r)\frac{dP_n}{d\theta}\right]e^{i\omega t}\end{aligned} \quad (2.2.3)$$

Here $P_n(\cos\theta)$ is a Legendre function. Substituting (2.2.2) into the equations of motion (2.1.1) and the boundary conditions (2.1.5) with respect to functions $U_{rn}(r)$, $U_{\theta n}(r)$ we obtain the following system of ordinary differential equations and boundary conditions:

$$\begin{aligned}&\left(U'_{rn}+\frac{2}{r}U_{rn}\right)'+\frac{1-2\nu}{2(1-\nu)}\left(\lambda^2-\frac{z^2-\frac{1}{4}}{r^2}\right)U_{rn}\\ &-\frac{1}{2(1-\nu)}\frac{1}{r}\left(z^2-\frac{1}{4}\right)\left(U'_{\theta n}+\frac{4\nu-3}{r}U_{\theta n}\right)=0\\ &\frac{1}{1-2\nu}\frac{1}{r}\left(U'_{rn}+\frac{4-4\nu}{r}U_{rn}\right)+U''_{\theta n}+\frac{2}{r}U'_{\theta n}\\ &+\left[\lambda^2-\frac{2(1-\nu)}{1-2\nu}\frac{z^2-\frac{1}{4}}{r^2}\right]U_{\theta n}=0\end{aligned} \quad (2.2.4)$$

where

$$n(n+1)=z^2-\frac{1}{4},\ \lambda^2=\frac{2\,(1+\nu)gR_0^2\omega^2}{E},$$

$R_0 = \frac{R_1 + R_2}{2}$ is the radius of the midsurface

$$\frac{2G}{1-2\nu}\left[(1-\nu)U'_{rn} + \frac{2\nu}{r}U_{rn} - \nu\left(z^2 - \frac{1}{4}\right)U_{\theta n}\right]_{r=R_s} = \sigma_{ns}$$
$$G\left[U'_{\theta n} - \frac{U_{\theta n}}{r} + \frac{U_{rn}}{r}\right]_{r=R_s} = \tau_{ns} \quad (2.2.5)$$

The primes denote derivatives with respect to r.

To solve the problems obtained here one may use different methods including numerical ones, for example, the method of orthogonal sweep by Godunov.

The described method of constructing heterogeneous solutions is rather versatile and is not dependent on different parameters of the shell including its thickness. However, if the relative thickness of the shell is sufficiently small and the load specified on the front surfaces is sufficiently smooth then for the construction of inhomogeneous solutions it is appropriate to use the first iteration process of the asymptotic method [1] which is less time consuming and makes it easier to achieve the ultimate goal. This method was used in [2] for the construction of inhomogeneous solutions of radially inhomogeneous spherical shells in the static case.

2.3 Construction of Homogeneous Solutions

By a homogeneous solution we mean a solution of equations of motion (2.1.1) satisfying the condition of zero stresses on the front surfaces

$$\sigma_r = 0, \tau_{r\theta} = 0 \quad \text{for } \rho = \rho_s \ (s = 1, 2). \quad (2.3.1)$$

The following boundary conditions are given on the remaining part of the boundary:

$$\sigma_\theta = Q_j(\rho)e^{i\omega t}, \tau_{r\theta} = \tau_j(\rho)e^{i\omega t} \text{ for } \theta = \theta_j \ (j = 1, 2). \quad (2.3.2)$$

Here $\rho = R_0^{-1}r$ is a dimensionless radial coordinate.

To construct the homogeneous solutions for spherical layer, the solutions of Eq. (2.1.1) will be sought in the form

$$[U_r, U_\theta] = \left[U(\rho)m(\theta), W(\rho)\frac{dm}{d\theta}\right]e^{i\omega t} \quad (2.3.3)$$

where the function is subject to the condition

$$\frac{d^2m}{d\theta^2} + ctg\theta\frac{dm}{d\theta} + \mu(\mu+1)m(\theta) = 0 \quad (2.3.4)$$

and the parameter μ is determined after the boundary conditions are satisfied on the surface $S_{1,2}$. Substituting (2.3.4) into (2.1.1) and employing the separation of variables for the pair of functions (U, W) we obtain the following system of ordinary differential equations:

$$\begin{aligned}
&\left(U' + \frac{2}{\rho}U\right)' + \frac{1-2v}{2(1-v)}\left(\lambda^2 - \frac{z^2 - \frac{1}{4}}{\rho^2}\right)U \\
&\quad - \frac{1}{2(1-v)}\frac{1}{\rho}\left(z^2 - \frac{1}{4}\right)\left[W' + \frac{4v-3}{r}W\right] = 0 \\
&\frac{1}{1-2v}\frac{1}{\rho}\left(U' + \frac{4-4v}{\rho}\right) + W'' + \frac{2}{\rho}W' \\
&\quad + \left[\lambda^2 - \frac{2(1-v)}{1-2v}\frac{1}{\rho^2}\left(z^2 - \frac{1}{4}\right)\right]W = 0, \mu(\mu+1) = z^2 - \frac{1}{4}.
\end{aligned} \tag{2.3.5}$$

Without going into further details, we present below the final solution of (2.3.5)

$$\begin{aligned}
U(\rho) &= \frac{1}{\sqrt{\rho}}\left[\alpha Z_z'(\alpha\rho) - \frac{1}{2\rho}Z_z(\alpha\rho) - \frac{1}{\rho}\left(z^2 - \frac{1}{4}\right)Z_z(\lambda\rho)\right] \\
W(\rho) &= \frac{1}{\sqrt{\rho}}\left[\frac{1}{\rho}Z_z(\alpha\rho) - \lambda Z_z'(\lambda\rho) - \frac{1}{2\rho}Z_z(\lambda\rho)\right]
\end{aligned} \tag{2.3.6}$$

Here

$$\alpha^2 = \frac{1-2v}{2(1-v)}\lambda^2, Z_z(x) = C_{1z}J_z(x) + C_{2z}Y_z(x)$$

$J_z(x)$, $Y_z(x)$ are the Bessel functions of the first and second kinds respectively, and C_{zi} $(i = 1, 2, 3, 4)$ are arbitrary constants.

If the homogeneous boundary conditions (2.3.1) are satisfied, we obtain the dispersion equation

$$\begin{aligned}
\Delta(z, \lambda, \rho_1, \rho_2) = &-32\pi^{-2}\left(z^2 - \frac{1}{4}\right)\varphi_1(z, \rho_1)(z, \rho_2) \\
&- F_1(\alpha, \rho_1, \rho_2)G_1(\lambda, \rho_1, \rho_2) + \left(z^2 - \frac{1}{4}\right) \\
&\times [F_2(\alpha, \rho_1, \rho_2)G_2(\lambda, \rho_1, \rho_2) + F_2(\alpha, \rho_2, \rho_1)G_2(\lambda, \rho_2, \rho_1)] \\
&- \left(z^2 - \frac{1}{4}\right)^2 F_3(\alpha, \rho_1, \rho_2)G_3(\lambda, \rho_1, \rho_2)
\end{aligned} \tag{2.3.7}$$

where

$$
\begin{aligned}
F_1(\alpha,\rho_1,\rho_2) = {} & \varphi_2(z,\rho_1)\varphi_2(z,\rho_2)L_z^{(0,0)}(\alpha\rho_1,\alpha\rho_2) \\
& - 4\alpha\rho_1\varphi_2(z,\rho_2)L_z^{(1,0)}(\alpha\rho_1,\alpha\rho_2) - 4\alpha\rho_2\varphi_2(z,\rho_1) \\
& \times L_z^{(0,1)}(\alpha\rho_1,\alpha\rho_2) + 16\alpha^2\rho_1\rho_2L_z^{(1,1)}(\alpha\rho_1,\alpha\rho_2) \\
G_1(\lambda,\rho_1,\rho_2) = {} & \varphi_3(z,\rho_1)\varphi_3(z,\rho_2)L_z^{(0,0)}(\lambda\rho_1,\lambda\rho_2) \\
& - 2\lambda\rho_1\varphi_3(z,\rho_2)L_z^{(1,0)}(\lambda\rho_1,\lambda\rho_2) - 2\lambda\rho_2\varphi_3(z,\rho_1) \\
& \times L_z^{(0,1)}(\lambda\rho_1,\lambda\rho_2) + 4\lambda^2\rho_1\rho_2L_z^{(1,1)}(\lambda\rho_1,\lambda\rho_2).
\end{aligned}
$$

The transcendental Eq. (2.3.7) has a countable set of roots z_k and the corresponding constants C_{1z_k}, C_{2z_k}, C_{3z_k}, C_{4z_k} are proportional to the cofactors of any row or column of the determinant of the system. As in the first chapter, choosing as a solution the cofactors of the first row, we obtain:

$$
\begin{aligned}
C_{1z_k} = {} & -\frac{4}{\pi}\left(z_k^2-\frac{1}{4}\right)\varphi_1(z_k,\rho_2)\Big[2\alpha\rho_1Y'_{z_k}(\alpha\rho_1)-3Y_{z_k}(\alpha\rho_1)\Big] \\
& + \Big[4\alpha\rho_2Y'_{z_k}(\alpha\rho_2)-\varphi_2(z_k,\rho_2)Y_{z_k}(\alpha\rho_2)\Big]G_1(\lambda,\rho_1,\rho_2) \\
& - \left(z_k^2-\frac{1}{4}\right)\Big[2\alpha\rho_2Y'_{z_k}(\alpha\rho_2)-3Y_{z_k}(\alpha\rho_2)\Big]G_2(\lambda,\rho_1,\rho_2) \\
C_{2z_k} = {} & -\frac{4}{\pi}\left(z_k^2-\frac{1}{4}\right)\varphi_1(z_k,\rho_2)\big[2\alpha\rho_1J'_z(\alpha\rho_1)-J_z(\alpha\rho_1)\big] \\
& + \Big[4\alpha\rho_2J'_{z_k}(\alpha\rho_2)-\varphi_2(z_k,\rho_2)J_{z_k}(\alpha\rho_2)\Big]G_1(\lambda,\rho_1,\rho_2) \\
& - \left(z_k^2-\frac{1}{4}\right)\big[2\alpha\rho_2J'_z(\alpha\rho_2)-3J_z(\alpha\rho_2)\big]G_2(\lambda,\rho_1,\rho_2) \\
C_{3z_k} = {} & \frac{4}{\pi}\varphi_1(z_k,\rho_1)\Big[2\lambda\rho_1Y'_{z_k}(\lambda\rho_1)-\varphi_3(z_k,\rho_1)Y_{z_k}(\lambda\rho_1)\Big] \\
& + \Big[2\lambda\rho_2Y'_{z_k}(\lambda\rho_2)-\varphi_3(z_k,\rho_2)Y_{z_k}(\lambda\rho_2)\Big]F_2(\alpha,\rho_2,\rho_1) \\
& \left(+\left(z_k^2-\frac{1}{4}\right)\Big[2\lambda\rho_2Y'_{z_k}(\lambda\rho_2)-3Y_{z_k}(\lambda\rho_2)\Big]F_2(\alpha,\rho_2,\rho_1\right)
\end{aligned}
\tag{2.3.8}
$$

$$
\begin{aligned}
F_2(\alpha,\rho_1,\rho_2) &= 3\varphi_2(z,\rho_1)L_z^{(0,0)}(\alpha\rho_1,\alpha\rho_2) \\
&\quad - 12\alpha\rho_1 L_z^{(1,0)}(\alpha\rho_1,\alpha\rho_2) - 2\alpha\rho_2\varphi_2(z,\rho_2) \\
&\quad \times L_z^{(0,1)}(\alpha\rho_1,\alpha\rho_2) + 8\alpha^2\rho_1\rho_2 L_z^{(1,1)}(\alpha\rho_1,\alpha\rho_2) \\
G_2(\lambda,\rho_1,\rho_2) &= 3\varphi_3(z,\rho_1)L_z^{(0,0)}(\lambda\rho_1,\lambda\rho_2) \\
&\quad - 6\lambda\rho_1 L_z^{(1,0)}(\lambda\rho_1,\lambda\rho_2) - 2\lambda\rho_2\varphi_3(z,\rho_1) \\
&\quad \times L_z^{(0,1)}(\lambda\rho_1,\lambda\rho_2) + 4\lambda^2\rho_1\rho_2 L_z^{(1,1)}(\lambda\rho_1,\lambda\rho_2) \\
F_3(\alpha,\rho_1,\rho_2) &= 9L_z^{(0,0)}(\alpha\rho_1,\alpha\rho_2) - 6\alpha\rho_1 L_z^{(1,0)}(\alpha\rho_1,\alpha\rho_2) \\
&\quad - 6\lambda\rho_2 L_z^{(0,1)}(\alpha\rho_1,\alpha\rho_2) + 4\alpha^2\rho_1\rho_2 L_z^{(1,1)}(\alpha\rho_1,\alpha\rho_2) \\
G_3(\lambda) &= F_3(\lambda) \\
L_z^{(s,l)}(\psi\rho_1,\psi\rho_2) &= J_z^{(s)}(\psi\rho_1)Y_z^{(l)}(\psi\rho_2) - J_z^{(j)}(\psi\rho_2)Y_z^{(s)}(\psi\rho_1); \\
\varphi_1(z,\rho) &= 2z^2 - 9/2 - \lambda^2\rho^2, \varphi_2(z,\rho) = 2z^2 + 3/2 - \lambda^2\rho^2, \\
\varphi_3(z,\rho) &= 2z^2 - 3/2 - \lambda^2\rho^2
\end{aligned}
$$

$$
\begin{aligned}
C_{4z_k} &= \frac{4}{\pi}\varphi_1(z_k,\rho_1)\left[2\lambda\rho_1 J'_{z_k}(\lambda\rho_1) - \varphi_3(z_k,\rho_1)J_{z_k}(\lambda\rho_1)\right] \\
&\quad + \left[2\lambda\rho_2 J'_{z_k}(\lambda\rho_2) - \varphi_3(z_k,\rho_2)J_{z_k}(\lambda\rho_2)\right]F_2(\alpha_5\rho_1,\rho_2) \\
&\quad + \left(z_k^2 - \frac{1}{4}\right)\left[2\lambda\rho_2 J'_{z_k}(\lambda\rho_2) - 3J_{z_k}(\lambda\rho)\right]F_3(\alpha,\rho_1,\rho_2).
\end{aligned}
$$

Substituting (2.3.8) into (2.3.6) and using the generalized Hooke's law we obtain

$$
\begin{aligned}
U_r &= \frac{1}{\sqrt{\rho}}\sum_{k=1}^{\infty} C_k U_{rk}(\rho)m_k(\theta)e^{i\omega t} \\
U_\theta &= \frac{1}{\sqrt{\rho}}\sum_{k=1}^{\infty} C_k U_{\theta k}(\rho)\frac{dm_k}{d\theta}e^{i\omega t} \\
\sigma_\theta &= \frac{G}{\rho^2\sqrt{\rho}}\sum_{k=1}^{\infty} C_k\left[Q_{\theta k}^{(1)}(\rho)ctg\theta\frac{dm_k}{d\theta} + Q_{\theta k}^{(2)}m_k(\theta)\right]e^{i\omega t} \\
\sigma_\varphi &= \frac{G}{\rho^2\sqrt{\rho}}\sum_{k=1}^{\infty} C_k\left[Q_{\varphi k}^{(1)}(\rho)m_k(\theta) + Q_{\varphi k}^{(2)}\frac{dm_k}{d\theta}\right]e^{i\omega t} \\
\sigma_r &= \frac{G}{\rho^2\sqrt{\rho}}\sum_{k=1}^{\infty} Q_{rk}(\rho)C_k m_k(\theta)e^{i\omega t} \\
\tau_{r\theta} &= \frac{G}{\rho^2\sqrt{\rho}}\sum_{k=1}^{\infty} C_k T_k(\rho)\frac{dm_k}{d\theta}(\theta)e^{i\omega t}
\end{aligned}
\tag{2.3.9}
$$

Here

$$
\begin{aligned}
U_{rk}(\rho) &= \alpha Z'_{z_k}(\alpha\rho) - \frac{1}{2\rho} Z_{z_k}(\alpha\rho) - \frac{1}{\rho}\left(z_k^2 - \frac{1}{4}\right) Z_{z_k}(\lambda\rho) \\
U_{\theta k}(\rho) &= \frac{1}{\rho} Z_{z_k}(\alpha\rho) - \lambda Z'_{z_k}(\lambda\rho) - \frac{1}{2\rho} Z_{z_k}(\lambda\rho) \\
Q_{\theta k}^{(1)} &= 2Z_{z_k}(\alpha\rho) - 2\lambda\rho Z'_{z_k}(\lambda\rho) - Z_{z_k}(\lambda\rho) \\
Q_{\theta k}^{(2)} &= 2\alpha\rho Z'_{z_k}(\alpha\rho) - \left(1 + \frac{\nu}{1-\nu}\lambda^2\rho^2\right) Z_{z_k}(\alpha\rho) - 2\left(Z_k^2 - \frac{1}{4}\right) Z_{z_k}(\lambda\rho) \\
Q_{\varphi k}^{(1)} &= 2\alpha\rho Z'_{z_k}(\alpha\rho) - \left(1 + \frac{\nu}{1-\nu}\lambda^2\rho^2\right) Z_{z_k}(\alpha\rho) - 2\left(Z_k^2 - \frac{1}{4}\right) Z_{z_k}(\lambda\rho) \\
Q_{\varphi k}^{(2)} &= 2Z_{z_k}(\alpha\rho) - 2\lambda\rho Z'_{z_k}(\lambda\rho) - Z_{z_k}(\lambda\rho) \\
Q_{rk}(\rho) &= \varphi_2(Z_k, \rho) Z_{z_k}(\alpha\rho) - 4\alpha\rho Z'_{z_k}(\alpha\rho) \\
&\quad + \left(Z_k^2 - \frac{1}{4}\right)\left[3Z_{z_k}(\lambda\rho) - 2\lambda\rho Z'_{z_k}(\lambda\rho)\right]
\end{aligned}
$$

$$
\begin{aligned}
T_k(\rho) &= -3Z_{z_k}(\alpha\rho) + 2\alpha\rho Z'_{z_k}(\alpha\rho) \\
&\quad - \varphi_3(Z_k, \rho) Z_{z_k}(\lambda\rho) + 2\lambda\rho Z_{z_k}(\lambda\rho)
\end{aligned}
$$

where C_k are arbitrary constants.

2.4 Asymptotic Analysis of the Dispersion Equation

The left-hand side of Eq. (2.3.7) as an integer function of the parameter Z has a countable set of zeros with an accumulation point at infinity. For an effective study of its roots we make some assumptions about the geometric parameters of the sphere. Assume that

$$\rho_1 = 1 - \varepsilon, \rho_2 = 1 + \varepsilon, 2\varepsilon = \frac{R_2 - R_1}{R_0} = \frac{2h}{R_0} \tag{2.4.1}$$

where ε is considered as a small parameter. Substituting (2.4.1) into Eq. (2.3.7), we obtain

$$D(z, \lambda_0, \varepsilon) = \Delta(z, \lambda, \rho_1, \rho_2) = 0. \tag{2.4.2}$$

The case $\lambda_0^2 = 1$ and $Z = \pm\frac{1}{2}(\mu = 0)$ are special cases and therefore considered separately.

Concerning zeros of $D(z, \lambda_0, \varepsilon)$ we formulate the following statement: the function $D(z, \lambda_0, \varepsilon)$ has three groups of zeros at a finite λ_0:

(a) the first group consists of two zeros given by $z_k = O(1)(k = 1, 2)$,
(b) the second group consists of four zeros, which are of the order of $O(\varepsilon^{-1/2})$,
(c) the third group contains a countable set of zeros, which are of the order of $O(\varepsilon^{-1})$.

We outline the proof of the statement. As before, we expand $D(z, \lambda_0, \varepsilon)$ into a series in the small parameter ε which results in

$$\begin{aligned} D(z, \lambda_0, \varepsilon) = & \frac{64}{3\pi^2}\left(\frac{1+v}{1-v}\right)^2 \lambda_0^4\varepsilon^2\Big\langle 12(1-v^2)(\lambda_0^2-1)z^2 \\ & + 27(1-v^2) + 9(1-v^2)(4v+1)\lambda_0^2 - 12(1-v^2)^2\lambda_0^4 \\ & + \{-4z^6 + [3 + 16v^2 + 2(1+v)(3-2v)\lambda_0^2]z^4 \\ & + \left[24v^2 - \frac{195}{4} + 4(1+v)(16-13v-12v^2)\lambda_0^2\right. \\ & \left. - 4(1+v)^2(4v^2-16v+11)\lambda_0^4\right]Z^2 + \frac{2241}{16} - 135v^2 \\ & + 30(1-v^2)\lambda_0^2\left[3\left(2v+\frac{1}{2}\right) - 2(1-v^2)\lambda_0^2\right]\Big\}\varepsilon^2 \\ & + \frac{1}{15}(-32z^8 + \ldots)\varepsilon^4 + \ldots\Big\rangle = 0 \end{aligned} \tag{2.4.3}$$

We seek z_k as an expansion of the form

$$z_k = z_{k0} + \varepsilon z_{k1} + \varepsilon^2 z_{k2} + \ldots \; (k = 1, 2). \tag{2.4.4}$$

After substituting (2.4.4) into (2.4.3) we obtain

$$\begin{aligned} z_{k0}^2 = & \left[4(\lambda_0^2-1)\right]^{-1}\left[4(1-v^2)\lambda_0^4 - 3(4v+1)\lambda_0^2 - 9\right] \\ z_{k1} = & \, 0 \\ z_{k2} = & \left[24(1-v^2)(\lambda_0^2-1)\right]^{-1} z_{k0}^{-1}\{4Z_{k0}^6 \\ & - [3 + 16v^2 + 2(1+v)(3-2v)\lambda_0^2]z_{k0}^4 \\ & + \left[\frac{195}{4} - 24v^2 + 4(1+v)(12v^2+13v-16)\lambda_0^2\right. \\ & \left. + 4(1+v)^2(4v^2-16v+11)\lambda_0^4\right]z_{k0}^2 + 135v^2 \\ & - \frac{2241}{16} + 30(1-v^2)\lambda_0^2\left[2(1-v^2)\lambda_0^2 - 3\left(2v+\frac{1}{2}\right)\right]\Big\} \end{aligned} \tag{2.4.5}$$

Formula (2.4.5) shows that in the cases $\lambda_0^2 < 1$ and $\lambda_0^2 > 3[8(1-v^2)]^{-1}\left[4v + 1 + \sqrt{17+8v}\right]$ we have two real zeros and when $1 < \lambda_0^2 < 3[8(1-v^2)]^{-1} \times \left[4v + 1 + \sqrt{17+8v}\right]$ we have two purely imaginary zeros. A penetrating solution corresponds to those group of zeros that are real.

It is easy to prove that all other zeros of the function $D(z, \lambda, \varepsilon)$ increase indefinitely when $\varepsilon \to 0$. They can be divided into two groups according to their behaviour as $\varepsilon \to 0$:

(1) $\varepsilon z_k \to 0$ as $\varepsilon \to 0$ 2) $\varepsilon z_k \to const$ as $\varepsilon \to 0$.

We first define those z_k, for which $\varepsilon z_k \to 0$ as $\varepsilon \to 0$. Seeking the roots $z_k (k = 3, 4, 5, 6)$ in terms of an expansion of the form

$$z_k = \frac{z_{k0}}{\sqrt{\varepsilon}} + z_{k1}\sqrt{\varepsilon} + \ldots (k = 3, 4, 5, 6) \tag{2.4.6}$$

and substituting in (2.4.3) we obtain

$$\begin{aligned} & z_{k0}^4 - 3(1 - \nu^2)(\lambda_0^2 - 1) = 0 \\ & z_{k1} = \left[40(\lambda_0^2 - 1)z_{k0}\right]^{-1}\left[(16\nu^2 + 5\nu - 11)\lambda_0^4 \right. \\ & \qquad \left. + (18\nu^2 + 25\nu + 32)\lambda_0^2 - 24\nu^2 - 1\right] \end{aligned} \tag{2.4.7}$$

From (2.4.7) it can be seen that when $\lambda_0^2 < 1$ we get four complex zeros. These zeros correspond to the solutions of the edge effect type. When $\lambda_0^2 < 1$ there are two real and two purely imaginary roots. The penetrating solutions correspond to real roots.

To construct the asymptotics of zeros of the third group we seek $z_n (n = k - 6,\ k = 7, 8, \ldots)$ as

$$z_n = \frac{\delta_n}{\varepsilon} + \lambda_0^2 O(\varepsilon) (n = 1, 2, \ldots). \tag{2.4.8}$$

Substituting (2.4.8) into the dispersion Eq. (2.3.7) and employing the asymptotic expansions of $J_z(x)$, $Y_z(x)$ for large z for δ_n we obtain the following equation:

$$sh^2 2\delta_n - 4\delta_n^2 = 0. \tag{2.4.9}$$

It should be noted that Eq. (2.4.9) coincides with the equation that determines the performance of St. Venant's boundary effects in the static of shells [3]. Since Eq. (2.4.9) has a countable set of zeros, then (2.3.7) also has a countable set of zeros for which $\varepsilon z_k \to const$ as $\varepsilon \to 0$.

Note that the zeros defined by (2.4.4), (2.4.6) and (2.4.8) as $\lambda_0 \to 0$ are completely transformed into the zeros found in [4].

Consider the case: (a) $z^2 - \frac{1}{4} = 0$ and (b) $\lambda_0^2 = 1$.

Case (a) corresponds to thickened resonances of a hollow sphere. It is of independent interest and will be discussed later.

In case (b) Eq. (2.4.3) takes the form:

$$
\begin{aligned}
D_0(z,\varepsilon) = \frac{64}{3\pi^2}\left(\frac{1+\nu}{1-\nu}\right)^2 \varepsilon^2 \Big\langle 12(1-\nu^2)(1+\nu)(\nu+2) \\
+ \big[-4z^6 + (14\nu^2+2\nu+9)z^4 - (16\nu^4+16\nu^3+8\nu^2 \\
+ 12\nu + \frac{115}{4}\Big)z^2 + \frac{2001}{16} + 180\nu + 60\nu^2 \\
-180\nu^3 - 60\nu^4\big] \times \varepsilon^2 + \frac{1}{15}(-32z^8 + \ldots)\varepsilon^4 + \ldots \Big\rangle = 0
\end{aligned}
\tag{2.4.10}
$$

Hence, we get the following sets of roots:

$$
z_q = \frac{1}{\sqrt[3]{\varepsilon}} z^*_{q0} + \sqrt[3]{\varepsilon} z^*_{q1} + \ldots (q = 1,2,3,4,5,6). \tag{2.4.11}
$$

On inserting the roots (2.4.11) into Eq. (2.4.10) we obtain

$$
\begin{aligned}
z^{*6}_{q0} - 3(1-\nu^2)(1+\nu)(\nu+2) = 0 \\
z^*_{q1} = \left(24 z^*_{q0}\right)^{-1} (12\nu^2 + 2\nu + 9)
\end{aligned}
\tag{2.4.12}
$$

As for the zeros defined by the formula (2.4.8), they remain valid in this case. Thus, in the case $\lambda_0^2 = 1$ we have six zeros (two of which are purely imaginary) increasing as $\varepsilon^{-1/3}$ as $\varepsilon \to 0$ and a countable set of zeros defined by the formula (2.4.9).

As in the case of a cylinder, we establish a connection of zeros defined by (2.4.11) with the zeros defined by the formulas (2.4.4) and (2.4.6). Studying the behaviour of the roots of Eq. (2.4.3) in a neighbourhood of $\lambda_0^2 = 1$ leads to the following result:

1. $\beta = 2\gamma,\ 0 < \gamma < \frac{1}{3} \lambda_0^2 - 1 = C_0 \varepsilon^{\alpha} (\alpha > 0)$

$$
\begin{aligned}
z_k &= z_{k0}\varepsilon^{-\frac{\beta}{2}} + z_{k1}\varepsilon^{\frac{\beta}{2}} + \ldots \quad 0 < \beta \le \frac{1}{2} \\
z_k &= z_{k0}\varepsilon^{-\frac{\beta}{2}} + z_{k1}\varepsilon^{2-\frac{7}{2}\beta} + \ldots, \quad \frac{1}{2} < \beta < \frac{2}{3} \\
z^2_{k0} &+ \frac{(1+\nu)(2+\nu)}{C_0} = 0, \\
z_{k1} &= -\frac{8\nu^2 + 12\nu - 5}{8 z_{k0}} \quad 0 < \beta < \frac{1}{2} \\
z_{k1} &= \left[24(1-\nu^2) C_0 z_{k0}\right]^{-1} \left[4 z^6_{k0} - 3(1-\nu^2) \right. \\
&\quad \left. \times (8\nu^2 + 12\nu - 5) C_0\right] \quad \beta = \frac{1}{2} \\
z_{k1} &= z^5_{k0}\left[6(1-\nu)^2 C_0\right]^{-1} \quad \frac{1}{2} < \beta < \frac{2}{3}
\end{aligned}
\tag{2.4.13}
$$

$$\begin{aligned} z_k &= z_{k0}\varepsilon^{-\frac{1}{3}} + z_{k1}\varepsilon^{\frac{1}{3}} + \ldots \beta = \frac{2}{3} \\ z_{k0}^6 &- 3(1-v^2)C_0 z_{k0}^2 - 3(1-v^2)(1+v)(2+v) = 0 \\ z_{k1} &= \left[z_{k0}^4 - (1-v^2)C_0\right](24 z_{k0})^{-1} \times \left[(12v^2+2v+9)\right. \\ &\quad \left. \times\, z_{k0}^4 + 3(1-v^2)(8v^2+12v-5)C_0\right] \end{aligned} \tag{2.4.14}$$

2. $\beta = 2 - 4\gamma \quad \frac{1}{2} < \beta < \frac{2}{3}$,

$$\begin{aligned} z_k &= z_{k0}\varepsilon^{\frac{\beta}{4}-\frac{1}{2}} + z_{k1}\varepsilon^{\frac{1}{2}-\frac{5\beta}{4}} + \ldots \\ z_{k0}^4 &- 3(1-v^2)C_0 = 0, \quad z_{k1} = -\frac{(1+v)(2+v)}{2 z_{k0} C_0} \end{aligned} \tag{2.4.15}$$

3. $\beta > 2\gamma \quad \gamma = \frac{1}{3}$

$$\begin{aligned} z_{k0}^6 &- 3(1-v^2)(1+v)(2+v) = 0 \\ z_{k1} &= \frac{(1-v^2)C_0}{2 z_{k0}^3} \end{aligned} \tag{2.4.16}$$

The comments about the cylinder given in the first chapter are also applicable to a sphere.

Until now, we have studied the asymptotic properties of the zeros Z of the dispersion equation under the assumption that the frequency parameter λ remains finite in the limit $\varepsilon \to 0$.

Consider now the case when λ increases infinitely as $\varepsilon \to 0$. We can show that all the zeros of the function given in (2.4.2) grow without bound when $\lambda \to \infty$ as $\varepsilon \to 0$. Here the following limiting cases are possible: (a) $\lambda\varepsilon \to 0$ as $\varepsilon \to 0$, (b) $\lambda\varepsilon \to const$ as $\varepsilon \to 0$, (c) $\lambda\varepsilon \to \infty$ as $\varepsilon \to 0$.

We first define that z_k in the limit $\lambda\varepsilon \to 0$ as $\varepsilon \to 0$. For this, we again use the decomposition (2.4.3). Assume that the leading asymptotic terms z_k, λ have the form:

$$\begin{aligned} z_k &= \chi_{k0}\varepsilon^{-\gamma}, \lambda_0 = \Lambda\varepsilon^{-\beta}, \chi_{k0} = O(1) \\ \Lambda &= O(1), 0 < \gamma < 1, 0 < \beta < 1 \end{aligned} \tag{2.4.17}$$

It is easy to prove that $\beta \le \gamma$.

We will individually consider the cases when $\gamma = \beta$ and $\beta < \gamma$.

In the first case, we seek z_k in the form

$$\begin{aligned} z_k &= z_{k0}\varepsilon^{-\gamma} + z_{k2}\varepsilon^{\gamma} + \ldots, \quad 0 < \gamma < \frac{1}{2} \\ z_k &= z_{k0}\varepsilon^{-\gamma} + z_{k2}\varepsilon^{2-3\gamma} + \ldots, \quad \frac{1}{2} \le \gamma < 1 \end{aligned} \tag{2.4.18}$$

Substituting (2.4.18) into (2.4.3) we obtain

$$
\begin{aligned}
z_{k0}^2 &= (1-\nu^2)\Lambda^2,\, z_{k2} = \frac{1-12\nu-4\nu^2}{8z_{k0}} \quad 0<\gamma<\frac{1}{2}\\
z_{k2} &= \frac{1-12\nu-4\nu^2}{8z_{k0}} + \frac{(1+\nu)^2(8\nu^2-31\nu+21)\Lambda^4}{12z_{k0}} \quad \gamma=\frac{1}{2}\\
z_{k2} &= \frac{(1+\nu)^2(8\nu^2-31\nu+21)\Lambda^4}{12z_{k0}} \quad \frac{1}{2}<\gamma<1
\end{aligned}
\tag{2.4.19}
$$

In the case $\beta<\gamma$ substituting (2.4.17) into (2.4.3) and keeping only the leading order terms for χ_{k0} we obtain the following limiting equation:

$$
\begin{aligned}
D = {}& \frac{64}{3\pi^2}\left(\frac{1+\nu}{1-\nu}\right)^2 \lambda_0^4\varepsilon^2\langle\left[12(1-\nu^2)\Lambda^2\chi_{k0}^2 + O(\varepsilon^{2\gamma-2\beta})\right]\\
& \times \varepsilon^{-2\gamma-2\beta} + \left\{-4\chi_{k0}^6 + O\left[\max\left(\varepsilon^{2\gamma-2\beta}, \varepsilon^{2-2\gamma}\right)\right]\right\}\varepsilon^{2-6\gamma} = 0
\end{aligned}
\tag{2.4.20}
$$

This implies that $\beta = 2\gamma - 1$. From the condition $\beta > 0$ we get that $\gamma > \frac{1}{2}$. Thus $\frac{1}{2}<\gamma<1$. From Eq. (2.4.20) we obtain

$$
\chi_{k0}^4 - 3(1-\nu^2)\Lambda^2 = 0 \tag{2.4.21}
$$

Writing λ_0 in the form $\lambda_0 = \Lambda\varepsilon^{1-2\gamma}$ we seek z_k in the form

$$
\begin{aligned}
z_k &= \varepsilon^{-\gamma}\left[z_{k0} + \varepsilon^{-2+4\gamma}z_{k1} + \ldots\right] \quad \frac{1}{2}<\gamma<\frac{2}{3}\\
z_k &= \varepsilon^{-\gamma}\left[z_{k0} + \varepsilon^{2-2\gamma}z_{k1} + \ldots\right] \quad \frac{2}{3}\le\gamma<1\\
z_{k0} &= \chi_{k0},\, z_{k1} = -\frac{z_{k0}}{4\Lambda^2}, \quad \frac{1}{2}<\gamma<\frac{2}{3}
\end{aligned}
\tag{2.4.22}
$$

$$
\begin{aligned}
z_{k1} &= \frac{1}{40z_{k0}}(1+\nu)(16\nu-11)\Lambda^2 - \frac{z_{k0}}{4\Lambda^2} \quad \gamma=\frac{2}{3}\\
z_{k1} &= \frac{1}{40z_{k0}}(1+\nu)(16\nu-11)\Lambda^2, \quad \frac{2}{3}<\gamma<1.
\end{aligned}
\tag{2.4.23}
$$

So, in this case, we have four zeros increasing as $\varepsilon^{-\gamma}$, two of which are real but two are purely imaginary. Real zeros correspond to the so-called irregular degeneration. It should be noted that, unlike the cylinder, ultra-low frequency vibrations are not available for the sphere.

In the cases $\lambda_0 = \Lambda\varepsilon^{-\gamma}$ and $\lambda_0 = \Lambda\varepsilon^{1-2\gamma}$ Eq. (2.4.9) remains valid.

To construct the asymptotics of the zeros of the second group of zeros ($z \sim \lambda, \lambda\varepsilon \to const$, $z\varepsilon \to const$ as $\varepsilon \to 0$) writing $\lambda = \frac{s}{\varepsilon}$ we seek z_n in the following form:

$$z_n = \frac{\delta_n}{\varepsilon} + O(\varepsilon). \tag{2.4.24}$$

After substituting (2.4.24) in the dispersion Eq. (2.3.7) and making use of the asymptotic expansions of functions $J_z(x)$, $Y_z(x)$ at large z and x for δ_n we get the following equation

$$\begin{aligned}&\left[\left(\delta_n^2 + \beta_n^2\right)^2 sh\alpha ch\beta_n - 4\alpha_n\beta_n\delta_n^2 sh\alpha_n ch\beta_n\right]\\ &\quad \times \left[\left(\delta_n^2 + \beta_n^2\right)^2 ch\alpha_n sh\beta_n - 4\alpha_n\beta_n\delta_n^2 sh\alpha_n ch\beta_n\right] = 0\end{aligned} \tag{2.4.25}$$

where

$$\alpha_n^2 = \delta_n^2 - \frac{1-2\nu}{2(1-\nu)}s^2, \beta_n^2 = \delta_n^2 - s^2.$$

For given λ, Eq. (2.4.25) defines a countable set of z_n. It should be noted that Eq. (2.4.25) coincides with the Rayleigh-Lamb dispersion equation for elastic layer [5] and is rather well studied in the literature.

Finally, in case (c), as in the first chapter, we denote $\mu_k\varepsilon$ by χ_k, and $\lambda\varepsilon$ by Y. Again, using the asymptotic expansions of the Bessel functions in the first term of the asymptotics, Eq. (2.3.7) can be written in the form

$$\begin{aligned}&\left[\left(x_k^2 + y^2\right)^2 sh\alpha_k ch\beta_k - 4\alpha_k\beta_k x_k^2 ch\alpha_k sh\beta_k\right]\\ &\quad \times \left[\left(x_k^2 + y^2\right)^2 ch\alpha_k sh\beta_k - 4\alpha_k\beta_k x_k^2 sh\alpha_k ch\beta_k\right] = 0\end{aligned} \tag{2.4.26}$$

where

$$\alpha_k^2 = x_k^2 - \frac{1-2\nu}{2(1-\nu)}y^2, \beta_k^2 = x_k^2 - y^2.$$

Thus, in the case of $\gamma > 1$ Eq. (2.4.25) is still valid.

2.5 Asymptotic Analysis of Homogeneous Solutions for a Spherical Shell

We introduce here an asymptotic analysis of homogeneous solutions for a spherical shell corresponding to different groups of roots of the dispersion equation.

The study will begin with the construction of systems of homogeneous solutions corresponding to the case $\lambda = O(1)$. As it has been shown before, in this case the dispersion equation has three groups of roots with different asymptotic properties. Therefore, the displacement vector and the stress tensor are divided into three parts respectively, each of which is determined by the type of the asymptotic expansion by ε. Supposing $\rho = 1 + \varepsilon\eta$, $-1 \leq \eta \leq 1$ and expanding them in the parameter ε, we obtain the following leading asymptotic terms (given only in terms of the amplitude values and the summation sign is omitted) from (2.3.9):

$$
\begin{aligned}
U_r^{(1)} &= R_0 C_k \left[4z_{k0}^2 + 12\nu - 13 - 2\left(1-\nu^2\right)\lambda_0^2 + O(\varepsilon)\right] m_k(\theta) \\
U_\theta^{(1)} &= R_0 C_k [4(3\nu - 1) + O(\varepsilon)] \frac{dm_k}{d\theta} \\
\sigma_\theta^{(1)} &= 2GC_k \left\{ 4(1-3\nu) ctg\theta \frac{dm_k}{d\theta} \right. \\
&\quad \left. + \left[8z_{k0}^2 - 12(1+\nu) - 2(1+\nu)^2 \lambda_0^2\right] m_k(\theta) + O(\varepsilon) \right\} (k = 1, 2) \\
\sigma_\varphi^{(1)} &= 2GC_k \left\{ 4(3\nu - 1) ctg\theta \frac{dm_k}{d\theta} \right. \\
&\quad \left. + \left[4(3\nu+1)z_{k0}^2 - (15\nu + 13) - 2(1+\nu)^2 \lambda_0^2\right] m_k(\theta) + O(\varepsilon) \right\} \\
\sigma_r &= O\left(\varepsilon^2\right), \tau_{r\theta} = O\left(\varepsilon^3\right).
\end{aligned}
$$

Thus, in this case, certain penetrating solutions correspond to the first group of zeros.

$$
\begin{aligned}
U_r^{(2)} &= -R_0 C_k \left[z_{k0}^2 + O(\varepsilon)\right] m_k(\theta) \\
U_\theta^{(2)} &= \varepsilon R_0 C_k \left[z_{k0}^2 \eta + O(\varepsilon)\right] \frac{dm_k}{d\theta} \\
\sigma_\theta^{(2)} &= -2GC_k \left[3(1+\nu)\left(\lambda_0^2 - 1\right)\eta m_k(\theta) + O(\varepsilon)\right] \\
\sigma_\varphi^{(2)} &= -2GC_k \left[3\nu(1+\nu)\left(\lambda_0^2 - 1\right)\eta m_k(\theta) + O(\varepsilon)\right] \\
\sigma_r^{(2)} &= O(\varepsilon), \tau_{r\theta} = O(\varepsilon) \\
&(k = 3, 4, 5, 6)
\end{aligned}
\tag{2.5.2}
$$

As noted earlier, in the case of $\lambda_0^2 < 1$ we get four damped solutions similar to the simple boundary effect in the static of shells. In case of $\lambda_0^2 > 1$ we get two damped solutions and two penetrating ones. Solutions corresponding to the third group of zeros have the character of the boundary layer and are localized at the side surfaces of the shell. At a distance from the conic sections $\theta = \theta_{j\,(j=1,2)}$, these solutions decay exponentially and in the first term the asymptotics coincides with a similar solution in static of shells. The following asymptotic formulas are obtained for these solutions:

$$
\begin{aligned}
U_r^{(3)} &= R_0\delta_k C_k \varepsilon\left[\left(ch\delta_k + \frac{2-2v}{\delta_k}sh\delta_k\right)sh\delta_k\eta\right.\\
&\quad \left.-\eta sh\delta_k ch\delta_k\eta + O(\varepsilon)\right]m_k(\theta)\\
U_\theta^{(3)} &= R_0 C_k \varepsilon^2\left[\left(ch\delta_k + \frac{2v-1}{\delta_k}\right)ch\delta_k\eta\right.\\
&\quad \left.-\eta sh\delta_k sh\delta_k\eta + O(\varepsilon)\right]\frac{dm_k}{d\theta}\\
\sigma_\theta^{(3)} &= 2GC_k[2vsh\delta_k ch\delta_k\eta m_k(\eta) + O(\varepsilon)]\\
\sigma_\varphi^{(3)} &= 2GC_k[2vsh\delta_k ch\delta_k\eta m_k(\theta) + O(\varepsilon)]\\
\sigma_r^{(3)} &= 2GC_k\delta_k[(\delta_k ch\delta_k + sh\delta_k)ch\delta_k\eta - \eta\delta_k sh\delta_k\eta + O(\varepsilon)]m_k(\theta)\\
\tau_{r\theta}^{(3)} &= 2GC_k\delta_k\varepsilon[ch\delta_k sh\delta_k\eta - \eta sh\delta_k ch\delta_k\eta + O(\varepsilon)]\frac{dm_k}{d\theta}\\
&sh\delta_k + 2\delta_k = 0 \quad (k = 7, 9, 11, \ldots)
\end{aligned}
\tag{2.5.3}
$$

$$
\begin{aligned}
U_r^{(3)} &= R_0 C_k \varepsilon\delta\left[\left(sh\delta_k + \frac{2-2v}{\delta_k}ch\delta_k\right)ch\delta_k\eta\right.\\
&\quad \left.-\eta\delta_k ch\delta_k sh\delta_k\eta + O(\varepsilon)\right]m_k(\theta)\\
U_\theta^{(3)} &= R_0 C_k \varepsilon^2\left[\left(sh\delta_k + \frac{2v-1}{\delta_k}ch\delta_k\right)sh\delta_k\eta\right.\\
&\quad \left.-\eta ch\delta_k ch\delta_k\eta + O(\varepsilon)\right]\frac{dm_k}{d\theta}\\
\sigma_\theta^{(3)} &= 2G\delta_k C_k[2vch\delta_k sh\delta_k\eta + O(\varepsilon)]m_k(\theta)\\
\sigma_\varphi^{(3)} &= 2G\delta_k C_k[2vch\delta_k sh\delta_k\eta m_k(\theta) + O(\varepsilon)]\\
\sigma_r^{(3)} &= 2G\delta_k C_k[(\delta_k sh\delta_k + ch\delta_k)sh\delta_k\eta\\
&\quad - \eta\delta_k ch\delta_k ch\delta_k\eta + O(\varepsilon)]m_k(\theta)\\
\tau_{r\theta}^{(3)} &= 2GC_k\delta_k\varepsilon[sh\delta_k ch\delta_k\eta - \eta ch\delta_k sh\delta_k\eta + O(\varepsilon)]\frac{dm_k}{d\theta}\\
&sh2\delta_k - 2\delta_k = 0 \quad (k = 8, 10, 12, \ldots).
\end{aligned}
\tag{2.5.4}
$$

We now give expressions for the displacements and stresses in the case of microwave vibrations of the sphere. Here we will individually consider the cases when (a) $\lambda\varepsilon \to 0$ as $\varepsilon \to 0$, (b) $\lambda\varepsilon \to const$ as $\varepsilon \to 0$.

In case (a) we have

$$
\begin{aligned}
U_r &= -R_0 C_k\left[\left(\frac{1}{4} - 3\nu - \nu^2\right)\varepsilon^{2\gamma} + O\left(\varepsilon^{2-2\gamma}\right)\right] m_k(\theta) \\
U_\theta &= R_0 C_k \varepsilon\left[\nu(1-\nu)\Lambda^2 + O\left(\varepsilon^{2\gamma}\right)\right]\frac{dm_k}{d\theta} \\
\sigma_\theta &= -2GC_k\varepsilon^{1-2\gamma}\left[\nu^2(1-\nu)^2\Lambda^4 m_k(\theta)\eta + O\left(\varepsilon^{2\gamma}\right)\right] \\
\sigma_\varphi &= 2GC_k\varepsilon^{1-2\gamma}\left\{\left[\nu(1+\nu)\left(1-\nu^2\right)\Lambda^4\eta + \nu(1+\nu)^2\Lambda^4\right]\right. \\
&\quad \left.\times\, m_k(\theta) + O\left(\varepsilon^{2\gamma}\right)\right\}, \quad 0<\gamma<\frac{1}{2} \\
U_r &= O\left(\varepsilon^{2-2\gamma}\right), U_\theta = O(\varepsilon), \sigma_\theta = O\left(\varepsilon^{1-2\gamma}\right), \\
\sigma_\varphi &= O\left(\varepsilon^{1-2\gamma}\right), \frac{1}{2} \le \gamma < 1 \quad (k = 1, 2) \\
\sigma_r &= O\left(\varepsilon^{2-3\gamma}\right), \tau_{r\theta} = O\left(\varepsilon^{3-4\gamma}\right), \\
\lambda_0 &= \Lambda\varepsilon^{-\gamma}, z = z_{k0}\varepsilon^{-\gamma}
\end{aligned}
\tag{2.5.5}
$$

$$
\begin{aligned}
U_r &= -R_0 C_k\left[z_{k0}^2 + O\left(\varepsilon^{4\gamma-2}\right)\right] m_k(\theta) \\
U_\theta &= R_0 C_k \varepsilon\left[z_{k0}^2\eta + O\left(\varepsilon^{4\gamma-2}\right)\right]\frac{dm_k}{d\theta} \quad \frac{1}{2}<\gamma<\frac{2}{3} \\
\sigma_\theta &= -2GC_k\varepsilon^{1-2\gamma}\left[3(1+\nu)\eta m_k(\theta) + O\left(\varepsilon^{4\gamma-2}\right)\right] \\
\sigma_\varphi &= 2GC_k\varepsilon^{1-2\gamma}\left[3\nu(1+\nu)\eta m_k(\theta) + O\left(\varepsilon^{4\gamma-2}\right)\right]
\end{aligned}
\tag{2.5.6.}
$$

$$
\begin{aligned}
U_r &= -R_0 C_k\left[z_{k0}^2 + O\left(\varepsilon^{2-2\gamma}\right)\right] m_k(\theta) \\
U_\theta &= R_0 C_k \varepsilon\left[z_{k0}^2\eta + O\left(\varepsilon^{2-2\gamma}\right)\right]\frac{dm_k}{d\theta} \\
\sigma_\theta &= -2GC_k\varepsilon^{1-2\gamma}\left[3(1+\nu)\eta m_k(\theta) + O\left(\varepsilon^{2-2\gamma}\right)\right] \\
\sigma_\varphi &= 2GC_k\varepsilon^{1-2\gamma}\left[3\nu(1+\nu)\eta m_k(\theta) + O\left(\varepsilon^{2-2\gamma}\right)\right] \\
\sigma_r &= O\left(\varepsilon^{4-4\gamma}\right), \tau_{r\theta} = O\left(\varepsilon^{3-4\gamma}\right), \quad \frac{1}{2}<\gamma<1 \\
\lambda &= O\left(\varepsilon^{1-2\gamma}\right), z = O\left(\varepsilon^{-\gamma}\right) \quad (k = 3, 4, 5, 6).
\end{aligned}
$$

Let us now present the displacements and stresses in case (b). Using the first term of the asymptotic expansion of Bessel functions for displacements and stresses in the first approximation, we find two classes of solutions, the first of which corresponds to zeros of the function

$$
\left(\delta_n^2 + \beta_n^2\right)^2 sh\alpha_n ch\beta_n - 4\alpha_n\beta_n\delta_n^2 ch\alpha_n sh\beta_n
$$

and the second to the zeros of the function

$$\left(\delta_n^2+\beta_n^2\right)^2 ch\alpha_n sh\beta_n-4\alpha_n\beta_n\delta_n^2 ch\beta_n sh\alpha_n$$

As in the case of a cylinder they are identical in structure and may be represented by the following expressions:

$$\begin{aligned}
U_r &= R_0\alpha_n C_n\varepsilon\Big[\left(\delta_n^2+\beta_n^2\right)^2 ch\beta_n ch\alpha_n\eta \\
&\quad -2\delta_n^2 ch\alpha_n ch\beta_n\eta+O(\varepsilon)\Big]m_n(\theta)\\
U_\theta &= R_0\alpha_n C_n\varepsilon^2\Big[\left(\delta_n^2+\beta_n^2\right)^2 ch\beta_n sh\alpha_n\eta \\
&\quad -2\alpha_n\beta_n ch\alpha_n sh\beta_n\eta+O(\varepsilon)\Big]\frac{dm_n}{d\theta}\\
\sigma_\theta &= G\frac{\nu}{1-\nu}\lambda_0^2 C_n\left[\left(\delta_n^2+\beta_n^2\right)ch\beta_n sh\alpha_n\eta m_n(\theta)+O(\varepsilon)\right]\\
\sigma_\varphi &= G\frac{\nu}{1-\nu}\lambda_0^2 C_n\left[\left(\delta_n^2+\beta_n^2\right)ch\beta_n sh\alpha_n\eta m_n(\theta)+O(\varepsilon)\right]\\
\sigma_r &= GC_n\Big[\left(\delta_n^2+\beta_n^2\right)^2 ch\beta_n sh\alpha_n\eta \\
&\quad -4\alpha_n\beta_n\delta_n^2 ch\alpha_n sh\beta_n\eta+O(\varepsilon)\Big]m_n(\theta)\\
\tau_{r\theta} &= GC_n\alpha_n\varepsilon\left(\delta_n^2+\beta_n^2\right)\big[ch\beta_n ch\alpha_n\eta \\
&\quad -ch\alpha_n sh\beta_n\eta+O(\varepsilon)\big]\frac{dm_n}{d\theta}\qquad (n=1,3,5,\ldots).
\end{aligned}\tag{2.5.7}$$

Here C_n are arbitrary constants.

Expressions for $n=2,4,6,\ldots$ are obtained from (2.5.7) by replacing chx with shx and shx with chx respectively.

Finally, we give the first term of the asymptotic expansions for the displacements and stresses in the case $\lambda_0^2=1$.

$$\begin{aligned}
U_r &= -R_0 B_k\left[z_{k0}^2+O\left(\varepsilon^{2/3}\right)\right]m_k(\theta)\\
U_\theta &= R_0 B_k\varepsilon\left[z_{k0}^2\eta+O\left(\varepsilon^{2/3}\right)\right]\frac{dm_k}{d\theta}\\
\sigma_\theta &= 2GB_k\varepsilon^{-2/3}\left[\frac{z_{k0}^4}{1-\nu}\eta m_k(\theta)+O\left(\varepsilon^{2/3}\right)\right]\\
\sigma_\varphi &= 2GB_k\varepsilon^{-2/3}\left[\frac{\nu}{1-\nu}z_{k0}^4\eta m_k(\theta)+O\left(\varepsilon^{2/3}\right)\right]\\
\sigma_r &= O(\varepsilon)\qquad \tau_{r\theta}=O(\varepsilon).
\end{aligned}\tag{2.5.8}$$

Here B_k are arbitrary constants.

As for the solutions defined by (2.5.3), (2.5.4) they should also be valid in this case.

Here we should pay attention to the following fact. The solution of Eq. (2.3.4), in general, can be written in terms of Legendre functions. However, as shown in [3], it is more convenient to use approximate methods. Here we should consider two cases separately:

(1) the shell does not contain any of the poles $0, \pi$,
(2) the shell comprises at least one of these poles.

In the first case for the approximate integration it is convenient to use asymptotic methods, as is thoroughly explained in [3].

Regarding the second case, here the asymptotic method of integration cannot give an approximate solution at any of the relative thickness ε of the shell. The fact is that the asymptotic approximations lose accuracy in the neighbourhood of the vertex $\theta = 0$. In this case it is necessary to select only those solutions of Eq. (2.3.4) which remain bounded for $\theta = 0$. These solutions are constructed in [6] where approximate methods are presented to calculate such solutions. Therefore, we will not consider them any further.

We point out that the homogeneous solutions satisfy the so-called generalized orthogonality conditions. According to (2.3.3), the homogeneous solutions have the form:

$$
\begin{aligned}
U_r &= U_k(r)m_k(\theta)e^{i\omega t} \\
U_\theta &= W_k(r)\frac{dm_k}{d\theta}e^{i\omega t} \\
\sigma_\theta &= 2G\left[\sigma_{1k}(r)m_k(\theta) - \sigma_{2k}(r)ctg\theta\frac{dm_k}{d\theta}\right]e^{i\omega t} \\
\tau_{r\theta} &= GT_k(r)\frac{dm_k}{d\theta}e^{i\omega t}
\end{aligned}
\tag{2.5.9}
$$

where

$$
\begin{aligned}
\sigma_{1k} &= \frac{1}{1-2\nu}\left[\nu U_k' + \frac{1-\nu}{r}U_k - \frac{1-\nu}{r}\left(z_k^2 - \frac{1}{4}\right)W_k\right] \\
\sigma_{2k} &= \frac{1}{r}W_k, \quad T_k = \frac{1}{r}U_k + W_k' - \frac{1}{r}W_k.
\end{aligned}
$$

Suppose, U_θ^i, U_r^i, σ_θ^i, $\tau_{r\theta}^i (i = 1, 2)$ are displacements and stresses of the first and second states. Then the following equality holds true for any θ by Batty's theorem

$$
\begin{aligned}
&\int_{R_1}^{R_2} \left(U_\theta^1 \delta_\theta^2 + U_r^1 \tau_{r\theta}^2\right) r \sin\theta dr \\
&= \int_{R_1}^{R_2} \left(U_\theta^2 \sigma_\theta^1 + U_r^2 \tau_{r\theta}^1\right) r \sin\theta dr.
\end{aligned}
\tag{2.5.10}
$$

Substituting (2.5.9) into (2.5.10), we obtain

$$\begin{aligned} & \cos\theta \frac{dm_k}{d\theta}\frac{dm_n}{d\theta}\int\limits_{R_1}^{R_2}(\sigma_{2k}W_n-\sigma_{2n}W_k)rdr \\ & + \sin\theta m_n\frac{dm_k}{d\theta}\int\limits_{R_1}^{R_2}(W_k\sigma_{1n}-U_nT_k)rdr \\ & + \sin\theta m_k\frac{dm_n}{d\theta}\int\limits_{R_1}^{R_2}(U_kT_n-W_n\sigma_{1k})rdr=0 \end{aligned} \tag{2.5.11}$$

Since equality (2.5.11) is valid for all θ it holds true only under the conditions

$$\int\limits_{R_1}^{R_2}(\sigma_{2k}W_n-\sigma_{2n}W_k)rdr=0 \tag{2.5.12}$$

$$\int\limits_{R_1}^{R_2}(W_k\sigma_{1n}-U_nT_k)rdr=0 \tag{2.5.13}$$

$$\int\limits_{R_1}^{R_2}(U_kT_n-W_n\sigma_{1k})rdr=0. \tag{2.5.14}$$

The relation (2.5.12) is satisfied identically, the conditions (2.5.13) and (2.5.14) are equivalent. Thus, we arrive at the equality

$$\int\limits_{R_1}^{R_2}(U_kT_n-W_n\sigma_{1k})rdr=0\ k\neq n. \tag{2.5.15}$$

From (2.5.15) we obtain the following orthogonality condition for functions $U_n(r)$, $W_n(r)$:

$$\begin{aligned} & \int\limits_{R_1}^{R_2}\left\{\left(\frac{1}{r}U_n+W_n'-\frac{1}{r}W_n\right)U_k-\frac{2}{1-2\nu}W_n\right. \\ & \left.\times\left[\nu U_k'+\frac{1-\nu}{r}U_k-\frac{1-\nu}{r}\left(z_k^2-\frac{1}{4}\right)W\right]\right\}rdr=0\ k\neq n. \end{aligned} \tag{2.5.16}$$

However, while satisfying the boundary conditions at the side surfaces of the shell, here the generalized conditions of orthogonality of homogeneous solutions, as in the case of a cylinder, does not permit a complete solution to the question of an accurate fulfilment of the boundary conditions on the side surfaces of the sphere. Apparently, in the general case, it is not possible to suggest anything here but the reduction to an infinite system of linear equations. However, under special conditions of the shell edge bearing the generalized orthogonality condition for homogeneous solutions allow us to represent the solution in the form of a series whose coefficients are determined exactly. Moreover, the condition (2.5.16) can be useful in solving the infinite systems of equations since it always allows to satisfy one of the boundary conditions on the side surfaces of a spherical layer accurately.

With the help of generalized orthogonality conditions, we can consider the following problem: let the front surfaces $r = R_{s\,(s=1,2)}$ be stress-free and the following mixed boundary conditions be given on the conical surfaces $\theta = \theta_{j\,(j=1,2)}$:

$$\begin{gathered} U_\theta(r,\theta,t) = a(r)e^{i\omega t}, \quad \tau_{r\theta} = \tau(\theta)e^{i\omega t}, \quad \theta = \theta_1 \\ U_\theta(r,\theta,t) = 0, \quad \tau_{r\theta} = 0, \quad \theta = \theta_2\,. \end{gathered} \tag{2.5.17}$$

The general solution of the Legendre Eq. (2.3.4) has the form

$$m\,(\theta) = A_n P_{z_n-1/2}(\cos\theta) + B_n Q_{z_n-1/2}(\cos\theta).$$

Then, to satisfy the boundary conditions (2.5.17) by means of the generalized orthogonality conditions for A_n and B_n we obtain the algebraic equations

$$\begin{gathered} A_n P^{(1)}_{z_n-1/2}(\cos\theta_1) + B_n Q^{(1)}_{z_n-1/2}(\cos\theta_1) = H_n \\ A_n P^{(1)}_{z_n-1/2}(\cos\theta_2) + B_n Q^{(1)}_{z_n-1/2}(\cos\theta_2) = 0 \end{gathered} \tag{2.5.18}$$

where

$$H_n = \Delta_H^{-1}\int_{r_1}^{r_2}\left\{\frac{2a(r)}{1-2\nu}\left[\nu U_n' + \frac{1-\nu}{r}U_n - \frac{1-\nu}{r}\left(z_n^2 - \frac{1}{4}\right)W_n\right] - \tau(r)U_n(r)\right\}rdr$$

$$\Delta_n = \int_{r_1}^{r_2}\left\{\frac{2W_n}{1-2\nu}\left[\nu U_n' + \frac{1-\nu}{r}U_n - \frac{1-\nu}{r}\left(z^2 - \frac{1}{4}\right)W_n\right]\right.$$
$$\left. - U_n\left(\frac{1}{r}U_n + W_n' - \frac{1}{r}W_n\right)\right\}rdr$$

The constants A_n and B_n are determined from the linear systems (2.5.18) for any n.

$$A_n = H_n \Delta^{-1} Q^{(1)}_{z_{n-1/2}}(\cos\theta_2), \quad B_n = H_n \Delta^{-1} P^{(1)}_{z_{n-1/2}}(\cos\theta_2)$$
$$\Delta = P^{(1)}_{z_{n-1/2}}(\cos\theta_1) - Q^{(1)}_{z_{n-1/2}}(\cos\theta_2) - P^{(1)}_{z_{n-1/2}}(\cos\theta_2) - Q^{(1)}_{z_{n-1/2}}(\cos\theta_1).$$

Note that this problem for an inhomogeneous spherical zone in the static case is considered in [2].

As noted above, usually generalized orthogonality conditions do not allow the satisfaction of the boundary conditions accurately on the side surfaces of a spherical zone. Here, to satisfy the boundary conditions on the side surfaces it is convenient to use Hamilton's variational principle. According to this principle, we can write

$$\delta H = \delta \int_{t_0}^{t} dt \int_{v} (T - U) dv + \int_{t_0}^{t} dt \oint_{s} \int_{r_1}^{r_2} (T_p \delta U_p) dr ds = 0. \qquad (2.5.19)$$

Here T is the density of kinetic energy and U is the density of the strain energy given, respectively, by

$$\begin{aligned} T &= \frac{1}{2} g \frac{\partial U_i}{\partial t} \frac{\partial U_i}{\partial t} \\ U &= \frac{1}{2} C_{ijkl} \varepsilon_{ij} \varepsilon_{kl} \end{aligned} \qquad (2.5.20)$$

Let $\theta = \theta_j$ be given at the following boundary conditions

$$\sigma_\theta = Q_j(r) e^{iwt}, \quad \tau_{r\theta} = \tau_j(r) e^{iwt}. \qquad (2.5.21)$$

Since homogeneous solutions satisfy the equations of motion (2.1.1) and the homogeneous boundary conditions (2.3.1), taking into account (2.5.21) and the variational principle stated above we obtain

$$\sum_{j=1}^{2} \int_{r_1}^{r_2} \left[(\sigma_\theta - Q_j) \delta U_\theta + (\tau_{r\theta} - \tau_j) \delta U_r \right]_{\theta=\theta_j} r dr = 0. \qquad (2.5.22)$$

The constants C_k may be determined by substituting (2.5.9) into (2.5.22) and considering the variations of C_k to be independent. On doing so, we obtain the infinite system of equations

$$\sum_{k=1}^{\infty} H^j_{nk} C_k = N^j_n \quad (n = 1, 2, \ldots) \qquad (2.5.23)$$

where

$$H_{nk}^{j} = \left[2m_k(\theta)\frac{dm_n}{d\theta}\int\limits_{r_1}^{r_2}\sigma_{1k}(r)W_n(r)rdr - 2ctg\theta\frac{dm_k}{d\theta}\frac{dm_n}{d\theta}\int\limits_{r_1}^{r_2}\sigma_{2k}(r)W_n(r)rdr + m_n(\theta)\frac{dm_k}{d\theta}\int\limits_{r_1}^{r_2}T_k(r)U_n(r)rd(r)\right]\theta=\theta_j \tag{2.5.24}$$

$$N_n^j = \left[\frac{dm_n}{d\theta}\int\limits_{r_1}^{r_2}Q_j(r)W_n(r)rdr + m_n(\theta)\int\limits_{r_1}^{r_2}T_j(r)U_n(r)rdr\right]\theta=\theta_j.$$

Using the smallness of the parameter of the shell's thin-walledness $\varepsilon = h/R_0$, we can construct an asymptotic solution of (2.5.23). To do this, we must substitute the asymptotic expressions for σ_θ, $\tau_{r\theta}$, U_r, U_θ corresponding to different groups of zeros of the dispersion equation in formula (2.5.24). In conclusion, we note that in the cases of $\lambda = O(1)$, $\lambda = O(\varepsilon^{-j})$, $\lambda = O(\varepsilon^{1-2j})$ $(0<j<1)$ the system (2.5.23) actually coincides with the infinite systems obtained in [3].

In the case $\lambda = O(\varepsilon^{-\gamma})$, $Z = O(\varepsilon^{-\gamma})$ $(\gamma \geq 1)$ we obtain an infinite system of algebraic equations generated by the dynamic problem of elasticity theory.

In conclusion, we note that all the arguments concerning the system (1.4.18) hold true for the system (2.5.23).

2.6 Dynamical Torsion of a Spherical Layer

Consider the torsional vibrations of a spherical zone when its front surfaces $r = r_s(s=1,2)$ are free and the following boundary conditions on the conical sections $\theta = \theta_j(\theta_1 < \theta_2)$ (θ_j are constant) are given:

$$\tau_{\theta\varphi} = \tau_j(r)e^{i\omega t} \quad \text{at} \quad \theta = \theta_j. \tag{2.6.1}$$

The solution of Eq. (2.1.2) will be sought in the following form

$$U_\varphi = U(\rho)\frac{dm}{d\theta}e^{i\omega t}. \tag{2.6.2}$$

Substituting (2.6.2) into (2.1.2) and taking into account (2.1.4), applying a separation of variables with respect to $m(\theta)$ and $U(\rho)$ we obtain the following boundary value problem:

$$\frac{d^2m}{d\theta^2} + ctg\theta\frac{dm}{d\theta} + \left(z^2 - \frac{1}{4}\right)m(\theta) = 0 \tag{2.6.3}$$

$$\left[2ctg\theta\frac{dm}{d\theta} + \left(z^2 - \frac{1}{4}\right)m(\theta)\right]_{\theta=\theta_k} = -\tau_k \tag{2.6.4}$$

$$U'' + \frac{2}{\rho}U' + \left(\lambda^2 - \frac{z^2 - \frac{1}{4}}{\rho^2}\right)U = 0 \tag{2.6.5}$$

$$\left(U' - \frac{1}{\rho}U\right)_{\rho=\rho_s} = 0. \tag{2.6.6}$$

In the meantime, let us take up the study of the spectral problem (2.6.5), (2.6.6). Performing the replacement $U(\rho) = h(\rho)/\rho$, the boundary value problem (2.6.5), (2.6.6) can be written in the form:

$$\begin{aligned} &\rho^2h''(\rho) + \lambda^2\rho^2h(\rho) = \gamma h(\rho) \\ &\left[h'(\rho) - \frac{2}{\rho}h(\rho)\right]_{\rho=\rho_s} = 0. \end{aligned} \tag{2.6.7}$$

Introducing the operator L as

$$Lh = \begin{cases} \rho^2h'' + \lambda^2\rho^2h \\ \left(h' - \frac{2}{\rho}h\right)_{\rho=\rho_s} = 0 \end{cases}$$

we may re-write the boundary value problem in the terms of the operator L as follows:

$$Lh = \gamma h \quad \gamma = z^2 - \frac{1}{4}.$$

We prove that the system of eigenfunctions of operator L forms a basis in the space $L_2(\rho_1, \rho_2)$. It is easy to show that the operator L is self-adjoint in the space $L_2\left[\rho_1, \rho_2 : \frac{1}{\rho^2}\right]$. Let us construct the Green's function $G(\rho, x)$ of L.

The general solution of the differential equation

$$h''(\rho) + \lambda^2h(\rho) = 0$$

has the form

$$h(\rho) = C_1 \cos \lambda\rho + C_2 \sin \lambda\rho. \tag{2.6.8}$$

In order to satisfy the boundary conditions on the surfaces we introduce two auxiliary functions. The first of these functions is defined as

$$\begin{aligned} h_1(\rho) = {} & \left(\lambda \cos \lambda\rho_1 - \frac{2 \sin \lambda\rho_1}{\rho_1}\right) \cos \lambda\rho \\ & + \left(\lambda \sin \lambda\rho_1 + \frac{2 \cos \lambda\rho_1}{\rho_1}\right) \sin \lambda\rho \end{aligned}$$

and satisfies the first boundary condition and the second function, which satisfies the second boundary condition, is given by

$$\begin{aligned} h_2(\rho) = {} & \left(\lambda \cos \lambda\rho_2 - \frac{2 \sin \lambda\rho_2}{\rho_2}\right) \cos \lambda\rho \\ & + \left(\lambda \sin \lambda\rho_2 - \frac{2 \cos \lambda\rho_2}{\rho_2}\right) \sin \lambda\rho. \end{aligned}$$

Therefore, the Green's function $G(\rho, x)$, for some $C_\lambda \neq 0$, has the form:

$$\begin{aligned} G(\rho, x) = {} & C_\lambda \left\{ \left[\left(\lambda \cos \lambda\rho_1 - \frac{2 \sin \lambda\rho_1}{\rho_1}\right) \cos \lambda\rho \right.\right. \\ & \left. + \left(\lambda \sin \lambda\rho_1 + \frac{2 \cos \lambda\rho_1}{\rho_1}\right) \sin \lambda\rho \right] \left[\left(\lambda \cos \lambda\rho_2 - \frac{2 \sin \lambda\rho_2}{\rho_2}\right) \right. \\ & \left.\left. \times \cos \lambda x + \left(\lambda \sin \lambda\rho_2 + \frac{2 \cos \lambda\rho_2}{\rho_2}\right) \sin \lambda x \right] \right\} \quad \rho_1 \leq \rho < x \\ G(\rho, x) = {} & C_\lambda \left\{ \left[\left(\lambda \cos \lambda\rho_1 - \frac{2 \sin \lambda\rho_1}{\rho_1}\right) \cos \lambda x \right.\right. \\ & \left. + \left(\lambda \sin \lambda\rho_1 - \frac{2 \cos \lambda\rho_1}{\rho_1}\right) \sin \lambda x \right] \left[\left(\lambda \cos \lambda\rho_2 - \frac{2 \sin \lambda\rho_2}{\rho_2}\right) \right. \\ & \left.\left. \times \cos \lambda\rho + \left(\lambda \sin \lambda\rho_2 - \frac{2 \cos \lambda\rho_2}{\rho_2}\right) \sin \lambda\rho \right] \right\} \quad x \leq \rho \leq \rho_2. \end{aligned}$$

Note that the Green's function $G(\rho, x)$ is symmetric, i.e. $G(\rho, x) = G(x, \rho)$. Suppose that λ is chosen such that zero is not an eigenvalue of the problem (2.6.7) equivalent to the eigenvalue problem for the homogeneous integral equation

$$h(\rho) = \gamma \int_{\rho_1}^{\rho_2} G(\rho, x) \frac{1}{x^2} h(x) dx, \quad h \in C[\rho_1, \rho_2]. \tag{2.6.9}$$

Turning to the unknown function $U(\rho) = h(\rho)/\rho$, we reduce the integral Eq. (2.6.9) to the equivalent form

$$U(\rho) = \gamma \int_{\rho_1}^{\rho_2} \frac{1}{x\rho} G(\rho, x) U(x) dx, . \tag{2.6.10}$$

The kernel $\frac{1}{\rho x} G(\rho, x) \neq 0$ is real, continuous and symmetric. Therefore, Hilbert-Schmidt theorem is applicable to the integral Eq. (2.6.10)—which asserts that the eigenfunctions $U_k(x)$ corresponding to the eigenvalue γ_k form an orthogonal system in $L_2(\rho_1, \rho_2)$ and this system is complete in $L_2(\rho_1, \rho_2)$ forming a basis in the same space.

Now consider the boundary value problem (2.6.3) and (2.6.4). The general solution of Eq. (2.6.3) can be written as

$$m_n(\theta) = C_{1n} P_{z_{n-1/2}}(\cos\theta) + C_{2n} Q_{z_{n-1/2}}(\cos\theta). \tag{2.6.11}$$

The constants C_{1n}, C_{2n} may be determined from the condition (2.6.4). To this end, we expand $\tau_k(\rho)$ by the eigenfunctions of the boundary value problem (2.6.5), (2.6.6)

$$\tau_k(\rho) = \sum_{n=1}^{\infty} a_{nk} U_n(\rho) \tag{2.6.12}$$

where

$$a_{nk} = \int_{\rho_1}^{\rho_2} \tau_k(\rho) \overline{U}_n(\rho) d\rho \left[\int_{\rho_1}^{\rho_2} U_n(\rho) \overline{U}_n(\rho) d\rho \right]^{-1}.$$

Satisfying the boundary conditions (2.6.4) leads to a linear algebraic system of second order with respect to the constants C_{1n}, C_{2n}. The solution of the system so constructed may be presented as follows:

$$\begin{aligned}
C_{1n} = \Delta_n^{-1} \Big\{ & a_{n2} \left[2ctg\theta_1 Q^{(1)}_{z_{n-1/2}}(\cos\theta_1) + \left(z_n^2 - \frac{1}{4} \right) Q_{z_{n-1/2}}(\cos\theta_1) \right] \\
& - a_{n1} \left[2ctg\theta_2 Q^{(1)}_{z_{n-1/2}}(\cos\theta_2) + \left(z_n^2 - \frac{1}{4} \right) Q_{z_{n-1/2}}(\cos\theta_2) \right] \Big\} \\
C_{n2} = \Delta_n^{-1} \Big\{ & a_{n1} \left[2ctg\theta_1 P^{(1)}_{z_{n-1/2}}(\cos\theta_1) + \left(z_n^2 - \frac{1}{4} \right) P_{z_{n-1/2}}(\cos\theta_1) \right] \\
& - a_{n2} \left[2ctg\theta_2 P^{(1)}_{z_{n-1/2}}(\cos\theta_2) + \left(z_n^2 - \frac{1}{4} \right) P_{z_{n-1/2}}(\cos\theta_2) \right] \Big\}
\end{aligned}$$

$$\begin{aligned}\Delta_n = &\left(z_n^2 - \frac{1}{4}\right)^2 D^{(0,0)}_{z_n-1/2}(\theta_1,\theta_2) + 2\left(z_n^2 - \frac{1}{4}\right) ctg\theta_1 \\ &\times D^{(1,0)}_{z_n-1/2}(\theta_1,\theta_2) + 2\left(z_n^2 - \frac{1}{4}\right) ctg\theta_2 D^{(0,1)}_{z_n-1/2}(\theta_1,\theta_2) \\ &+ 4ctg\theta_1 ctg\theta_2 D^{(1,1)}_{z_n-1/2}(\theta_1,\theta_2)\end{aligned} \tag{2.6.13}$$

$$D^{(s,l)}_z(\varphi,\psi) = P^{(s)}_z(\cos\varphi)Q^{(l)}_z(\cos\psi) - P^{(l)}_z(\cos\psi)Q^{(l)}_z(\cos\psi)$$

$(s, l = 0; 1)$.

Let us investigate the asymptotic behaviour of the solution of the problem on vibrations of a spherical zone.

The general solution of Eq. (2.6.5) has the form:

$$U = \frac{1}{\sqrt{\rho}}[C_{1z}J_z(\lambda\rho) + C_{2z}Y_z(\lambda\rho)].$$

Satisfying the homogeneous boundary conditions (2.6.6), we obtain the dispersion equation

$$\begin{aligned}\Delta(z,\lambda,\rho_1,\rho_2) = &9L^{(0,0)}_z(\lambda\rho_1,\lambda\rho_2) - 6\lambda\rho_1 L^{(1,0)}_z(\lambda\rho_1,\lambda\rho_2) \\ &- 6\lambda\rho_2 L^{(0,1)}_z(\lambda\rho_1,\lambda\rho_2) + 4\lambda^2\rho_1\rho_2 L^{(1,1)}_z(\lambda\rho_1,\lambda\rho_2) = 0\end{aligned} \tag{2.6.14}$$

Equation (2.6.14) has a countable set of zeros which correspond to the homogeneous solutions

$$\begin{aligned}U_\varphi &= R_0 \sum_{k=1}^{\infty} C_k U_k(\rho) \frac{dm}{d\theta} e^{i\omega t} \\ \tau_{r\varphi} &= G \sum_{k=1}^{\infty} C_k \tau_{rk}(\rho) \frac{dm_k}{d\theta} e^{i\omega t} \\ \tau_{\theta\varphi} &= G \sum_{k=1}^{\infty} C_k \tau_{\theta k}(\rho) \left[2ctg\theta \frac{dm_k}{d\theta} + \left(z_k^2 - \frac{1}{4}\right) m_k\right] e^{i\omega t}\end{aligned} \tag{2.6.15}$$

where C_k are arbitrary constants, and

$$\begin{aligned}U_k(\rho) &= \frac{1}{\sqrt{\rho}}\left[3L^{(0,0)}_{z_k}(\lambda\rho,\lambda\rho_2) - 2\lambda\rho_2 L^{(0,1)}_{z_k}(\lambda\rho,\lambda\rho_2)\right] \\ \tau_{rk}(\rho) &= \frac{1}{\rho\sqrt{\rho}}\left[9L^{(0,0)}_{z_k}(\lambda\rho,\lambda\rho_2) - 6\lambda\rho L^{(1,0)}_{z_k}(\lambda\rho,\lambda\rho_2)\right. \\ &\quad \left. - 6\lambda\rho_2 L^{(0,1)}_{z_k}(\lambda\rho,\lambda\rho_2) + 4\lambda^2\rho\rho_2 L^{(1,1)}_{z_k}(\lambda\rho,\lambda\rho_2)\right] \\ \tau_{\theta k}(\rho) &= \frac{1}{\rho\sqrt{\rho}}\left[3L^{(0,0)}_{z_k}(\lambda\rho,\lambda\rho_2) - 2\lambda\rho_2 L^{(0,1)}_{z_k}(\lambda\rho,\lambda\rho_2)\right]\end{aligned}$$

We now turn our attention to the study of the roots of Eq. (2.6.14).

For any finite $\lambda[\lambda = O(1)]$ as $\varepsilon \to 0$, the function $D(z, \lambda, \varepsilon) = \Delta(z, \lambda, \rho_1, \rho_2)$ has two zeros due to the asymptotic properties $z_k = O(1)$ as $\varepsilon \to 0$ $(k = 1, 2)$. To prove this, we expand $D(z, \lambda, \varepsilon)$ in a series by ε and obtain

$$\begin{aligned} D(z, \lambda, \varepsilon) = & -4\pi^{-1}\varepsilon\{4z^2 - 9 - 4\lambda^2 \\ & + \frac{1}{3}\left[8z^4 - \left(14 + 16\lambda^2\right)z^2 + 8\lambda^4 - 6\lambda^2 - 9\right]\varepsilon^2 \\ & + \frac{1}{15}\left[8z^6 + 2\left(11 - 12\lambda^2\right)z^4 + \left(24\lambda^4 - 76\lambda^2 - 78\right)z^2 \right. \\ & \left. -27 - 12\lambda^2 + 22\lambda^4 - 8\lambda^6\right]\varepsilon^4 + \ldots\} = 0. \end{aligned} \tag{2.6.16}$$

We seek z_k in terms of the expansion

$$z_k = z_{k0} + \varepsilon^2 z_{k2} + \ldots \quad (k = 1, 2). \tag{2.6.17}$$

Substituting (2.6.17) into (2.6.16) we obtain

$$z_{k0}^2 = 9/4 + \lambda^2, \ z_{k2} = \frac{5}{6}\frac{\lambda^2}{z_{k0}}. \tag{2.6.18}$$

Consider the case where λ increases infinitely with $\varepsilon \to 0$. Here we will consider separately the following limiting cases: (a) $\lambda\varepsilon \to 0$ as $\varepsilon \to 0$ and (b) $\lambda\varepsilon \to const$ as $\varepsilon \to 0$.

We first define such z_k when $\lambda\varepsilon \to 0$, $\varepsilon \to 0$ we assume that the principal terms of the asymptotic behaviour of z_k and λ have the form:

$$\begin{aligned} & z_k = z_{k0}\varepsilon^{-\beta}, \lambda = \lambda_0\varepsilon^{-q}, z_{k0} = O(1), \\ & \lambda_0 = O(1), \ 0 < \beta < 1, \ 0 < q < 1. \end{aligned} \tag{2.6.19}$$

Substituting (2.6.19) into (2.6.16), from the condition of consistency of the built asymptotic process we find that only the case $q = \beta$ is possible.

Assuming $\lambda = \lambda_0\varepsilon^{-\beta}$, we seek z_k as

$$z_k = z_{k0}\varepsilon^{-\beta} + z_{k2}\varepsilon^{\beta} + \ldots \quad 0 < \beta < 1 \tag{2.6.20}$$

Substituting the expansions of z_k into (2.6.16) we obtain

$$z_{k0}^2 = \lambda_0^2, \quad z_{k2} = \frac{9}{8}z_{k0}^{-1}. \tag{2.6.21}$$

We consider the case $\lambda\varepsilon \to 0$ with $\varepsilon \to 0$ as well as $\lambda\varepsilon \to const$ with $\varepsilon \to 0$. We seek z_n $(n = k - 4)$ in the following form

$$z_n = i\left[\frac{\delta_n}{\varepsilon} + O\left(\varepsilon^{1-\beta}\right)\right], \lambda = \lambda_0\varepsilon^{-\beta}, 0 \le \beta < 1. \tag{2.6.22}$$

If the asymptotic relations (2.6.22) are inserted into Eq. (2.6.14) and asymptotic expansions of the functions $J_z(x)$, $y_z(x)$ for δ_n are employed we get

$$\sin 2\delta_n = 0. \tag{2.6.23}$$

This equation completely agrees with the equation that determines the performance of Saint-Venant's boundary effects in the vortex theory of thick plates [7]. To construct the asymptotics of the zeros of the second group $\lambda\varepsilon \to const$ $(\lambda \sim z)$, $z\varepsilon \to const$ as $\varepsilon \to 0$, defining $\lambda = \lambda_0\varepsilon^{-1}$ we seek z_p $(p = 1, 2, \ldots, p \ne k)$ as follows:

$$z_p = \frac{\gamma_p}{\varepsilon} + O(\varepsilon). \tag{2.6.24}$$

As mentioned above, after substituting (2.6.24) into (2.6.14) and converting it by means of asymptotic expansions of $J_z(x)$, $Y_z(x)$ for large values of the index and the argument γ_p we obtain the equation

$$\sin 2\sqrt{\lambda_0^2 - \gamma_p^2} = 0. \tag{2.6.25}$$

For given λ, Eq. (2.6.25) defines a countable set of z_p.

We now present an asymptotic formula for U_φ, $\tau_{r\varphi}$, $\tau_{\theta\varphi}$. Assuming $\rho = 1 + \varepsilon\eta$, $-1 \le \eta \le 1$ and expanding in the small parameter ε, we obtain, from (2.6.15), the following leading order terms of the respective asymptotic expansions:

$$\begin{aligned}
U_\varphi^{(1)} &= R_0\sum_{k=1}^{2} C_k[2 + (2\eta - 3)\varepsilon + \ldots]\frac{dm_k}{d\theta}e^{i\omega t} \\
\tau_{\theta\varphi}^{(1)} &= G\sum_{k=1}^{2} C_k[-2 + 3\varepsilon + \ldots]\frac{dm_k}{d\theta}e^{i\omega t} \\
\tau_{r\varphi}^{(1)} &= O\left(\varepsilon^2\right)
\end{aligned} \tag{2.6.26}$$

$$\begin{aligned}
U_\varphi^{(2)} &= R_0\sum_{k=3}^{4} C_k[2 + O(\varepsilon)]\frac{dm_k}{d\theta}e^{i\omega t} \\
\tau_{\theta\varphi}^{(2)} &= G\sum_{k=3}^{4} C_k[-2 + O(\varepsilon)]\frac{dm_k}{d\theta}e^{i\omega t} \\
\tau_{r\varphi}^{(2)} &= O(\varepsilon) \quad 0 < \beta \le \frac{1}{2} \\
\tau_{r\varphi}^{(2)} &= O\left(\varepsilon^{2-2\beta}\right) \quad \frac{1}{2} < \beta < 1.
\end{aligned} \tag{2.6.27}$$

Let us note that $\tau_{r\varphi}$ has the same order as $\tau_{\theta\varphi}$ at $\beta = \frac{3}{4}$. $\tau_{r\varphi}$ is principal at $\beta > \frac{3}{4}$.

As noted above, the solution corresponding to the roots of Eq. (2.6.23) in the first term of the asymptotics coincides with the solution of Saint-Venant's boundary effects in the theory of thick plates and, therefore, is not included here.

Finally, in the case of microwave vibrations we get

$$\begin{aligned} U_{\varphi}^{(3)} &= R_0\varepsilon \sum_{k=1}^{\infty} B_k \cos\left[\sqrt{\lambda_0^2 - \gamma_k^2}(\eta - 1)\right] \frac{dm_k}{d\theta} e^{i\omega t} \\ \tau_{\theta\varphi}^{(3)} &= G\varepsilon \sum_{k=1}^{\infty} B_k \cos\left[\sqrt{\lambda_0^2 - \gamma_k^2}(\eta - 1)\right] \frac{dm_k}{d\theta} e^{i\omega t} \\ \tau_{r\varphi}^{(3)} &= G \sum_{p=1}^{\infty} B_p \sqrt{\lambda_0^2 - \gamma_k^2} \sin\left[\sqrt{\lambda_0^2 - \gamma_k^2}(\eta - 1)\right] m_k e^{i\omega t}. \end{aligned} \tag{2.6.28}$$

As seen from (2.6.28), in the case of microwave vibrations $\tau_{r\varphi}$ is a basic stress which is ignored in the applied theory of shells.

As already pointed out, the case $z^2 = \frac{1}{4}$ is special and corresponds to thickened resonances. In this case, the problem (2.6.5), (2.6.6) takes the form:

$$\begin{aligned} & U'' + \frac{2}{\rho} U' + \lambda^2 U = 0 \\ & \left(U' - \frac{1}{\rho} U\right)_{\rho = \rho_s} = 0 \end{aligned} \tag{2.6.29}$$

The general solution of (2.6.29) has the following form:

$$U(\rho) = \frac{1}{\rho}(C_1 \cos \lambda\rho + C_2 \sin \lambda\rho).$$

On using the boundary conditions, we obtain:

$$\begin{aligned} D_0(\lambda, \varepsilon) = \frac{1}{\rho_1\rho_2} & \left[(\lambda^2 + 4 - \lambda^2\varepsilon^2) \sin 2\lambda\varepsilon \right. \\ & \left. - 4\lambda\varepsilon \cos 2\lambda\varepsilon\right] = 0 \end{aligned} \tag{2.6.30}$$

Equation (2.6.30) defines a countable set of frequencies of thickened resonances.

2.7 Non-axisymmetric Dynamic Problems of Elasticity Theory for a Hollow Sphere

In this section, we consider non-axisymmetric dynamic problem of elasticity theory for a hollow sphere. Due to spherical symmetry, the general boundary value problem is divided into two parts one of which coincides exactly with the boundary value problem for axisymmetric vibrations of a hollow sphere, whereas the second one describes the vortex motion of a hollow sphere (it coincides with the boundary value problem for purely torsional vibrations of a hollow sphere).

Consider a spherical layer referred to in a spherical coordinate system

$$r_1 \le r \le r_2,\ \theta_1(\varphi) \le \theta \le \theta_2(\varphi),\ 0 \le \varphi \le 2\pi. \tag{2.7.1}$$

The equations of motion in terms of displacements have the form:

$$\begin{aligned}
&G\left\{\Delta U_r - \frac{2}{r^2}\left[U_r + \frac{1}{\sin\theta}\frac{\partial}{\partial\theta}(\sin\theta U_\theta) + \frac{1}{\sin\theta}\frac{\partial U_\varphi}{\partial\varphi}\right]\right\}\\
&+(\lambda+G)\frac{\partial}{\partial r}\left[\frac{1}{r^2}\frac{\partial}{\partial r}(rU_r) + \frac{1}{r\sin\theta}\frac{\partial}{\partial\theta}(\sin U_\theta)\right.\\
&\left.+\frac{1}{r\sin\theta}\frac{\partial U_\varphi}{\partial\varphi}\right] = g\frac{\partial^2 U_r}{\partial t^2}\\
&G\left\{\Delta U_\theta + \frac{2}{r^2}\left[\frac{\partial U_r}{\partial\theta} - \frac{1}{2\sin^2\theta}U_\theta - \frac{\cos\theta}{\sin\theta}\frac{\partial U_\varphi}{\partial\varphi}\right]\right\}\\
&+(\lambda+G)\frac{1}{r}\frac{\partial}{\partial\theta}\left[\frac{1}{r^2}\frac{\partial}{\partial r}(r^2U_r) + \frac{1}{r\sin\theta\partial\theta}(\sin U_\theta)\right.\\
&\left.\frac{1}{r\sin\theta}\frac{\partial U_\varphi}{\partial\varphi}\right] = g\frac{\partial^2 U_\theta}{\partial t^2}\\
&G\left\{\Delta U_\varphi + \frac{2}{r^2\sin\theta}\left[\frac{\partial U_r}{\partial\varphi} + ctg\theta\frac{\partial U_\varphi}{\partial\varphi} - \frac{\partial U_\varphi}{2\sin\theta}\right]\right\}\\
&+\frac{(\lambda+G)}{r\sin\theta}\frac{\partial}{\partial\varphi}\left[\frac{1}{r^2}\frac{\partial}{\partial r}(r^2U_r) + \frac{1}{r\sin\theta}\frac{\partial}{\partial\theta}(\sin\theta U_\theta)\right.\\
&\left.\frac{1}{r\sin\theta}\frac{\partial U_\varphi}{\partial\varphi}\right] = g\frac{\partial^2 U_\varphi}{\partial t^2}
\end{aligned} \tag{2.7.2}$$

It is assumed that the load is acting on the layer from the side of face surfaces:

$$\begin{aligned}
\sigma_r &= q_r^{(k)}(\theta,\varphi)e^{i\omega t}, \quad \tau_{r\theta} = q_{r\theta}^{(k)}(\theta,\varphi)e^{i\omega t}\\
\tau_{r\varphi} &= q_{r\varphi}^{(k)}(\theta,\varphi)e^{i\omega t} \quad \text{for} \quad r = r_k \quad (k=1,2)
\end{aligned} \tag{2.7.3}$$

and on the rest of the boundary the following boundary conditions are given:

$$\begin{aligned}
\sigma_\theta &= Q_\theta^{(s)}(r,\varphi)e^{i\omega t}, \quad \tau_{r\theta} = Q_{r\theta}^{(s)}(r,\varphi)e^{i\omega t}\\
\tau_{\theta\varphi} &= Q_{\theta\varphi}^{(s)}(r,\varphi)e^{i\omega t} \quad \text{for} \quad \theta = \theta_s \quad (s=1,2)
\end{aligned} \tag{2.7.4}$$

Using the results of [2, 8], a two-dimensional vector field (U_θ, U_φ) is represented in the form

$$U_\theta = \frac{\partial \Phi}{\partial \theta} + \frac{1}{\sin\theta}\frac{\partial F}{\partial \varphi}, \quad U_\varphi = \frac{1}{\sin\theta}\frac{\partial \Phi}{\partial \theta} - \frac{\partial F}{\partial \varphi}. \tag{2.7.5}$$

Substituting (2.7.5) into the equations of motion (2.7.2) and the boundary conditions (2.5.3) respectively, we obtain

$$L_1(U_r, \Phi) = 0 \tag{2.7.6}$$

$$\frac{\partial}{\partial \theta} L_2(U_r, \Phi) + \frac{1}{\sin\theta}\frac{\partial}{\partial \varphi} L_3(F) = 0 \tag{2.7.7}$$

$$\frac{1}{\sin\theta}\frac{\partial}{\partial \varphi} L_2(U_r, \Phi) + \frac{\partial}{\partial \theta} L_3(F) = 0 \tag{2.7.8}$$

$$M_1(U_r, \Phi)\big|_{r=r_k} = q_r^{(k)}(\theta, \varphi)e^{i\omega t} \tag{2.7.9}$$

$$\left[\frac{\partial}{\partial \theta} M_2(U_r, \Phi) + \frac{1}{\sin\theta}\frac{\partial}{\partial \varphi} M_3(F)\right]_{r=r_k} = q_{r\theta}^{(k)}(\theta, \varphi)e^{i\omega t} \tag{2.7.10}$$

$$\left[\frac{1}{\sin\theta}\frac{\partial}{\partial \varphi} M_2(U_r, \Phi) - \frac{\partial}{\partial \theta} M_3(\psi)\right]_{r=r_k} = q_{r\varphi}^{(k)}(\theta, \varphi)e^{i\omega t} \tag{2.7.11}$$

$$\begin{aligned}
L_1(U_r, \Phi) &= \frac{2(1-\nu)}{1-2\nu}\left(\frac{\partial^2 U_r}{\partial r^2} + \frac{2}{r}\frac{\partial U_r}{\partial r} - \frac{2}{r^2}U_r\right) \\
&\quad + \frac{1}{r^2}\Delta_0 U_r + \frac{1}{1-2\nu}\left(\frac{1}{r}\frac{\partial}{\partial r} + \frac{4\nu-3}{r^2}\right)\Delta_0 \Phi - gG^{-1}\frac{\partial^2 U_r}{\partial t^2} \\
L_2(U_r, \Phi) &= \frac{1}{1-2\nu}\frac{1}{r}\left[\frac{\partial U_r}{\partial r} + \frac{4-4\nu}{r}U_r\right] \\
&\quad + \frac{2(1-\nu)}{1-2\nu}\frac{1}{r^2}\Delta_0 \Phi + \frac{\partial^2 \Phi}{\partial r^2}\frac{\varepsilon}{r}\frac{\partial \Phi}{\partial r} + gG^{-1}\frac{\partial^2 \Phi}{\partial t^2} \\
L_3(F) &= \frac{\partial^2 F}{\partial r^2} + \frac{2}{r}\frac{\partial F}{\partial r} + \frac{1}{r^2}\Delta_0 F + \frac{2}{r}\frac{\partial \Phi}{\partial r} - gG^{-1}\frac{\partial^2 F}{\partial t^2} \\
\Delta_0 &= \frac{\partial^2}{\partial \theta^2} + ctg\theta\frac{\partial}{\partial \theta} + \frac{1}{\sin^2\theta}\frac{\partial^2}{\partial \varphi^2} \\
M_1(U_r, \Phi) &= \frac{2G}{1-2\nu}\left[(1-\nu)\frac{\partial U_r}{\partial r} + \frac{\nu}{2}(2U_r + \Delta_0 \Phi)\right] \\
M_2(U_r, \Phi) &= G\left(\frac{U_r}{r} + \frac{\partial \Phi}{\partial r} - \frac{1}{r}\Phi\right) \\
M_3(F) &= G\left(\frac{\partial F}{\partial r} - \frac{1}{r}F\right).
\end{aligned}$$

The relations (2.7.7), (2.7.8) are identically satisfied if we suggest that

$$\begin{aligned} L_2(U_r, \Phi) &= -\frac{\partial \chi(r,\theta,\varphi,t)}{\partial \varphi} \\ L_3(F) &= \left(\frac{1}{\sin\theta}\right)^{-1} \frac{\partial \chi(r,\theta,\varphi,t)}{\partial \varphi} \end{aligned} \tag{2.7.12}$$

where $\chi(r,\theta,\varphi,t)$ satisfies the equation

$$\Delta_0 \chi(r,\theta,\varphi,t) = 0 \tag{2.7.13}$$

Now we represent $\left[q_{r\theta}^{(k)}, \quad q_{r\varphi}^{(k)}\right]$ in the form

$$\begin{aligned} q_{r\theta}^{(k)} &= \frac{\partial q_2^{(k)}}{\partial \theta} + \frac{1}{\sin\theta}\frac{\partial q_3^{(k)}}{\partial \varphi}, \\ q_{r\varphi}^{(k)} &= \frac{1}{\sin\theta}\frac{\partial q_2^{(k)}}{\partial \varphi} - \frac{\partial q_3^{(k)}}{\partial \theta} \end{aligned} \tag{2.7.14}$$

Then, the original boundary value problem (2.7.2)–(2.7.3) splits into two

$$L_1(U_r, \varPhi) = 0, \quad L_2(U_r, \varPhi) = -\frac{\partial \chi}{\partial \varphi} \tag{2.7.15}$$

$$\begin{aligned} \left[M_1(U_r, \varPhi)\right]_{r=r_k} &= q_r^{(k)}, \\ \left[M_2(U_r, \varPhi)\right]_{r=r_k} &= q_2^{(k)} - \partial \frac{e^{(k)}}{\partial \varphi} \end{aligned} \tag{2.7.16}$$

$$L_3(F) = \sin\theta \frac{\partial \chi}{\partial \theta} \tag{2.7.17}$$

$$\left[M_3(F)\right]_{r=r_k} = q_3^{(k)} + \sin\theta \frac{\partial e^{(k)}}{\partial \theta} \tag{2.7.18}$$

$e^{(k)}(\theta,\varphi)$—are arbitrary functions satisfying the equation
$\Delta_0 e^{(k)}(\theta,\varphi) = 0.$

The validity of representation (2.7.5) is discussed thoroughly in [2] and there is no need to discuss it here. It is also shown that, without loss of generality, one can always suppose that

$$\chi = 0,\ e^{(k)=0}.$$

To construct the homogeneous solutions for a spherical layer in (2.7.15)–(2.7.18), we assume that $q_r^{(k)} = q_2^{(k)} = q_3^{(k)} = 0$. Thus, we finally obtain

$$L_1(U_r, \Phi) = 0,\ L_2(U_r, \Phi) = 0$$

$$\begin{aligned} &[M_1(U_r, \Phi)]_{r=r_k} = 0, \\ &[M_2(U_r, \Phi)]_{r=r_k} = 0 \end{aligned} \tag{2.7.19}$$

$$L_3(F) = 0,\ [M_3(F)]_{r=r_k} = 0. \tag{2.7.20}$$

The solution of (2.7.19) and (2.7.20) is sought in the form

$$\begin{aligned} U_r &= a(r)T(\theta, \varphi)\, e^{i\omega t} \\ \Phi &= b(r)T(\theta, \varphi)\, e^{i\omega t} \\ F &= \psi(r)\frac{\partial T}{\partial \varphi} e^{i\omega t}, \end{aligned} \tag{2.7.21}$$

where $T(\theta, \varphi)$ is subject to the condition

$$\Delta_0 T + \left(z^2 - \frac{1}{4}\right) T = 0. \tag{2.7.22}$$

Substituting (2.7.21) into (2.7.19), (2.7.20) and taking into account (2.7.22), on separating the variables we respectively obtain

$$\begin{aligned} &\left(a' + \frac{2}{r}a\right)' + \frac{1-2v}{2(1-v)}\left(\lambda^2 - \frac{z^2 - \frac{1}{4}}{r^2}\right)a \\ &+ \frac{1}{2(1-v)}\frac{1}{r}\left(z^2 - \frac{1}{4}\right)\left[b' + \frac{4v-3}{r}b\right] = 0 \end{aligned} \tag{2.7.23}$$

$$\begin{aligned} &\frac{1}{1-2v}\left(\frac{1}{r}a' + \frac{4-4v}{r^2}a\right) + b'' + \frac{2}{r}b' + \left[\lambda^2 - \frac{2(1-v)}{1-2v}\frac{z^2 - \frac{1}{4}}{r^2}\right]b = 0 \\ &\left[(1-v)a' + \frac{2v}{r}a - \frac{v}{2}\left(z^2 - \frac{1}{4}\right)b\right]_{r=r_k} = 0 \\ &\left[\frac{a}{r} + b' - \frac{1}{r}b\right]_{r=r_k} = 0 \end{aligned} \tag{2.7.24}$$

$$\psi'' + \frac{2}{r}\psi' + \left(\lambda^2 - \frac{z^2 - \frac{1}{4}}{r^2}\right)\psi = 0 \tag{2.7.25}$$

$$\left[\psi' - \frac{1}{r}\psi\right]_{r=r_k} = 0. \quad (2.7.26)$$

The problem (2.7.23)–(2.7.24) describes the potential motion of a shell and completely coincides with the boundary value problem of axisymmetric vibrations of a spherical zone. However, the problem (2.7.25)–(2.7.26) describes the vortex motion of a shell and coincides with the boundary value problem of purely torsional vibrations of a spherical zone.

It should be noted that such a partition of vibrations could be obtained on the basis of general concepts of solving the dynamic problem of elasticity theory by means of Bessel and Legendre functions. However, this method would be more cumbersome and would never allow a direct relationship of problems (2.7.23)–(2.7.24), (2.7.25)–(2.7.26) with the boundary value problems of axisymmetric and torsional vibrations of a shell, respectively. Representation (2.7.5) leads to the dismemberment of the stress–strain state. Index "1" at the top will denote a part of the displacements and stresses corresponding to the potential problem, index "2" will correspond to the vortex one.

$$\begin{aligned}
U_r^{(1)} &= \frac{C_1}{\sqrt{r}}\left[\alpha Z_z'(\alpha r) - \frac{1}{2r}Z_z(\alpha r)\right.\\
&\left.- \frac{1}{r}\left(z^2 - \frac{1}{4}\right)Z_z(\lambda r)\right]T(\theta,\varphi)e^{i\omega t}\\
U_\theta^{(1)} &= \frac{C_1}{\sqrt{r}}\left[\frac{1}{r}Z_z(\alpha r) - \lambda Z_z'(\lambda r) - \frac{1}{2r}Z_z(\lambda r)\right]\frac{\partial T}{\partial \theta}e^{i\omega t}\\
U_\varphi^{(1)} &= \frac{C_1}{\sqrt{r}}\frac{1}{\sin\theta}\left[\frac{1}{r}Z_z(\alpha r) - \lambda Z_z'(\lambda r) - \frac{1}{2r}Z_z(\lambda r)\right]\frac{\partial T}{\partial \theta}e^{i\omega t}
\end{aligned} \quad (2.7.27)$$

$$\begin{aligned}
\sigma_\theta^{(1)} &= \frac{GC_1}{r^2\sqrt{r}}\left\{\left[2\alpha Z_z'(\alpha r) - \left(1 + \frac{\nu}{1-\nu}\right)\lambda^2 r^2\right)Z_z(\alpha r)\right.\\
&\left.-2\left(z^2 - \frac{1}{4}\right)Z_z(\lambda r)\right]T(\theta,\varphi) + [2Z_z(\alpha r)\\
&\left.-2\lambda r Z_z'(\lambda r) - Z_z(\lambda r)]\frac{\partial^2 T}{\partial \theta^2}\right\}e^{i\omega t}\\
\sigma_\varphi^{(1)} &= \frac{GC_1}{r^2\sqrt{r}}\left\{\left[2\alpha r Z_z'(\alpha r) - \left(1 + \frac{\nu}{1-\nu}\lambda^2 r^2\right)Z_z(\alpha r)\right.\right.\\
&\left.-2\left(z^2 - \frac{1}{4}\right)Z_z(\lambda r)\right]T(\theta,\varphi) + [2Z_z(\alpha r)\\
&\left.-2\lambda r Z_z'(\lambda r) - Z_z(\lambda r)]\left(\frac{\partial T}{\partial \theta} + \frac{1}{\sin^2\theta}\frac{\partial^2 T}{\partial \varphi^2}\right)\right\}e^{i\omega t}\\
\sigma_r^{(1)} &= \frac{GC_1}{r^2\sqrt{r}}\left\{\varphi_2(r,z)Z_z(\alpha r) - 4\alpha r Z_z'(\alpha r)\right.\\
&\left.+\left(z^2 - \frac{1}{4}\right)[3Z_z(\lambda r) - 2\lambda r Z_z'(\lambda r)]\right\}Te^{i\omega t}
\end{aligned} \quad (2.7.28)$$

$$\begin{aligned}
\tau_{r\varphi}^{(1)} &= \frac{GC_1}{r^2\sqrt{r}}\left[2\alpha r Z_z'(\alpha r) - 3Z_z(\alpha r)\right.\\
&\quad \left. + 2\lambda r Z_z'(\lambda r) - \varphi_3(z,r) Z_z(\lambda r)\right]\frac{1}{\sin\theta}\frac{\partial T}{\partial\varphi}e^{i\omega t}\\
\tau_{r\theta}^{(1)} &= \frac{GC_1}{r^2\sqrt{r}}\left[2\alpha r Z_z'(\alpha r) - 3Z_z(\alpha r)\right.\\
&\quad \left. + 2\lambda r Z_z'(\lambda r) - \varphi_3(z,r) Z_z(\lambda r)\right]\frac{\partial T}{\partial\theta}e^{i\omega t}\\
\tau_{\theta\varphi}^{(1)} &= \frac{GC_1}{r^2\sqrt{r}}\left[-2Z_z(\alpha r) + 2\lambda r Z_z'(\lambda r)\right.\\
&\quad \left. + Z_z(\lambda r)\right]\frac{ctg\theta}{\sin\theta}\frac{\partial T}{\partial\varphi}e^{i\omega t}
\end{aligned} \tag{2.7.29}$$

$$\begin{aligned}
U_r^{(2)} &= 0, \sigma_r^{(2)} = 0\\
\sigma_\theta^{(2)} &= \frac{GC_2}{r\sqrt{r}}Z_z(\lambda r)\frac{\partial}{\partial\theta}\left(\frac{1}{\sin\theta}\frac{\partial T}{\partial\varphi}\right)e^{i\omega t}\\
\sigma_\varphi^{(2)} &= \frac{GC_2}{r\sqrt{r}}Z_z(\lambda r)\frac{1}{\sin\theta}\frac{\partial}{\partial\varphi}\left(ctg\theta T - \frac{\partial T}{\partial\varphi}\right)e^{i\omega t}\\
\tau_{r\varphi}^{(2)} &= \frac{GC_2}{2r\sqrt{r}}\left[3Z_z(\lambda r) - 2\lambda r Z_z'(\lambda r)\right]\frac{\partial T}{\partial\theta}e^{i\omega t}\\
\tau_{r\theta}^{(2)} &= \frac{GC_2}{2r\sqrt{r}}\left[3Z_z(\lambda r) - 2\lambda r Z_z'(\lambda r)\right]\frac{1}{\sin\theta}\frac{\partial T}{\partial\varphi}e^{i\omega t}\\
\tau_{\theta\varphi}^{(2)} &= \frac{GC_2}{r\sqrt{r}}Z_z(\lambda r)\left[2ctg\theta\frac{\partial T}{\partial\theta} + \left(z^2 - \frac{1}{4}\right)^T\right]e^{i\omega t}\\
U_\theta^{(2)} &= \frac{C_2}{\sqrt{r}}Z_z(\lambda r)\frac{1}{\sin\theta}\frac{\partial^2 T}{\partial\varphi^2}e^{i\omega t}\\
U_\varphi^{(2)} &= -\frac{C_2}{\sqrt{r}}Z_z(\lambda r)\frac{\partial^2 T}{\partial\varphi\partial\theta}e^{i\omega t}.
\end{aligned} \tag{2.7.30}$$

The notation is the same as in the previous sections.

Thus, the total dynamic problem of elasticity theory for a spherical layer is divided into two. However, the solutions to these two problems are related through the boundary conditions on the side surface. Therefore, while satisfying the boundary conditions on the side surface there are some difficulties mainly related to the non-orthogonality of homogeneous solutions. As shown in the preceding paragraphs, the potential solutions of the problem have the property of generalized orthogonality, solutions of the vortex problem are orthogonal. The solutions of different groups, on the contrary, do not possess these properties. That's why we have the question of satisfying boundary conditions on the side surface using homogeneous solutions. In the case of $\theta_j = const$, the reduction of boundary value problems to the solution of infinite systems of algebraic equations is carried out in the same way as in the axisymmetric case.

Let us consider the case of $\theta_j = \theta_j(\varphi)$. For simplicity, consider a spherical layer with one boundary $\theta = \theta_1(\varphi)\,(\theta_1 \le \theta \le \pi)$. In the plane of the variables $\varphi,\ \theta$ we introduce the system of local dimensionless coordinates [3], associated with the side contour layer s, n, r (Fig. 2.2). Define parametrically the boundary

$$\varphi = \varphi_1(s), \quad \theta = \theta_1(s).$$

The coordinates of points on the plane $\varphi,\ \theta$ are then expressed in terms of n and s by the following formulas:

$$\varphi(s,n) = \varphi_1(s) + nl(s), \quad l(s) = \frac{d\theta_1}{ds}$$
$$\theta(s,n) = \theta_1(s) + nm(s), \quad m(s) = -\frac{d\varphi_1}{ds}$$

where $l(s)$, $m(s)$ are direction cosines of the normal to the boundary.

We need to pass to local coordinates $n,\ s$ in the expressions for displacements and stresses, that is, to find expressions for σ_n, σ_s, τ_{ns}, U_n, U_s etc. We write down all the components of the displacement vector and the stress tensor in the new variables using well-known formulas of the theory of elasticity

$$\begin{aligned}
\sigma_n &= \sigma_{\theta\theta} m^2 + \sigma_{\varphi\varphi} l^2 + 2\tau_{\theta\varphi} ml \\
\sigma_s &= \sigma_{\theta\theta} l^2 + \sigma_{\varphi\varphi} m^2 - 2\tau_{\theta\varphi} ml \\
\sigma_r &= \sigma_r \\
\tau_{ns} &= \left(\sigma_{\varphi\varphi} - \sigma_{\theta\theta}\right) ml + \tau_{\theta\varphi}\left(m^2 - l^2\right) \\
\tau_{nr} &= \tau_{r\theta m} + \tau_{\theta\varphi l} \\
\tau_{sr} &= m\tau_{r\varphi} - l\tau_{r\theta} \\
U_n &= U_\theta l + U_\varphi m, \ U_s = -U_\theta m + U_\varphi l.
\end{aligned} \tag{2.7.31}$$

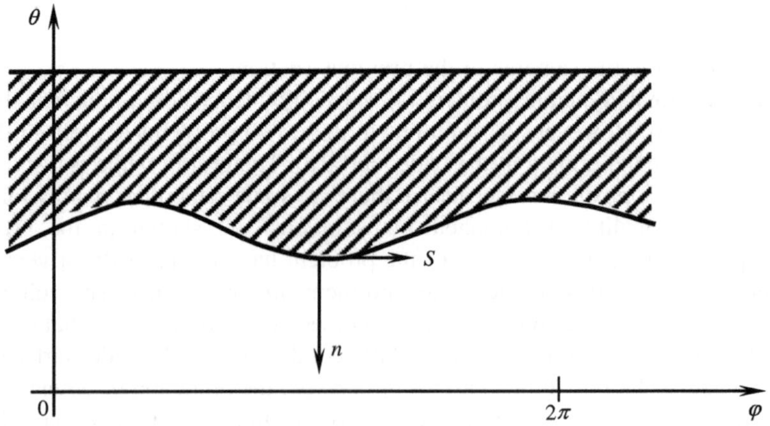

Fig. 2.2 Local coordinate system for a sphere

The relationships between the derivative functions in the two coordinate systems are established through the equations

$$\frac{dl}{ds} = -\frac{m(s)}{R(s)}, \frac{dm}{ds} = \frac{l(s)}{R(s)}.$$

Here, $R(s)$ is the dimensionless radius of curvature of the curve $\theta = \theta_1(\varphi)$ in the plane of variables φ, θ.

$$\frac{\partial\theta}{\partial s} = -m\left(1+\frac{n}{R}\right) = -mH$$
$$\frac{\partial\varphi}{\partial s} = l\left(1+\frac{n}{R}\right) = lH,\ H = 1+\frac{n}{R}$$

Using the derived relations, we obtain

$$\begin{aligned}
&\frac{\partial}{\partial\theta} = \frac{l}{H}\frac{\partial}{\partial s} + m\frac{\partial}{\partial n},\quad \frac{\partial}{\partial\varphi} = -\frac{m}{H}\frac{\partial}{\partial s} + l\frac{\partial}{\partial n}\\
&\frac{\partial^2}{\partial\theta^2} = \frac{l^2}{H^2}\frac{\partial^2}{\partial s^2} + \frac{2lm}{H}\frac{\partial^2}{\partial s\partial n} + m^2\frac{\partial^2}{\partial n^2}\\
&\quad + \left[\frac{1}{\sin^2\theta}ml^2\left(2m' - \frac{m}{H}H'\right) + \frac{l}{H}\left(l^2 + \frac{ctg\theta}{\sin\theta}m^2\right)\right]\frac{\partial}{\partial s}\\
&\quad + \left[\frac{lm}{H}\left(mm' - \frac{1}{\sin^2\theta}ll'\right) + ml^2\left(1 - \frac{ctg\theta}{\sin\theta}\right)\right]\frac{\partial}{\partial n}\Bigg\}\\
&\frac{\partial^2}{\partial\varphi^2} = \frac{m^2}{H^2}\frac{\partial^2}{\partial s^2} - \frac{2lm}{H}\frac{\partial^2}{\partial s\partial n} + l^2\frac{\partial^2}{\partial n^2}\\
&\quad + \frac{m}{H}\left(2m' - \frac{m}{H}H'\right)\frac{\partial}{\partial s} - \frac{m}{mH}l'\frac{\partial}{\partial n}.
\end{aligned} \tag{2.7.32}$$

Now substituting the expressions for the stresses and displacements (2.7.27), (2.7.28), (2.7.29) and (2.7.30) into (2.7.31) and taking into account (2.7.32), we obtain the stresses and displacements of vortex and potential solutions in the local coordinate system as:

$$\begin{aligned}
\sigma_s^{(1)} &= \frac{GC_1}{r^2\sqrt{r}}\left\langle\left[2\alpha r Z_z'(\alpha r) - \left(1 + \frac{\nu}{1-\nu}\lambda^2 r^2\right)Z_z(\alpha r)\right.\right.\\
&\quad \left. - 2\left(z^2 - \frac{1}{4}\right)Z_z(\lambda r)\right]T(s,n) + \left[2Z_z(\alpha r) - 2\lambda Z_z'(\lambda r) - Z_z(\lambda r)\right]\\
&\quad \times\left\{\left(1 + \frac{m^4}{\sin^2\theta}\right)\frac{1}{H^2}\frac{\partial^2 T}{\partial s^2} + \frac{2lm}{H}\left(l^2 - \frac{m^2}{\sin^2\theta}\right)\frac{\partial T}{\partial s\partial n}\right.\\
&\quad + \left(1 + \frac{1}{\sin^2\theta}\right)l^2 m^2\frac{\partial^2 T}{\partial n^2} + \left[\frac{l^3}{H^2}\left(2l' - \frac{l}{H}H'\right)\right.\\
&\quad \left. + \frac{1}{\sin^2\theta}\frac{m^3}{H^2}\left(2m' - \frac{m}{H}H'\right) + \frac{lm^2}{H}\left(1 - \frac{ctg\theta}{\sin\theta}\right)\right]\frac{\partial T}{\partial s}
\end{aligned}$$

$$+\left[\frac{1}{H}\left(l^3 m' - \frac{m^3 l'}{\sin^2\theta}\right) + m\left(m^2 + \frac{ctg\theta}{\sin\theta}l^2\right)\right]\frac{\partial T}{\partial n}\bigg\}\bigg\rangle e^{i\omega t}\tau_{ns}^{(1)}$$

$$= \frac{GC_1}{r^2\sqrt{r}}\langle[2Z_z(\alpha r) - 2\lambda r Z_z'(\lambda r) - Z_z(\lambda r)]$$

$$\times\left\{\frac{lm}{H^2}\left(\frac{m^2}{\sin^2\theta} - l^2\right)\frac{\partial^2 T}{\partial s^2} - \frac{2m^2 l^2}{H}\left(1 + \frac{1}{\sin^2\theta}\right)\frac{\partial^2 T}{\partial s\partial n}\right.$$

$$+ lm\left(\frac{l^2}{\sin^2\theta} - m^2\right)\frac{\partial T}{\partial n^2} + \left[\frac{1}{\sin^2\theta}\frac{ml}{H^2}\left(2m' - \frac{m}{H}H'\right)\right.$$

$$\left. - \frac{ml}{H}\left(2l' - \frac{l}{H}H'\right) + \frac{l^2}{H} + \frac{ctg\theta}{\sin\theta}\frac{m^2 - l^2}{H}\right]m\frac{\partial T}{\partial s}$$

$$+\left[m^2 - \frac{m}{H}\left(lm' + \frac{ml'}{\sin^2\theta}\right) + \frac{ctg\theta}{\sin\theta}(l^2 - m^2)\right]l\frac{\partial T}{\partial n}\bigg\rangle e^{i\omega t}$$

$$\sigma_n^{(1)} = \frac{GC_1}{r^2\sqrt{r}}\bigg\langle\bigg[2\alpha r Z_z'(\alpha r) - \left(1 + \frac{\nu}{1-\nu}\lambda^2 r^2\right)Z_z(\alpha r)$$

$$- 2\left(z^2 - \frac{1}{4}\right)Z_z(\lambda r)\bigg]T(s,n) + [2Z_z(\alpha r) - 2\lambda Z_z'(\lambda r) - Z_z(\lambda r)]$$

$$\times\left\{\left(1 + \frac{1}{\sin^2\theta}\right)\frac{m^2 l^2}{H^2}\frac{\partial^2 T}{\partial s^2} + \frac{2lm}{H}\left(m^2 - \frac{l^2}{\sin^2\theta}\right)\frac{\partial^2 T}{\partial s\partial n}\right.$$

$$+\left(m^4 + \frac{l^4}{\sin^2\theta}\right)\frac{\partial^2 T}{\partial n^2} + \left[\frac{lm^2}{H^2}\left(2l' - \frac{l}{H}H'\right)\right.$$

$$\left. + \frac{l}{H^2}\left(2l' - \frac{l}{H}H'\right)\frac{\partial T}{\partial s} - \frac{l}{H}m'\frac{\partial T}{\partial n}\right\} e^{i\omega t}$$

$$\tau_{sr}^{(1)} = -\frac{GC_1}{r^2\sqrt{r}}\langle[2\alpha r Z_z'(\alpha r) - 3Z_z(\alpha r)$$

$$+ 2\lambda r Z_z'(\lambda r) - \varphi_3(z,r)Z_z(\lambda r)]\left[\left(\frac{1}{H}l^2 + \frac{m^2}{\sin\theta}\right)\frac{\partial T}{\partial s}\right.$$

$$\left. + \left(1 - \frac{1}{\sin\theta}\right)ml\frac{\partial T}{\partial n}\right]e^{i\omega t}$$

$$\tau_{nr}^{(1)} = \frac{GC_1}{r^2\sqrt{r}}\{[2\alpha r Z_z'(\alpha r) - 3Z_z(\alpha r)$$

$$+ 2\lambda r Z_z'(\lambda r) - \varphi_3(z,r)Z_z(\lambda r)]\left(\frac{lm}{H}\frac{\partial T}{\partial s} + m^2\frac{\partial T}{\partial n}\right)$$

$$- [2Z_z(\alpha r) - 2\lambda r Z_z'(\lambda r) - Z_z(\lambda r)]\frac{ctg\theta}{\sin\theta}$$

$$\times\left(-\frac{ml}{H}\frac{\partial T}{\partial s} + l^2\frac{\partial T}{\partial n}\right)\bigg\}e^{i\omega t}$$

$$U_n^{(1)} = \frac{C_1}{\sqrt{r}}\left[\frac{1}{r}Z_z(\alpha r) - \lambda Z_z'(\lambda r) - \frac{1}{2r}Z_z(\lambda r)\right]$$
$$\times\left(\frac{l^2}{H} - \frac{m^2}{\sin\theta}\right)\frac{\partial T}{\partial s} + \left(1 + \frac{1}{\sin\theta}\right)ml\frac{\partial T}{\partial n}\Bigg] e^{i\omega t}$$
$$U_s^{(1)} = \frac{C_1}{\sqrt{r}}\left[\frac{1}{r}Z_z(\alpha r) - \lambda Z_z'(\lambda r) - \frac{1}{2r}Z_z(\lambda r)\right]$$
$$\times\left[-\left(1 + \frac{1}{\sin\theta}\right)ml\frac{\partial T}{\partial s} + \left(m^2 + \frac{l^2}{\sin\theta}\right)\frac{\partial T}{\partial n}\right] e^{i\omega t}$$

$$\sigma_n^{(2)} = \frac{GC_1}{r\sqrt{r}}Z_z(\lambda r)\left\{\frac{ml(m^2 - l^2)}{\sin\theta}\left(\frac{\partial^2 T}{\partial n^2} - \frac{1}{H^2}\frac{\partial^2 T}{\partial s^2}\right)\right.$$
$$-\frac{1}{H\sin\theta}\frac{\partial^2 T}{\partial s\partial n} + \left[\frac{ml^2}{H^2\sin\theta}\left(2l' - \frac{l}{H}H'\right) - \frac{lm^2}{H^2\sin\theta}\right.$$
$$\times\left(2m' - \frac{m}{H}H'\right) - \frac{1}{RH^2\sin\theta} + \frac{ctg\theta}{\sin\theta}\frac{m(m^2 - l^2)}{H}$$
$$\left.+2ctg\theta\frac{ml^2}{H}\right]\frac{\partial T}{\partial s} + \left[\frac{1}{\sin\theta}\frac{ml}{H}(lm' - ml') - \frac{ctg\theta}{\sin\theta}\right.$$
$$\left.\times l(m^2 - l^2) + 2ctg\theta lm^2\right]\frac{\partial T}{\partial n} + \left(z^2 - \frac{1}{4}\right)mlT(s,n)\Bigg\} e^{i\omega t}$$
$$\tau_{ns}^{(2)} = \frac{2GC_2}{r\sqrt{r}}Z_z(\lambda r)\left\{-\frac{2lm^3}{H^2\sin\theta}\frac{\partial^2 T}{\partial s^2} + \frac{4m^2l^2}{H\sin\theta}\frac{\partial^2 T}{\partial s\partial n}\right.$$
$$-\frac{2ml^3}{\sin\theta}\frac{\partial^2 T}{\partial n^2} + \left[-\frac{2lm^2}{H^2\sin\theta}\left(2m' - \frac{m}{H}H'\right) - \frac{2ctg\theta}{\sin\theta}\frac{lm^2}{H}\right.$$
$$\left.+\frac{l}{H}(m^2 - l^2)ctg\theta\right]\frac{\partial T}{\partial s} + \left[-\frac{2lm^2l'}{H\sin\theta} + 2ml^2\frac{ctg\theta}{\sin\theta}\right.$$
$$\left.+m(m^2 - l^2)ctg\theta\right]\frac{\partial T}{\partial n} + \frac{1}{2}\left(z^2 - \frac{1}{4}\right)(m^2 - l^2)T(s,t)\Bigg\} e^{i\omega t}$$
$$\tau_{nr}^{(2)} = \frac{GC_2}{r\sqrt{r}}\left\{\left[\frac{3}{2}Z_z(\lambda r) - \lambda r Z_z'(\lambda r)\right]\left(\frac{lm}{H}\frac{\partial T}{\partial s} + m^2\frac{\partial T}{\partial n}\right)\right.$$
$$\left.+Z_z(\lambda r)\left[2ctg\theta\left(-\frac{ml}{H}\frac{\partial T}{\partial s}\right) + \left(z^2 - \frac{1}{4}\right)lT(s,n)\right]\right\} e^{i\omega t}$$
$$\tau_{sr}^{(2)} = -\frac{GC_2}{2r\sqrt{r}}\left[3Z_z(\lambda r) - 2\lambda r Z_z'(\lambda r)\right]\left[\left(1 + \frac{1}{\sin\theta}\right)\right.$$
$$\left.\times\frac{ml}{H}\frac{\partial T}{\partial s} + \left(m^2 - \frac{l^2}{\sin\theta}\right)\frac{\partial T}{\partial n}\right] e^{i\omega t}$$

$$U_n^{(2)} = \frac{C_2}{\sqrt{r}} Z_z(\lambda r)\left\{\frac{1}{\sin\theta}\frac{lm^2}{H^2}\frac{\partial^2 T}{\partial s^2} - \frac{2ml^2}{H\sin\theta}\frac{\partial^2 T}{\partial s \partial n} + \frac{l^3}{\sin\theta}\frac{\partial^2 T}{\partial n^2} + \left[\frac{ml}{H\sin\theta}\left(2m' - \frac{m}{H}H'\right) - \frac{lm}{H}\right]\frac{\partial T}{\partial s} - \left(\frac{ml}{H\sin\theta}l' + m^2\right)\frac{\partial T}{\partial n}\right]e^{i\omega t}$$

$$U_s^{(2)} = -\frac{C_2}{\sqrt{r}} Z_z(\lambda r)\left\{\frac{1}{\sin\theta}\frac{m^3}{H^3}\frac{\partial^2 T}{\partial s^2} - \frac{2lm^2}{H\sin\theta}\frac{\partial^2 T}{\partial s \partial n} + \frac{ml^2}{\sin\theta}\frac{\partial^2 T}{\partial n^2} + \left[\frac{m^2}{H\sin\theta}\left(2m' - \frac{m}{H}H'\right) - \frac{l^2}{H}\right]\frac{\partial T}{\partial s} + \left(lm - \frac{lm}{H}l'\right)\frac{\partial T}{\partial n}\right]e^{i\omega t} \tag{2.7.33}$$

$$\sigma_n = \sigma_n^{(1)} + \sigma_n^{(2)}, \quad \sigma_s = \sigma_s^{(1)} + \sigma_s^{(2)}$$

$$\tau_{ns} = \tau_{ns}^{(1)} + \tau_{ns}^{(2)}, \quad \tau_{nr} = \tau_{nr}^{(1)} + \tau_{nr}^{(2)}$$

$$\sigma_r = \sigma_r^{(1)}, \quad U_r = U_r^{(1)}$$

$$U_n = U_n^{(1)} + U_n^{(2)}, \quad U_s = U_s^{(1)} + U_s^{(2)}.$$

It is well-known that the solution of (2.7.22) is given in terms of the functions $T(\theta,\varphi)\cos n\varphi K_{z-1/2}^n(\theta)$ where

$$K_z^n(\theta) = A_z P_z^n(\cos\theta) + B_z Q_z^n(\cos\theta)$$

and A_z and B_z are arbitrary constants together with the associated Legendre functions P_z^n, Q_z^n of order n.

However, it is sometimes convenient to use approximate approaches. Therefore, in the case $\theta = \theta_j(\varphi)$ it is advisable to write this equation in the variables n and s. Using (2.7.32), we obtain

$$I_1(s,n)\frac{\partial^2 T}{\partial n^2} - 2ctg\theta\frac{ml}{H}\frac{\partial^2 T}{\partial s \partial n} + I_2(s,n)\frac{\partial^2 T}{\partial s^2} + [I_3(s,n) + mctg\theta]\frac{\partial T}{\partial n} + I_4(s,n)\frac{\partial T}{\partial s} + \left(z^2 - \frac{1}{4}\right)T(s,n) \tag{2.7.34}$$

where

$$I_1(s,n) = m^2 + \frac{l^2}{\sin^2\theta}, I_2(s,n) = \frac{1}{H^2}\left(l^2 + \frac{m^2}{\sin^2\theta}\right)$$
$$I_3(s,n) = \frac{1}{H}\left(lm' - \frac{ml'}{\sin^2\theta}\right)$$
$$I_4(s,n) = \frac{1}{H}\left[lctg\theta + \frac{m}{H\sin^2\theta}\left(2m' - \frac{m}{H}H'\right) + \frac{l}{H}\left(2l' - \frac{l}{H}H'\right)\right].$$

Equation (2.7.34) is studied by asymptotic methods in [2].

If we now use the asymptotic behaviour of the function $T(s,n)$ and the results of the Sects. 2.5–2.6, it is easy to obtain simple asymptotic formulas for the stresses and displacements which we will not consider here.

Now we formulate a variational principle to reduce the boundary value problems to solving infinite systems of linear algebraic equations in the case $\theta_j = \theta_j(\varphi)$.

We assume that the following system of strains is specified at each point on the side surface $[\theta = \theta_1(\varphi)]$ of the layer:

$$\sigma_n^0(s,r)e^{i\omega t}, \tau_{ns}^0(s,r)e^{i\omega t}, \tau_{nr}^0(s,r)e^{i\omega t}$$

which, of course, are also expressed in terms of $Q_\theta e^{i\omega t}$, $Q_{r\theta}e^{i\omega t}$, $Q_{\theta\varphi}e^{i\omega t}$.

To satisfy the boundary conditions on the side surface, similar to the axisymmetric case, we use the variational principle of Hamilton. Since the displacements (2.7.27) and (2.7.29) and the corresponding strains (2.7.28), (2.7.30) are the exact solutions of the equations of motion, from the variational principle we obtain

$$\int_{\partial s}^{0}\int_{r_1}^{r_2}\left[(\sigma_n^* - \sigma_n^0)\delta U_n^* + (\tau_{ns}^* - \tau_{ns}^0)\delta U_s^* + (\tau_{nr}^* - \tau_{nr}^0)\delta U_r^*\right] rdrds = 0. \quad (2.7.35)$$

Here ∂s is the boundary contour, $\sigma_n^*, \ldots, U_n^*$ are amplitude values of the corresponding quantities.

From (2.7.35) we obtain an infinite system of algebraic equations.

References

1. Goldenveizer, A.L.: An approximate theory of bending shells by means of asymptotic integration of equations of elasticity theory. J. Appl. Math. Mech. **27**(4), 593–608 (1963)
2. Boyev, N.V.: Asymptotic analysis of three-dimensional stress-strain state of a radially inhomogeneous spherical shell: Diss. of candi. of math-phys sci, 177 p. Rostov-on-Don (1981)

3. Wilensky, T.V., Vorovich, I.I.: Asymptotic behavior of the solution of the problem of elasticity theory for a spherical shell of small thickness. J. Appl. Math. Mech. **30**(2), 278–295 (1966)
4. Vasilenko, A.T.: To the assessment of some assumptions of the theory of shells. Reports of Ukrainian Academy of Sciences, № 4, pp. 306–309 (1978)
5. Hrinchenko, V.T., Myaleshka, V.V.: Harmonic vibrations and waves in elastic bodies, 283 p. Naukova Dumka, Kiev. (1981)
6. Lurie, A.I.: Statics of thin elastic shells, p. 252. Gostekhizdat, Moscow (1947)
7. Aksentyan, O.C., Vorovich, I.I.: Stress state a of plate with small thinness. PMM **27**(6), 1057–1074 (1963)
8. Puro, A.E.: Separation of equations of the theory of elasticity in the radial inhomogeneity. J. Appl. Math. Mech. **38**(6), 1139–1144 (1974)

Chapter 3
Free Vibrations of Isotropic Hollow Cylinder and Closed Hollow Sphere

Abstract Construction of an asymptotic process for finding the frequencies of free axisymmetric vibrations of a hollow cylinder and a closed hollow sphere is presented. It is well known that, even in relatively simple cases the analysis of the frequency equations is quite challenging and yet it is essential to determine all the frequencies in a certain frequency range. The author's approaches allow to devise algorithms that are able to capture all natural frequencies in a given interval which undoubtedly represents a scientific and practical value. Comparison of the results obtained by Kirchhoff-Love's and Timoshenko's theories with the results obtained by the three-dimensional elasticity theory are illustrated.

In the study of forced vibrations and other quasi-stationary processes for finite solids it is important to know the natural frequencies and the corresponding forms of vibrations. Determination of natural frequencies of shells has a long history which may be attributed to classical problems of mathematical physics. However, the results presented in [1], where proceeding from the classical linear two-dimensional equations of shell theory the authors examine the frequency spectra of free vibrations of shells, show that the relevant mathematical problems are not always simple even within the classical theory of shells. The difficulties that arise here are connected with the fact that the equations with variable coefficients in a certain frequency range have the so-called branching points in the domain of integration. In the neighbourhood of these points the usual asymptotic expansions become unusable. In the paper mentioned above the asymptotic method of solving the problem of free vibrations of shells is developed for the case when there are branching points in the domain of integration. It is not possible to list all the activities related to the free vibrations of shells, however the initial state of the problem is described in the monograph by Oniashvili [2]; further recent results are given in Gontkevich [3], Birger and Panovko [4]. Therefore, here we restrict ourselves with only certain information in [5] Goldenveyzer's asymptotic method which is employed within the three-dimensional elasticity theory for the analysis of vibrations of the elastic layer. Bolotin's asymptotic method is widely used in the framework of two-dimensional theories and is also used in the analysis of natural

M. F. Mekhtiev, *Vibrations of Hollow Elastic Bodies*,
Advanced Structured Materials 88,
https://doi.org/10.1007/978-3-319-74354-7_3

vibrations of the plates on the basis of equations of elasticity theory [6]. In [7] Vorovich's variant of asymptotic method is successfully used in the study of dynamic elasticity problems for a plate. In particular, effective relationships are obtained for the natural frequencies of plates. In [8] the eigenfunction method is employed to investigate both the features of the frequency spectra and the natural modes of vibrations of the end plates. A complete review of literature on the methods of solving dynamic boundary value problems of elasticity theory is also presented in the same work. The natural vibrations of a plate under shear deformations are investigated in [9].

However, the very important question of determining the vibration frequencies of the shells examined by three-dimensional theory of elasticity and comparison of the results with those obtained by the applied theories remains open. The present chapter is dedicated to this issue. The algorithm for solving the problem of forced vibrations developed in the previous chapters allows us to find the natural frequencies within a certain accuracy, since, in the general case, the equation that determines the natural frequencies is actually equivalent to the vanishing of the determinant of the infinite systems. Here, we may take the following steps: a wide interval containing the natural frequency is fixed. The subsequent fragmentation of this range enables one to find both the natural frequency and the corresponding waveform characteristics with a high accuracy. However, as noted in [8], when approaching the natural frequency, the accuracy of the boundary conditions deteriorates at a fixed order of a finite system. This imposes a restriction on the division of the frequency interval. In a sense, this restricts the degree of accuracy of the determination of the natural frequency at the selected order of the finite system. On the other hand, the determination of the interval containing the natural frequency itself is kind of a daunting task.

It will be shown below that, it is not mandatory to resort to infinite systems to determine the natural frequencies.

In all cases of practical importance, it is possible to construct an asymptotic process for finding the natural frequencies without resorting to infinite systems. This chapter provides an asymptotic process for finding the frequencies of free axisymmetric vibrations of an isotropic hollow cylinder and a closed hollow sphere. The asymptotic process is presented in details for a cylinder with free side surfaces and simply supported at the ends and a closed hollow sphere with free face surfaces.

These problems can be considered as the model ones, since the study of the asymptotic process for other types of boundary conditions presents no fundamental difficulties.

3.1 Free Vibrations of an Isotropic Hollow Cylinder

We consider the axisymmetric problem of natural vibrations of an isotropic hollow cylinder with the following boundary conditions:

$$\sigma_r = 0, \tau_{rz} = 0, \tau_{r\varphi} = 0 \quad \text{when } r = R_n \ (n = 1, 2) \tag{3.1.1}$$

$$U_r = 0, \sigma_z = 0, \tau_{z\varphi} = 0 \quad \text{when } z = \pm l \tag{3.1.2}$$

The solution is sought in the form

$$\begin{aligned} U_r &= U(\rho)\sin p\xi e^{i\omega t}, \quad U_z = W(\rho)\cos p\xi e^{i\omega t} \\ U_\varphi &= \upsilon(\rho)\cos p\xi e^{i\omega t}, \quad p = \frac{\pi}{l}k \end{aligned} \tag{3.1.3}$$

Satisfying the system of Lame's differential equations, the boundary conditions (3.1.2) and the results of the first chapter, we obtain the following equations for $U(\rho)$, $W(\rho)$ and $\upsilon(\rho)$:

$$\begin{aligned} U(\rho) &= -\alpha J_1(\alpha\rho)C_1 - \alpha Y_1(\alpha\rho)C_2 + pJ_1(\gamma\rho)C_3 + pY_1(\gamma\rho)C_4 \\ W(\rho) &= pJ_0(\alpha\rho)C_1 + pY_0(\alpha\rho)C_2 + \gamma J_0(\gamma\rho)C_3 + \gamma Y_0(\gamma\rho)C_4 \\ \upsilon(\rho) &= b_1 J_1(\gamma\rho) + b_2 Y_1(\gamma\rho) \\ \alpha^2 &= \frac{1}{2}\frac{1-2v}{1-v}\lambda^2 - p^2, \quad \gamma^2 = \lambda^2 - p^2. \end{aligned} \tag{3.1.4}$$

Taking into account the boundary conditions (3.1.1), we obtain a system of linear algebraic equations with respect to the constants $C_i (i = 1, 2, 3, 4)$ and b_k $(k = 1, 2)$. The condition for the existence of nontrivial solutions of the system so obtained leads to the frequency equations for λ^2 given as

$$\begin{aligned} \Delta_1(\lambda^2, \rho, \rho_1, \rho_2) = {} & -8\pi^{-2}\rho_1^{-1}\rho_2^{-1}p^2\delta^4 \\ & - 4^{-1}\rho_1^{-1}\rho_2^{-1}\lambda^4\alpha^2 L_{11}(\alpha)L_{11}(\gamma) + 2^{-1}\rho_1^{-1}\lambda^2\alpha^2 p^2\gamma \\ & \times L_{10}(\gamma)L_{11}(\alpha) + 2^{-1}\rho_2^{-1}\lambda^2\alpha^2 p^2\gamma L_{01}(\gamma)L_{11}(\alpha) \\ & + 2^{-1}\rho_1^{-1}\lambda^2\alpha\delta^4 L_{10}(\alpha)L_{11}(\gamma) + 2^{-1}\rho_2^{-1}\lambda^2\alpha\delta^4 L_{01}(\alpha)L_{11}(\gamma) \\ & - \alpha^2\gamma^2 p^4 L_{00}(\gamma)L_{11}(\alpha) - \delta^8 L_{00}(\alpha)L_{11}(\gamma) - \alpha\gamma p^4\delta^4 \\ & \times [L_{01}(\alpha)L_{11}(\gamma) + L_{01}(\gamma)L_{10}(\alpha)] = 0 \end{aligned} \tag{3.1.5}$$

$$\begin{aligned} \Delta_2(\lambda^2, \rho, \rho_1, \rho_2) &= \gamma^2 L_{00}(\gamma) - \rho_1^{-1}\gamma L_{10}(\gamma) \\ &\quad - \rho_2^{-1}\gamma L_{01}(\gamma) + \rho_1^{-1}\rho_2^{-1}L_{11}(\gamma) \\ \delta^2 &= \frac{1}{2}\lambda^2 - p^2 \end{aligned} \tag{3.1.6}$$

The left sides of Eqs. (3.1.5) and (3.1.6), as a transcendental entire function of parameter λ^2, has a countable set of zeros with an accumulation point at infinity the asymptotic behaviour of which will be considered below.

3.2 Analysis of the Frequency Equation and Vibration Forms of a Cylinder

We analyse the zeros of frequency Eqs. (3.1.5) and (3.1.6). To this end, as in the first chapter, we make assumptions about the geometric parameters of the cylinder. We propose that

$$\rho_1 = 1 - \varepsilon, \quad \rho_2 = 1 + \varepsilon, \quad 2\varepsilon = \frac{2h}{R_0} \tag{3.2.1}$$

where ε is taken as a small parameter. Substituting (3.2.1) into (3.1.5) and (3.1.6), respectively, we obtain

$$D_1\left(\lambda^2, p, \varepsilon\right) = \Delta_1\left(\lambda^2, p, \rho_1, \rho_2\right) = 0 \tag{3.2.2}$$

$$D_2\left(\lambda^2, p, \varepsilon\right) = \Delta_2\left(\lambda^2, p, \rho_1, \rho_2\right) = 0 \tag{3.2.3}$$

Equation (3.2.3) will be discussed later. Let us start with the study of the zeros of the function $D_1\left(\lambda^2, p, \varepsilon\right)$. We will consider the case $p = 0$-separately.

We can formulate the following statement with respect to the zeros of: $D_1\left(\lambda^2, p, \varepsilon\right)$: for any finite $p\left(p = O\left(\varepsilon^\beta\right),\ \beta \geq 0 \text{ at } \varepsilon \to 0\right)$ the function $D_1\left(\lambda^2, p, \varepsilon\right)$ has a finite number of zeros with the following asymptotic properties:

$$\Lambda_k = O(\varepsilon^q), \quad q \geq 0, \quad \Lambda^2 = \frac{m_0 \omega^2 R_0^2}{E}$$

We outline the proof of this statement. For this purpose, we expand the function $D_1\left(\lambda^2, p, \varepsilon\right)$ into a series in terms of the small variable ε;

$$\begin{aligned} D_1\left(\lambda^2, p, \varepsilon\right) = {} & 16(1+\nu)^2(1-\nu)^{-2}\Lambda^4\varepsilon^2\Big\langle -\left(1-\nu^2\right)^2\Lambda^4 \\ & + (1-\nu^2)p^2\Lambda^2 + (1-\nu^2)\Lambda^2 - (1-\nu^2)p^2 \\ & + \frac{1}{3}\big\{-p^6 + \left[2(1+\nu)(3-2\nu)\Lambda^2 - 4\left(1-\nu^2\right)\right]p^4 \\ & - \Big[(1+\nu)^2\left(4\nu^2 - 16\nu + 11\right)\Lambda^4 + 2(1+\nu)\left(2\nu^2 + 9\nu - 9\right)\Lambda^2 \\ & + 9\left(1-\nu^2\right)\Big]p^2 + 2\left(1-\nu^2\right)(1+\nu)^2(3-4\nu)\Lambda^6 - 2(1+\nu)^2 \\ & \times \left(6\nu^2 - 14\nu + 7\right)\Lambda^4 + 9\left(1-\nu^2\right)\Lambda^2\big\}\varepsilon^2 \\ & + 1/45\left(-8p^8 + \cdots\right)\varepsilon^4 + \cdots \Big\rangle = 0. \end{aligned} \tag{3.2.4}$$

Assume that the leading terms of the asymptotic behaviour of Λ_k and P have the form

$$\Lambda_k = \Lambda_{k0}\varepsilon^q, \quad p = p_0\varepsilon^\beta, \quad \Lambda_{k0} = O(1), \quad p_0 = O(1) \\ q \geq 0, \ \beta \geq 0. \tag{3.2.5}$$

Substituting (3.2.5) into (3.2.4), from the consistency condition of the constructed asymptotic process we see that there can only be the case $q = 0$ and $q = \beta$.

Note that here and henceforth the main interval of variation of parameters q and β will be sometimes split into sub-intervals, since depending on which interval contains q and β the zeros of $D_1\left(\lambda^2, p, \varepsilon\right)$ have different asymptotic representations. The range of variation of parameters q and β is shown in Fig. 3.1.

In the first case ($q = 0$, $p = p_0\varepsilon^\beta$, $\beta > 0$) we search Λ_k ($k = 1$) in the form of

$$\begin{aligned} \Lambda_k &= \Lambda_{k0} + \Lambda_{k2}\varepsilon^{2\beta} + \cdots \quad 0<\beta<1 \\ \Lambda_k &= \Lambda_{k0} + \Lambda_{k2}\varepsilon^{2} + \cdots \quad \beta \geq 1 \end{aligned} \tag{3.2.6}$$

On substituting (3.2.6) into (3.2.4) we get

$$\begin{aligned} \Lambda_{k0} &- \frac{1}{\sqrt{1-\nu^2}}, \quad \Lambda_{k2} - \frac{\nu^2 p_0^2}{2\sqrt{1-\nu^2}} && 0<\beta<1 \\ \Lambda_{k2} &= \frac{1}{2\sqrt{1-\nu^2}}\left[\nu^2 p_0^2 + \frac{3\nu+1}{3(1-\nu)}\right] && \beta = 1 \\ \Lambda_{k2} &= \frac{3\nu+1}{6(1-\nu)\sqrt{1-\nu^2}} && \beta > 1 \end{aligned} \tag{3.2.7}$$

In the second case ($q = \beta$) we seek Λ_k ($k = 2$) as

$$\Lambda_k = \Lambda_{k0}\varepsilon^\beta + \Lambda_{k2}\varepsilon^{3\beta} + \cdots \tag{3.2.8}$$

Then from (3.2.4) we obtain

$$\Lambda_{k0} = p_0, \ \Lambda_{k2} = -\frac{\nu^2 p_0^3}{2} \tag{3.2.9}$$

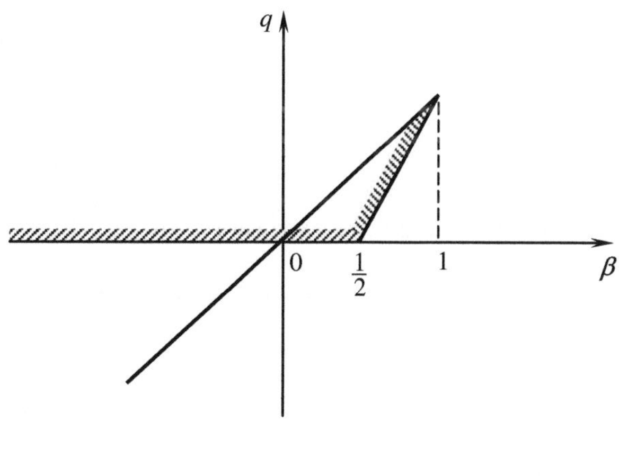

Fig. 3.1 Domain of variation of the parameter of frequency for a cylinder

These frequencies are the so-called ultra-low frequency vibrations.

Finally, consider the case when $q = \beta = 0$. Then we seek Λ_k in the following form

$$\begin{aligned}
&\Lambda_k = \Lambda_{k0} + \varepsilon^2 \Lambda_{k2} + \cdots \quad (k = 1,2) \\
&\Lambda_{k0}^2 = 2^{-1}\left(1 - \nu^2\right)^{-1}\left[p^2 + 1 - (-1)^k \sqrt{\left(p^2 - 1\right)^2 + 4\nu^2 p^2}\right]
\end{aligned} \tag{3.2.10}$$

$$\begin{aligned}
\Lambda_{k2} = {} & 6^{-1}\Lambda_{k0}^{-1}\left(1 - \nu^2\right)^{-1}\left[p^2 + 1 - 2\left(1 - \nu^2\right)\Lambda_{k0}^2\right]^{-1} \\
& \times \left\{p^6 + \left[4\left(1 - \nu^2\right) - 2(1+\nu)(3 - 2\nu)\Lambda_{k0}^2\right]p^4\right. \\
& + \left[(1+\nu)^2\left(4\nu^2 - 16\nu + 11\right)\Lambda_{k0}^4 + 2(1+\nu)\right. \\
& \times \left.\left(2\nu^2 + 9\nu - 9\right)\Lambda_{k0}^2 + 9\left(1 - \nu^2\right)\right]p^2 \\
& - 2(1+\nu)^2\left(1 - \nu^2\right)(3 - 4\nu)\Lambda_{k0}^6 + 2(1+\nu)^2 \\
& \times \left.\left(6\nu^2 - 14\nu + 7\right)\Lambda_{k0}^4 - 9\left(1 - \nu^2\right)\Lambda_{k0}^2\right\}
\end{aligned} \tag{3.2.11}$$

We have thus shown that for fixed finite p there are two natural frequencies.

Consider the case where p increases infinitely with $\varepsilon \to 0$. Here we consider only the following limiting cases: $p\varepsilon \to 0$ when $\varepsilon \to 0$; $p\varepsilon \to const$ when $\varepsilon \to 0$.

We first define those Λ_k $(k = 1,2)$ in the limiting case $p\varepsilon \to 0$ at $\varepsilon \to 0$. To do this, we again use the expansions (3.2.4). Assume that the leading terms of the asymptotics Λ_k and P have the form:

$$\begin{aligned}
&\Lambda_k = \Lambda_{k0}\varepsilon^{-q}, \quad p = p_0\varepsilon^{-\beta}, \quad \Lambda_{k0} = O(1) \\
&p_0 = O(1), \quad 0 \le q < 1, \quad 0 < \beta < 1
\end{aligned} \tag{3.2.12}$$

It is easy to prove that $q \le \beta$. Here we separately consider the cases when $q = 0$ and $q = \beta$, $q \ne 0$.

In the first case from the expansion (3.2.4) we get that $0 < \beta < \frac{1}{2}$. The case $\beta = \frac{1}{2}$ is considered separately. Let us seek Λ_k $(k = 1)$ in the form

$$\Lambda_k = \Lambda_{k0} + \Lambda_{k2}\varepsilon^{2\beta} + \cdots, \quad 0 < \beta \le \frac{1}{3} \tag{3.2.13}$$

$$\Lambda_k = \Lambda_{k0} + \Lambda_{k2}\varepsilon^{2-4\beta} + \cdots, \quad \frac{1}{3} < \beta < \frac{1}{2} \tag{3.2.14}$$

On substituting these expansions into (3.2.4), we obtain

$$\begin{aligned}
&\Lambda_{k0} = 1, \Lambda_{k2} = -\tfrac{\nu^2}{2p_0^2} \quad 0 < \beta < \tfrac{1}{3} \\
&\Lambda_{k2} = \tfrac{P_0^4}{6(1-\nu^2)} - \tfrac{\nu^2}{2p_0^2} \quad \beta = \tfrac{1}{3}
\end{aligned} \tag{3.2.15}$$

$$\Lambda_{k2} = \frac{p_0^4}{6(1-\nu^2)}, \quad \frac{1}{3} < \beta < \frac{1}{2} \tag{3.2.16}$$

In the case $q = 0$, $\beta = \frac{1}{2}$, we obtain

$$\begin{aligned}
\Lambda_k &= \Lambda_{k0} + \varepsilon\Lambda_{k1} + \cdots \quad (k = 2)\\
\Lambda_{k0} &= \sqrt{1 + 3^{-1}(1-\nu^2)^{-1}P_0^4}\\
\Lambda_{k1} &= \frac{\Lambda_{k0}}{10}\left[\frac{12 - 17\nu^2}{p_0^2} - \frac{12(1-\nu^2)}{\Lambda_{k0}^2 p_0^2} + \frac{7\nu - 17}{3(1-\nu)}p_0^2\right]
\end{aligned} \tag{3.2.17}$$

Similarly, in the case of $q = \beta$ we obtain the following from (3.2.4):

$$\begin{aligned}
\Lambda_k &= \Lambda_{k0}\varepsilon^{-\beta} + \Lambda_{k1}\varepsilon^{\beta} + \cdots, & 0 < \beta \le \tfrac{1}{2} \quad (k = 1)\\
\Lambda_k &= \Lambda_{k0}\varepsilon^{-\beta} + \Lambda_{k1}\varepsilon^{2-3\beta} + \cdots, & \tfrac{1}{2} < \beta < 1\\
\Lambda_{k0} &= \tfrac{P_0}{\sqrt{1-\nu^2}}, \Lambda_{k1} = \tfrac{\nu^2}{2\sqrt{1-\nu^2}}\tfrac{1}{p_0} & 0 < \beta < \tfrac{1}{2}\\
\Lambda_{k1} &= \tfrac{\nu^2}{2\sqrt{1-\nu^2}}\left[\tfrac{1}{p_0} - \tfrac{p_0^3}{3(1-\nu)^2}\right] & \beta = \tfrac{1}{2}\\
\Lambda_{k1} &= -\tfrac{\nu^2 p_0^3}{6(1-\nu)^2\sqrt{1-\nu^2}} & \tfrac{1}{2} < \beta < 1
\end{aligned} \tag{3.2.18}$$

In the case $q \ne 0$, $q < \beta$, substituting (3.2.12) into (3.2.4) and keeping only the leading terms, we obtain the following limiting equation for λ_k:

$$\begin{aligned}
D_1(\lambda^2, p, \varepsilon) &= 16\left(\frac{1+\nu}{1-\nu}\right)^2 \varepsilon^{\alpha_1}\Lambda_{k0}^4\left\langle\left[(1-\nu^2)\Lambda_{k0}^2 p_0^2 + O(\varepsilon^{\alpha_2})\right]\varepsilon^{\alpha_3}\right.\\
&\quad \left. + \frac{1}{3}\left\{-p_0^6 + O[\max(\varepsilon^{\alpha_4}, \varepsilon^{\alpha_5})]\right\}\varepsilon^{\alpha_6}\right\rangle = 0\\
\alpha_1 &= 2 - 4q, \alpha_2 = 2(\beta - q), \alpha_3 = -2(q + \beta)\\
\alpha_4 &= 4 - 2\beta, \alpha_5 = 2 - 2q - 2\beta, \alpha_6 = 2 - 6\beta
\end{aligned} \tag{3.2.19}$$

Hence, we get $q = 2\beta - 1$. And from the condition $q > 0$ we have $\beta > \frac{1}{2}$. Thus $\frac{1}{2} < \beta < 1$. Now we seek Λ_k in the form

$$\begin{aligned}
\Lambda_k &= \Lambda_{k0}\varepsilon^{1-2\beta} + \Lambda_{k2}\varepsilon^{2\beta-1} + \cdots, & \tfrac{1}{2} < \beta < \tfrac{2}{3}\\
\Lambda_k &= \Lambda_{k0}\varepsilon^{1-2\beta} + \Lambda_{k2}\varepsilon^{3-4\beta} + \cdots, & \tfrac{2}{3} \le \beta < 1
\end{aligned} \tag{3.2.20}$$

Substituting the expansions (3.2.20) in (3.2.4), we obtain

$$\begin{aligned}
\Lambda_{k0} &= \tfrac{p_0^2}{\sqrt{3(1-\nu^2)}}, \quad \Lambda_{k2} = \tfrac{\sqrt{3(1-\nu^2)}}{2p_0^2} & \tfrac{1}{2} < \beta < \tfrac{2}{3}\\
\Lambda_{k2} &= \tfrac{\sqrt{3(1-\nu^2)}}{2p_0^2} + \tfrac{(7\nu-17)p_0^4}{30(1-\nu)\sqrt{3(1-\nu^2)}} & \beta = \tfrac{2}{3}\\
\Lambda_{k2} &= \tfrac{7\nu-17}{30(1-\nu)\sqrt{3(1-\nu^2)}}p_0^4 & \tfrac{2}{3} < \beta < 1
\end{aligned} \tag{3.2.21}$$

Consider the second case where $p\varepsilon \to const$ for $\varepsilon \to 0$; ($\lambda\varepsilon \to const$ as $\varepsilon \to 0$). Let λ_n ($n = k-2, k = 3, 4, \ldots$) be sought in the form

$$\lambda_n = \frac{\delta_n}{\varepsilon} + O(\varepsilon), p = p_0\varepsilon^{-1}, (n = 1, 2, \ldots). \tag{3.2.22}$$

On substituting (3.2.22) into (3.1.5) and utilizing the asymptotic expansions of the functions $J_\nu(x)$, $Y_\nu(x)$ for large values of their arguments we obtain the equation for δ:

$$\begin{aligned} &\left[\left(\delta_n^2 - 2p_0^2\right)^2 \sin\alpha_n \cos\gamma_n + 4\alpha_n\gamma_n p_0^2 \cos\alpha_n \sin\gamma_n\right] \\ &\quad\times \left[\left(\delta_n^2 - 2p_0^2\right)^2 \cos\alpha_n \sin\gamma_n + 4\alpha_n\gamma_n p_0^2 \sin\alpha_n \cos\gamma_n\right] = 0 \\ &\alpha_n^2 = \frac{1-2\nu}{2(1-\nu)}\delta_n^2 - p_0^2, \gamma_n^2 = \delta_n^2 - p_0^2 \end{aligned} \tag{3.2.23}$$

At a given p_0, the transcendental Eq. (3.2.23) defines a countable set of λ_n. It should be noted that Eq. (3.2.23) actually coincides with the Rayleigh-Lamb frequency equation for an elastic layer [8].

In principle, there is a case particular case:

$$p = p_0\varepsilon^{-1}, \lambda = O(1) \quad (p \gg \lambda) \text{ as } \varepsilon \to 0 \tag{3.2.24}$$

In this case,

$$\begin{aligned} \alpha = i\alpha^*, \gamma = i\gamma^*, J_\nu(\alpha x) &= J_\nu(i\alpha^* x) \\ Y_\nu(\alpha x) = Y_\nu(i\alpha^* x), J_\nu(\gamma x) &= J_\nu(i\gamma^* x), \\ Y_\nu(\gamma x) = Y_\nu(i\gamma^* x), \gamma^* &= \sqrt{p^2 - \lambda^2} \\ \alpha^* &= \sqrt{p^2 - 2^{-1}(1-2\nu)\lambda^2/(1-\nu)} \end{aligned} \tag{3.2.25}$$

After substituting (3.2.24), (3.2.25) into (3.1.5) and referring to the asymptotic expansions of the functions $J_\nu(x), Y_\nu(x)$ for large values of their arguments we obtain the following expression in the first term of the asymptotic behaviour of $D_1(\lambda^2, p, \varepsilon)$:

$$D_1(\lambda^2, p, \varepsilon) = sh^2 p_0 - 4p_0^2 + O(\varepsilon) \tag{3.2.26}$$

A real parameter p cannot be a solution of the equation $sh^2 2p_0 - 4p_0^2 = 0$, as this equation has only complex zeros. Therefore, in this case $D_1(\lambda^2, p, \varepsilon) \neq 0$ and there is no free vibration. The shell can, of course, only perform forced vibrations.

As mentioned above, the case $p = 0$ is a particular one. In this case the boundary value problem is divided into two parts:

$$U_r = a_0(\rho)e^{i\omega t}, U_z \equiv 0 \quad (\tau_{rz} \equiv 0), \tag{3.2.27}$$

$$U_r \equiv 0, U_z = b_0(\rho)e^{i\omega t} \quad \left(\sigma_r = \sigma_\phi = \sigma_z \equiv 0\right). \tag{3.2.28}$$

Substituting (3.2.27) and (3.2.28) into the Lame's equation and the boundary conditions (3.1.1), respectively, we obtain

$$\begin{gathered} a_0'' + \tfrac{1}{\rho} a_0' + \left(\alpha_0^2 - \tfrac{1}{\rho^2}\right) a_0 = 0 \\ \alpha_0^2 = \tfrac{1-2v}{2(1-v)} \lambda^2 \end{gathered} \tag{3.2.29}$$

$$(1-v)a_0' + \frac{v}{\rho} a_0 = 0 \text{ in } \rho = \rho_s \quad (s = 1, 2) \tag{3.2.30}$$

$$b_0'' + \frac{1}{\rho} b_0' + \lambda^2 b_0 = 0 \tag{3.2.31}$$

$$b_0' = 0 \text{ in } \rho = \rho_s \tag{3.2.32}$$

The general solutions of (3.2.29) and (3.2.31) have the form:

$$\begin{aligned} a_0 &= C_1 J_1(\alpha_0 \rho) + C_2 Y_1(\alpha_0 \rho), \\ b_0 &= D_1 J_0(\lambda \rho) + D_2 Y_0(\lambda \rho) \end{aligned} \tag{3.2.33}$$

Fulfilling the boundary conditions (3.2.30) and (3.2.32), we obtain the frequency equations of the corresponding problems:

$$\begin{aligned} D_1^0(\alpha_0, \varepsilon) = {} & \alpha_0^2 L_{11}(\alpha_0)\rho_1^{-1}\rho_2^{-1} - \rho_1^{-1}\alpha_0\lambda^2 L_{10}(\alpha_0) \\ & - \rho_2^{-1}\alpha_0\lambda^2 L_{01}(\alpha_0) + 4^{-1}\lambda^4 L_{00}(\alpha_0) = 0 \end{aligned} \tag{3.2.34}$$

$$D_2^0(\lambda, \varepsilon) = L_{11}(\lambda) = 0 \tag{3.2.35}$$

It can be shown that Eq. (3.2.35) at the limit $\varepsilon \to 0$ has a bounded zero $\lambda = 0$ to which there corresponds the solution $b_0 = 1$ describing the motion of the shell as a solid body.

Equation (3.2.34) as $\varepsilon \to 0$ has one bounded zero with the following asymptotic property:

$$\Lambda_k = \Lambda_{k0} + \varepsilon^2 \Lambda_{k2} + \cdots \tag{3.2.36}$$

From (3.2.34) we obtain

$$\Lambda_{k0} = \tfrac{1}{\sqrt{1-v^2}}, \quad \Lambda_{k2} = \tfrac{3v+1}{6(1-v)\sqrt{1-v^2}} \tag{3.2.37}$$

Let us seek $\lambda_n^{(i)}$ in an expansion of the form

$$\lambda_n^{(i)} = \frac{S_n^{(i)}}{\varepsilon} + O(\varepsilon) \tag{3.2.38}$$

where $i = 1$ corresponds to Eq. (3.2.34) and $i = 2$ corresponds to Eq. (3.2.35). On substituting (3.2.38) into (3.2.34) and (3.2.35) we obtain

$$S_n^{(i)} = \sqrt{\frac{2(1-\nu)}{1-2\nu}\frac{n\pi}{2}} \tag{3.2.39}$$

Following the terminology of [8], these frequencies will be called locking frequencies. The solutions corresponding to the frequencies determined from formulas (3.2.36) and the first one of (3.2.39) describe the radial vibrations and the frequencies determined from the second formula of (3.2.39) describe purely longitudinal vibrations.

Note that the dispersion Eq. (3.2.2) has another group of zeros with the following asymptotic properties:

$$\lambda = s\varepsilon^{-1}, \quad p = p_0\varepsilon^{-\beta}, \; 0<\beta<1. \tag{3.2.40}$$

It is easy to see that in this case, in the first term of the asymptotic expansion we obtain the frequency defined by (3.2.36) and (3.2.38). Note that in the case $\lambda\varepsilon \to \infty$, $p\varepsilon \to \infty$ as $\varepsilon \to 0$ Eq. (3.2.39) remains valid.

Now we turn to the study of the roots of Eq. (3.2.3). Using the results of Sect. 1.6, we represent the function $D_2(\lambda, p, \varepsilon)$ in the form:

$$D_2(\lambda, p, \varepsilon) = \gamma^2 D_2^0(\lambda, p, \varepsilon). \tag{3.2.41}$$

It is easy to show that $D_2^0(\lambda, p, \varepsilon) \neq 0$ in the cases $\lambda = O(1)$ and $\lambda\varepsilon \to 0$ in the limit $\varepsilon \to 0$. This implies that the function γ^2 is a double zero of the function $D_2(\lambda, p, \varepsilon)$:

$$\gamma^2 = \lambda^2 - p^2 = 0, \quad \lambda^2 = p^2. \tag{3.2.42}$$

The remaining zeros of the function $D_2(\lambda, p, \varepsilon)$ grow infinitely as $\varepsilon \to 0$. They can be divided into two groups according to their behaviour as $\varepsilon \to 0$:

(a) $\varepsilon\lambda_k \to const, \quad \varepsilon \to 0;$
(b) $\varepsilon\lambda_k \to \infty, \quad \varepsilon \to 0.$

To construct this group of zeros by means of the asymptotic expansions of the Bessel functions for large values of their arguments, Eq. (3.2.3) can be written at the leading order of the asymptotic behaviour in the following form:

$$D_2(\lambda, p, \varepsilon) = \sin 2\gamma\varepsilon + O(\varepsilon). \tag{3.2.43}$$

Hence, specifying $p = p_0\varepsilon^{-\beta}$, $\beta \geq 1$, we seek

$$\lambda_k = \lambda_{k0}\varepsilon^{-\beta} + O(\varepsilon^{\beta})$$

From (3.2.43) we obtain

$$\lambda_{k0}^2 = \tfrac{k^2\pi^2}{4} + p_0^2 \quad \beta = 1 \quad (k = 1, 2, \ldots) \tag{3.2.44}$$

$$\lambda_{k0}^2 = \frac{k^2\pi^2}{4}\varepsilon^{2\beta-2} + p_0^2, \quad \beta > 1, \quad k = 0(\varepsilon^{2-2\beta}) \tag{3.2.45}$$

As mentioned above, the case $P = 0$ corresponds to the cylinder thickness variations. In this case, from (3.2.44) and (3.2.45), we obtain, respectively,

$$\begin{aligned} \lambda_{k0}^2 &= \frac{k^2\pi^2}{4} \quad (k = 1, 2, \ldots) \\ \lambda_{k0}^2 &= \frac{k^2\pi^2}{4}\varepsilon^{2\beta-2}, \quad k = O(\varepsilon^{2-2\beta}) \end{aligned} \tag{3.2.46}$$

Now for a comparison we present here the analysis of the frequency equations obtained by Kirchhoff-Love and Timoshenko theories. Frequency equations obtained in Kirchhoff-Love's theory have the form:

$$\begin{aligned} &(1-\nu^2)^2\Lambda^4 - (1-\nu^2)\Lambda^2p^2 - (1-\nu^2)\Lambda^2 \\ &+ (1-\nu^2)p^2 + \frac{1}{3}\left[p^6 - (1-\nu^2)\Lambda^2p^4\right]\varepsilon^2 = 0 \end{aligned} \tag{3.2.47}$$

$$\lambda^2 - p^2 + \frac{3}{4}p^2\varepsilon^2 = 0 \tag{3.2.48}$$

The following groups of zeros may be obtained from (3.2.48):

1.
$$\begin{aligned} &\Lambda_k = \Lambda_{k0} + \Lambda_{k2}\varepsilon^{2\beta} + \cdots \quad \beta > 0 \quad (k = 1) \\ &\Lambda_{k0} = \frac{1}{\sqrt{1-\nu^2}}, \quad \Lambda_{k2} = \frac{\nu^2p_0^2}{2\sqrt{1-\nu^2}} \end{aligned} \tag{3.2.49}$$

2.
$$\begin{aligned} &\Lambda_k = \Lambda_{k0}\varepsilon^{\beta} + \Lambda_{k2}\varepsilon^{3\beta} + \cdots \quad (k = 2) \\ &\Lambda_{k0} = p_0, \quad \Lambda_{k2} = \frac{\nu^2p_0^3}{2} \end{aligned} \tag{3.2.50}$$

3.
$$\Lambda_k = \Lambda_{k0} + \Lambda_{k2}\varepsilon^2 + \cdots \quad (k = 1, 2)$$
$$\Lambda_{k0}^2 = \frac{(1-\nu)^{-1}}{2}\left[p^2 + 1 - (-1)^k\sqrt{(p^2-1) + 4\nu^2 p^2}\right] \tag{3.2.51}$$
$$\Lambda_{k2} = \frac{1}{6(1-\nu^2)\Lambda_{k0}}\left[p^2 + 1 - 2(1-\nu^2)\Lambda_{k0}^2\right]^{-1}\left[p^6 - (1-\nu^2)\Lambda_{k0}^2 p^4\right]$$

4.
$$\Lambda_k = \Lambda_{k0} + \Lambda_{k2}\varepsilon^{2\beta} + \cdots \quad 0 < \beta \leq \tfrac{1}{3}$$
$$\Lambda_k = \Lambda_{k0} + \Lambda_{k2}\varepsilon^{2-4\beta} + \cdots \tfrac{1}{3} < \beta < \tfrac{1}{2}$$
$$\Lambda_{k0} = 1,\ \Lambda_{k2} = -\frac{\nu^2}{2P_0^2} \quad 0 < \beta < \tfrac{1}{3} \tag{3.2.52}$$
$$\Lambda_{k2} = \frac{p_0^4}{6(1-\nu^2)} - \frac{\nu^2}{2P_0^2} \quad \beta = \tfrac{1}{3}$$
$$\Lambda_{k2} = \frac{p_0^4}{6(1-\nu^2)} \quad \tfrac{1}{3} < \beta < \tfrac{1}{2}$$

5.
$$\Lambda_k = \Lambda_{k0} + \varepsilon\Lambda_{k1} + \cdots \quad (k = 2)$$
$$\Lambda_{k0} = \sqrt{1 + 3^{-1}(1-\nu^2)^{-1}p_0^4} \tag{3.2.53}$$
$$\Lambda_{k2} = -\frac{\nu^2}{2\Lambda_{k0}}\left[\frac{1}{p_0^2} + \frac{p_0^2}{3(1-\nu^2)}\right]$$

6.
$$\Lambda_k = \Lambda_{k0}\varepsilon^{-\beta} + \Lambda_{k1}\varepsilon^{\beta} + \cdots \quad (0 < \beta < 1)$$
$$\Lambda_{k0} = \frac{p_0}{\sqrt{1-\nu^2}}, \quad \Lambda_{k2} = \frac{\nu^2}{2\sqrt{1-\nu^2}}\frac{1}{p_0} \tag{3.2.54}$$

7.
$$\Lambda_k = \Lambda_{k0}\varepsilon^{1-2\beta} + \Lambda_{k2}\varepsilon^{2\beta-1} + \cdots \quad \frac{1}{2} < \beta < \frac{2}{3}$$
$$\Lambda_k = \Lambda_{k0}\varepsilon^{1-2\beta} + \Lambda_{k1}\varepsilon^{3-4\beta} + \cdots \quad \frac{2}{3} \leq \beta < 1$$
$$\Lambda_{k0} = \frac{p_0^2}{\sqrt{3(1-\nu^2)}}, \quad \Lambda_{k2} = \frac{\sqrt{3(1-\nu^2)}}{2p_0^2} \quad \frac{1}{2} < \beta \leq \frac{2}{3} \tag{3.2.55}$$
$$\Lambda_{k2} = 0 \quad \frac{2}{3} < \beta < 1$$

Similarly, from (3.2.48), we obtain

$$\lambda_k = \lambda_{k0} + \lambda_{k2}\varepsilon^2 + \cdots$$
$$\lambda_{k0}^2 = p^2, \quad \lambda_{k2} = -\frac{2}{3}\frac{p^2}{\lambda_{k0}} \tag{3.2.56}$$

Frequency equations in the Timoshenko type theory are of the form:

$$\begin{aligned}&K_1^2\left[(1-\nu^2)^2\Lambda^4-(1-\nu^2)(p^2+1)\Lambda^2+(1-\nu^2)p^2\right]\\&+\frac{1}{3}\{K_1^2p^6+(1-\nu^2)(1-\Lambda^2-2K_1^2\Lambda^2)p^4\\&+\left[(1-\nu^2)^2K_1^2\Lambda^4+2(1-\nu^2)^2\Lambda^4+(1-\nu^2)(\nu^2-2)\Lambda^2\right]p^2\\&+(1-\nu^2)^2\Lambda^4[1-(1-\nu^2)\Lambda^2]\}\varepsilon^2=0\end{aligned}\tag{3.2.57}$$

$$\gamma^2-p^2=0.\tag{3.2.58}$$

Equation (3.2.58) describes the torsional vibrations of a cylindrical shell based on Timoshenko type theory and it completely coincides with Eq. (3.2.42).

From (2.3.57) we obtain the following groups of zeros:

1.
$$\begin{aligned}&\Lambda_k=\Lambda_{k0}+\Lambda_{k2}\varepsilon^{2\beta}+\cdots\ (k=1)\ (q=0,\quad p=p_0\varepsilon^\beta)\\&\Lambda_{k0}=\frac{1}{\sqrt{1-\nu^2}},\quad \Lambda_{k2}=\frac{\nu^2}{2}\frac{p_0^2}{\sqrt{1-\nu^2}}\ \beta>0\end{aligned}\tag{3.2.59}$$

Comparing (3.2.49) and (2.3.59) with the exact expansion (3.2.6) shows that in the case of $0<\beta<1$ Kirchhoff-Love theory and Timoshenko-type theory correctly identify the two terms of the expansion of the roots of the frequency equation.

For $\beta\geq 1$ the applied theory of shells not only incorrectly identifies the second term but also distorts the order of the second term in the expansions.

2.
$$\begin{aligned}&\Lambda_k=\Lambda_{k0}\varepsilon^\beta+\Lambda_{k2}\varepsilon^{3\beta}+\cdots\quad (k=2)\quad p=p_0\varepsilon^\beta\\&\Lambda_{k0}=p_0,\quad \Lambda_{k2}=-\frac{\nu^2}{2}p_0^3\end{aligned}\tag{3.2.60}$$

In this case, Kirchhoff-Love theory and Timoshenko-type theory correctly identify the two terms of the expansion of the roots of the frequency equation.

3.
$$\begin{aligned}&\Lambda_k=\Lambda_{k0}+\Lambda_{k2}\varepsilon^2+\cdots(k=1,2)\quad p=O(1)\\&\Lambda_{k0}^2=2^{-1}(1-\nu^2)^{-1}\left[p^2+1-(-1)^k\sqrt{(p^2-1)+4\nu^2p^2}\right]\\&\Lambda_{k2}=6^{-1}(1-\nu^2)^{-1}\Lambda_{k0}^{-1}K_1^{-2}[p^2+1-2(1-\nu^2)\Lambda_{k0}^2]^{-1}\\&\quad\times\{K_1^2p^6+(1-\nu^2)(1-\Lambda_{k0}^2-2K_1^2\Lambda_{k0}^2)p^4\\&\quad+\left[(1-\nu^2)^2K_1^2\Lambda_{k0}^4+2(1-\nu^2)^2\Lambda_{k0}^4+(1-\nu^2)(\nu^2-2)\Lambda_{k0}^2\right]p^2\\&\quad+(1-\nu^2)^2\Lambda_{k0}^4[1-(1-\nu^2)\Lambda_{k0}^2]\}.\end{aligned}\tag{3.2.61}$$

The comparison of (3.2.51) and (2.3.61) with the exact expansion (3.2.10) shows that the applied theory of shells is true only to the first approximation.

4.
$$\begin{aligned}
&\Lambda_k = \Lambda_{k0} + \Lambda_{k2}\varepsilon^{2\beta} + \cdots \quad <\beta \le \tfrac{1}{3} \quad (k=1)\\
&\Lambda_k = \Lambda_{k0} + \Lambda_{k2}\varepsilon^{2-4\beta} + \cdots \quad \tfrac{1}{3} < \beta < \tfrac{1}{2} \quad p = p_0\varepsilon^{\beta}\\
&\Lambda_{k0} = 1, \quad \Lambda_{k2} = -\tfrac{\nu^2}{2P_0^2} \quad 0 < \beta < \tfrac{1}{3}\\
&\Lambda_{k2} = -\tfrac{\nu^2}{2p_0^2} + \tfrac{p_0^4}{6(1-\nu^2)} \quad \beta = \tfrac{1}{3}\\
&\Lambda_{k2} = \tfrac{p_0^4}{6(1-\nu^2)} \quad \tfrac{1}{3} < \beta < \tfrac{1}{2}.
\end{aligned} \tag{3.2.62}$$

In this case the applied theory of shells correctly identifies the two terms of the expansion

$$\begin{aligned}
&\Lambda_k = \Lambda_{k0} + \varepsilon\Lambda_{k1} + \cdots \left(q=0,\ \beta = \tfrac{1}{2}\right) \quad (k=2)\\
&\Lambda_{k0} = \sqrt{1 + 3^{-1}(1-\nu^2)^{-1}P_0^4},\\
&\Lambda_{k2} = -2^{-1}(1-\nu^2)^{-1}\Lambda_{k0}^{-1}\left[\tfrac{\nu^2(1-\nu^2)}{p_0^2} + \tfrac{p_0^2}{3} + \tfrac{p_0^6}{9}\left(1 + \tfrac{1}{K_1^2}\right)\right]
\end{aligned} \tag{3.2.63}$$

The comparison of (3.2.53) and (3.2.63) with the exact expansion (3.2.17) shows that the first terms of the expansions are the same but subsequent terms differ considerably:

5.
$$\begin{aligned}
&\Lambda_k = \Lambda_{k0}\varepsilon^{-\beta} + \Lambda_{k1}\varepsilon^{\beta} + \ldots \quad 0 < \beta < 1 \quad (k=1)\\
&\Lambda_{k0} = \frac{p_0}{\sqrt{1-\nu^2}}, \quad \Lambda_{k2} = \frac{\nu^2}{2\sqrt{1-\nu^2}}\frac{1}{p_0}
\end{aligned} \tag{3.2.64}$$

Here in the interval $0 < \beta < \frac{1}{2}$ the applied theory correctly identifies two terms of the expansions. In the interval $\frac{1}{2} \le \beta < 1$ the applied theory of shells even incorrectly defines the order of the second term of the expansions:

$$\begin{aligned}
&\Lambda_k = \Lambda_{k0}\varepsilon^{1-2\beta} + \Lambda_{k1}\varepsilon^{2\beta-1} + \ldots \quad \tfrac{1}{2} < \beta < \tfrac{2}{3} \quad (k=2)\\
&\Lambda_k = \Lambda_{k0}\varepsilon^{1-2\beta} + \Lambda_{k1}\varepsilon^{3-4\beta} + \ldots \quad \tfrac{2}{3} \le \beta < 1\\
&\Lambda_{k2} = \frac{P_0^2}{\sqrt{3(1-\nu^2)}}, \quad \Lambda_{k2} = \frac{\sqrt{3(1-\nu^2)}}{2p_0^2} \quad \tfrac{1}{2} < \beta < \tfrac{2}{3}\\
&\Lambda_{k2} = \frac{\sqrt{3(1-\nu^2)}}{2p_0^2} - \frac{P_0^4}{2\cdot 3\sqrt{3(1-\nu^2)}}\left(1 + \tfrac{1}{K_1^2}\right) \quad \beta = \tfrac{2}{3}\\
&\Lambda_{k2} = -\frac{p_0^4}{2\cdot 3\sqrt{3(1-\nu^2)}}\left(1 + \tfrac{1}{K_1^2}\right) \quad \tfrac{2}{3} < \beta < 1
\end{aligned} \tag{3.2.65}$$

Assuming that $K^2 = \frac{5}{6-\nu}$ in these formulas we obtain the two terms of the expansion defined by three-dimensional theory elasticity.

Note that unlike Kirchhoff-Love theory, in the domain $\frac{2}{3} \leq \beta < 1$ Timoshenko type theory correctly determines the order of the second term of the expansions of the zeros of the frequency equation.

In the case of torsional vibrations, comparing (3.2.56) with the expansion (3.2.42), we find that the comparison is appropriate only in the first term of the asymptotic expansions. As for the natural vibration frequencies defined by (3.2.23), (3.2.38), (3.2.44), (3.2.45), (3.2.46), these are not available in the applied theory of shells.

Thus, here, we have two natural frequencies in the first term of the asymptotic expansions coinciding with the frequencies determined by the applied theory of shells and a countable set of frequencies that are not available in the applied theory of shells. Such an assertion is true with respect to the frequency of torsional vibrations.

In the foregoing discussion, the roots of the frequency equation were investigated. The frequencies of the free axisymmetric vibrations of a hollow cylinder were obtained. An asymptotic representation of the forms of natural vibrations corresponding to different frequencies is given below at the leading term of the asymptotic expansions. These vibrations may be obtained through expanding (3.1.4) in terms of the small parameter ε. Only the amplitude values of the displacement components are given. We first present the forms of natural vibrations corresponding to ultra-low shell vibrations. According to formulas (3.2.6) and (3.2.8) we respectively have

$$\begin{aligned} U &= C_k \left\{ -1 + \begin{bmatrix} O(\varepsilon^{2\beta}), & 2\beta < 1 \\ O(\varepsilon), & 2\beta \geq 1 \end{bmatrix} \right\} \sin\left(p_0 \varepsilon^{\beta} \xi\right) \\ W &= C_k \varepsilon^{2\beta} \left[-\nu p_0^2 + O(\varepsilon) \right] \cos\left(p_0 \varepsilon^{\beta} \xi\right) \quad k = 1 \end{aligned} \tag{3.2.66}$$

$$\begin{aligned} U &= C_k \left[\nu^2 p_0^2 + O(\varepsilon) \right] \sin\left(p \varepsilon^{\beta} \xi\right) \quad k = 2 \\ W &= C_k \left[-\nu p_0^2 + O(\varepsilon) \right] \cos\left(p_0 \varepsilon^{\beta} \xi\right) \end{aligned} \tag{3.2.67}$$

In the case $\lambda_k = \lambda_{k0} + \varepsilon^2 \lambda_{k2} + \ldots$ we have

$$\begin{aligned} U &= C_k \left[p^2 - 1 + (-1)^k \sqrt{(p^2 - 1) + 4\nu p^2} + O(\varepsilon) \right] \sin p\xi \\ W &= C_k \left[-2\nu p^2 + O(\varepsilon) \right] \cos p\xi \quad k = 1, 2. \end{aligned} \tag{3.2.68}$$

For $p = p_0 \varepsilon^{-\beta}$, $q = 0$, $0 < \beta < \frac{1}{2}$ we obtain

$$\begin{aligned} U &= C_k \left[p_0^2 + O\left(\varepsilon^{2\beta}\right) \right] \sin\left(p_0 \varepsilon^{-\beta} \xi\right) \quad (k = 1) \\ W &= C_k \left[-\nu p_0^2 + O\left(\varepsilon^{1-2\beta}\right) \right] \cos\left(p_0 \varepsilon^{-\beta} \xi\right) \end{aligned} \tag{3.2.69}$$

Similarly, when $q = 0$, $\beta = \frac{1}{2}$ we get

$$
\begin{aligned}
U &= C_k\left[p_0^2 + O(\varepsilon)\right]\sin\left(\tfrac{p_0}{\sqrt{\varepsilon}}\xi\right),\\
W &= C_k\left[-\nu p_0^2 + p_0^4\eta + O(\varepsilon)\right]\cos\left(\tfrac{p_0}{\sqrt{\varepsilon}}\xi\right) \quad (k=2).
\end{aligned}
\tag{3.2.70}
$$

The natural forms corresponding to the frequencies defined by formulas (3.2.18) have the form

$$
\begin{aligned}
U &= C_k\left[-\nu^2 + O\left(\varepsilon^{1-2\beta}\right)\right]\sin\left(p_0\varepsilon^{-\beta}\xi\right) \quad 0<\beta<\frac{1}{2}\\
W &= C_k\left[-\nu p_0^2\varepsilon^{-2\beta} + O\left(\varepsilon^{1-4\beta}\right)\right]\cos\left(p_0\varepsilon^{-\beta}\xi\right)
\end{aligned}
\tag{3.2.71}
$$

$$
\begin{aligned}
U &= C_k\left[-\nu^2\left(1+2p_0^2\right) - \frac{\nu(1-2\nu)}{1-\nu}p_0^2\eta + O(\varepsilon)\right]\sin\left(\frac{p_0}{\sqrt{\varepsilon}}\xi\right)\\
W &= C_k\left[-\nu p_0^2\left(1+\frac{1}{1-\nu}p_0^2\right)\varepsilon^{-1} + O(\varepsilon)\right]\cos\left(\frac{p_0}{\sqrt{\varepsilon}}\xi\right)
\end{aligned}
\tag{3.2.72}
$$

$$
\begin{aligned}
U &= C_k\left[-\nu^2\left(1+\frac{p_0^4}{3-3\nu^2}\right)\varepsilon + O\left(\varepsilon^{1+2\beta}\right)\right]\sin\left(p_0\varepsilon^{-\beta}\xi\right)\\
W &= C_k\left[-\frac{\nu}{1-\nu}p_0^4 + O\left(\varepsilon^{1+2\beta}\right)\right]\cos\left(p_0\varepsilon^{-\beta}\xi\right)
\end{aligned}
\tag{3.2.73}
$$

In the case $p = p_0\varepsilon^{-\beta}$, $\lambda_k = \lambda_{k0}\varepsilon^{1-2\beta}$, $\frac{1}{2}<\beta<1$ we have

$$
\begin{aligned}
U &= C_k\left[p_0^2 + O(\varepsilon)\right]\sin\left(p_0\varepsilon^{-\beta}\xi\right)\\
W &= C_k\left[p_0^4\eta + O\left(\varepsilon^{2-2\beta}\right)\right]\varepsilon^{1-2\beta}\cos\left(p_0\varepsilon^{-\beta}\xi\right)
\end{aligned}
\tag{3.2.74}
$$

The frequencies determined from Rayleigh-Lamb's equation correspond to the following vibration forms

$$
\begin{aligned}
U = &-\alpha_{k0}C_k\left[\left(\delta_n^2 - 2p_0^2\right)\sin\gamma_{k0}\sin\alpha_{k0}\eta\right.\\
&\left.+ 2p_0^2\sin\alpha_{k0}\sin\gamma_{k0}\eta + O(\varepsilon)\right]\sin\left(\frac{p_0}{\varepsilon}\xi\right)\\
W = &\,p_0\left[\left(\delta_k^2 - 2p_0^2\right)\sin\gamma_{k0}\cos\alpha_{k0}\eta\right.\\
&\left.- 2\alpha_{k0}\gamma_{k0}\sin\alpha_{k0}\cos\gamma_{k0}\eta + O(\varepsilon)\right]\cos\left(\frac{p_0}{\varepsilon}\xi\right)
\end{aligned}
\tag{3.2.75}
$$

The expressions for $k = 2, 4, 6, \ldots$, are obtained from (3.2.75) by replacing $\cos x$ with $\sin x$ and $\sin x$ with $\cos x$ respectively.

According to the classification of vibrations shown in [1], in Fig. 3.1 the dashed line corresponds to the quasitransverse vibrations whereas the solid line corresponds to quasitangential vibrations of a shell; the line from (1/2, 0) to (1,1) corresponds to quasitransverse vibrations with high variability. In the remaining of the (β, q) plane there are no solutions corresponding to the two-dimensional theory of shells.

3.3 Axisymmetric Free Vibrations of a Hollow Sphere

In [10] an attempt was made to study the propagation of elastic waves in a hollow sphere in the three-dimensional theory of elasticity. However, in the study of the frequency equation some degenerate cases are considered which do not provide complete information on the frequency spectrum. In the second chapter forced vibrations of a spherical layer were investigated by the method of homogeneous solutions. Depending on the frequency of the driving forces a possible form of wave formation is studied.

In this section, we present an asymptotic process for finding the frequencies of free axisymmetric vibrations of a hollow sphere. The asymptotic process is constructed in detail for a closed sphere with free face surfaces. A comparison of the results obtained in Kirchhoff-Love theory with the results obtained by three-dimensional elasticity theory is given here.

1. For a closed hollow sphere we write the equations of motion in vector form as

$$\frac{1}{1-2v} graddiv\bar{U} + \Delta\bar{U} = gG^{-1}\frac{\partial^2\bar{U}}{\partial t^2} \tag{3.3.1}$$

Assume that the faces of the boundary are stress-free, i.e.

$$\sigma_r = 0,\ \tau_{r\theta} = 0 \quad \text{at } r = R_s\ (s = 1, 2) \tag{3.3.2}$$

Equation (3.3.1) may be represented, by means of the results of the second chapter, in the form

$$[U_r, U_\theta] = \left[U(\rho)P_n(\cos\theta),\quad W(\rho)\frac{dP_n}{d\theta}(\cos\theta)\right]e^{i\omega t} \tag{3.3.3}$$

where $P_n(\cos\theta)$ is a Legendre function of the first kind, n is an integer, and

$$\begin{aligned} U(\rho) &= \frac{1}{\sqrt{\rho}}\left[\alpha Z_z'(\alpha\rho) - \frac{1}{2\rho}Z_z(\lambda\rho) - \frac{1}{\rho}\left(z^2 - \frac{1}{4}\right)Z_z(\lambda\rho)\right] \\ W(\rho) &= \frac{1}{\sqrt{\rho}}\left[\frac{1}{\rho}Z_z(\alpha\rho) - \lambda Z_2'(\lambda\rho) - \frac{1}{2\rho}Z_z(\lambda\rho)\right] \\ z^2 &= n(n+1) + 1/4, \\ Z_z(x) &= C_1J_z(x) + C_2Y_z(x) \end{aligned} \tag{3.3.4}$$

Employing the homogeneous boundary conditions (3.3.2), we obtain the equation for the frequency λ:

$$\begin{aligned}\Delta(z,\lambda,\rho_1,\rho_2) &= 32\pi^{-2}\left(z^2-\frac{1}{4}\right)\varphi_1(z,\rho_1)\varphi_1(z,\rho_2)\\ &- F_1(\alpha,z,\rho_1,\rho_2)G_1(\lambda,z,\rho_1,\rho_2)\\ &+\left(z^2-\frac{1}{4}\right)[F_2(\alpha,z,\rho_1,\rho_2)G_2(\lambda,z,\rho_1,\rho_2)\\ &+F_2(\alpha,z,\rho_2,\rho_1)G_2(\lambda,z,\rho_2,\rho_1)]-\left(z^2-\frac{1}{4}\right)\\ &\times F_3(\alpha,z,\rho_1,\rho_2)G_3(\lambda,z,\rho_1,\rho_2)=0.\end{aligned} \tag{3.3.5}$$

Equation (3.3.5) is actually the same as Eq. (2.3.7) with the only difference that here the parameter z is purely real.

2. The zeros of (3.3.5) can be found numerically. However, in the case of thin shells the asymptotic method is more efficient and allows us to specify any frequency to any desired accuracy. Therefore, let us conduct an asymptotic analysis of the frequency equation. For this purpose, as before, we set

$$\rho_1 = 1-\varepsilon,\quad \rho_2 = 1+\varepsilon \tag{3.3.6}$$

Substituting (3.3.6) into (3.3.5), we obtain

$$D(z,\lambda,\varepsilon) = \Delta(z,\lambda,\rho_1,\rho_2) = 0 \tag{3.3.7}$$

$z^2 = \frac{1}{4}$ is a particular case and will be considered separately.

For any finite z [$z = O(1)$ at $\varepsilon \to 0$] the function $D(z,\lambda,\varepsilon)$ has two zeros with the following asymptotic properties: $\Lambda_k = O(1)$, $(k = 1,2)$ $\left(\Lambda^2 = \frac{9R_0^2\omega^2}{E}\right)$.

In order to obtain the proof, we use the expansion of $D(z,\lambda,\varepsilon)$ into a series in the small parameter ε given in (2.4.3). On doing so we obtain

$$\begin{aligned}D(z,\lambda,\varepsilon) &= \frac{64}{3\pi^2}\left(\frac{1+\nu}{1-\nu}\right)^2\varepsilon^2\Big\langle -12(1-\nu^2)^2\Lambda^4+12(1-\nu^2)\Lambda^2z^2\\ &+9(1-\nu^2)(4\nu+1)\Lambda^2-12(1-\nu^2)^2z^2+27(1-\nu^2)\\ &+\{-4z^6+[3+16\nu^2+2(1+\nu)(3-2\nu)\Lambda^2]\,z^4\\ &+\Big[24\nu^2-\frac{195}{4}+4(1+\nu)(16-13\nu-12\nu^2)\Lambda^2\\ &-4(1+\nu)^{-2}(4\nu^2+16\nu+11)\Lambda^4\Big]z^2+\frac{2241}{16}-135\nu^2+30(1-\nu^2)\Lambda^2\\ &\times\Big[3\left(2\nu+\frac{1}{2}\right)-2(1-\nu^2)\Lambda^2\Big]\Big\}\varepsilon^2+\frac{1}{15}(-32z^8+\cdots)\varepsilon^4+\cdots\Big\rangle=0\end{aligned} \tag{3.3.8}$$

We now seek the roots Λ_k in the form of the expansion

$$\Lambda_k = \Lambda_{k0} + \varepsilon^2 \Lambda_{k2} + \cdots. \tag{3.3.9}$$

Substituting (3.3.9) into (3.3.8) we obtain

$$\begin{aligned}
\Lambda_{k0}^2 &= 2^{-3}\left(1-v^2\right)^{-1}\Big[4z^2+12v+3+(-1)^k \\
&\quad\times\sqrt{16z^4+(64v^2+96v-40)z^2+153+72v}\Big] \\
\Lambda_{k2} &= 6^{-1}\left(1-v^2\right)^{-1}\Lambda_{k0}^{-1}\left[8\left(1-v^2\right)\Lambda_{k0}^2-4z^2-12v-3\right]^{-1} \\
&\quad\times\Big\{-4z^6+\left[3+16v^2+2(1+v)(3-2v)\Lambda_{k0}^2\right]\cdot z^4 \\
&\quad+\Big[24v^2-\frac{195}{4}+4(1+v)\left(16-13v-12v^2\right)\Lambda_{k0}^2 \\
&\quad-4(1+v)^{-2}\left(4v^2-16v+11\right)\Lambda_{k0}^4\Big]z^2+\frac{2241}{16}-135v^2 \\
&\quad+30\left(1-v^2\right)\Lambda_{k0}^2\left[3\left(2v+\frac{1}{2}\right)-2\left(1-v^2\right)\Lambda_{k0}^2\right]\Big\}
\end{aligned} \tag{3.3.10}$$

Thus, we have proved that at fixed finite z there are two natural frequencies.

Let's consider the case when z increases without bound as $\varepsilon \to 0$. Here we will consider separately the following limiting cases:

(a) $z\varepsilon \to 0$ as $\varepsilon \to 0$; $z\varepsilon \to const$ as $\varepsilon \to 0$.

We first define those Λ_k in the limit $z\varepsilon \to 0$ as $\varepsilon \to 0$. Assume that the leading terms of the asymptotic behaviour of Λ_k and Z have the form:

$$\begin{aligned}
&\Lambda_k = \Lambda_{k0}\varepsilon^{-q},\ z = z_0\varepsilon^{-\beta}, \quad \Lambda_{k0} = O(1) \\
&z_0 = O(1), \quad 0 \le q < 1, \quad 0 < \beta < 1
\end{aligned} \tag{3.3.11}$$

Substituting (3.3.9) into (3.3.8), from the consistency condition of the asymptotic process we find that the only possible cases here are $q = 0$ and $q \le \beta$.

Note that, sometimes, the basic interval of variation of the parameters q and β will be split into sub-intervals, since, depending on the range of q and β, the zeros of the function $D(z, \lambda, \varepsilon)$ have different asymptotic representations. The range of the variation of parameters q and β is indicated in Fig. 3.2. We consider the cases when $q = 0$ and $q = \beta$ separately.

In the first case, from the expansion (3.3.8), it is obtained that $0 < \beta < \frac{1}{2}$. The case $\beta = \frac{1}{2}$ is considered separately. We seek Λ_k $(k = 1)$ in the form

$$\begin{aligned}
&\Lambda_k = \Lambda_{k0} + \Lambda_{k2}\varepsilon^{2\beta} + \cdots, \quad 0 < \beta \le \tfrac{1}{3} \\
&\Lambda_k = \Lambda_{k0} + \Lambda_{k2}\varepsilon^{2-4\beta} + \cdots, \quad \tfrac{1}{3} < \beta < \tfrac{1}{2}
\end{aligned} \tag{3.3.12}$$

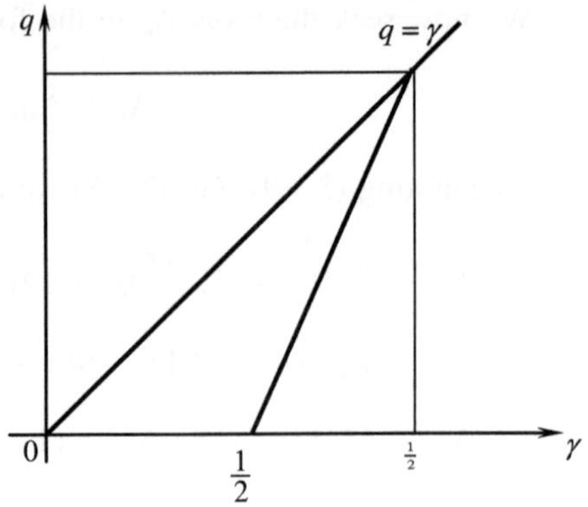

Fig. 3.2 Domain of variation of the parameter of frequency for a sphere

Substituting expansions into (3.3.8) we obtain

$$
\begin{aligned}
\Lambda_{k0} = 1, \quad \Lambda_{k2} &= -\frac{(1+\nu)(2+\nu)}{2z_0^2} \quad 0<\beta<\frac{1}{3} \\
\Lambda_{k2} &= \frac{z_0^4}{6(1-\nu^2)} - \frac{(1+\nu)(2+\nu)}{2z_0^2} \quad \beta = \frac{1}{3} \\
\Lambda_{k2} &= \frac{z_0^4}{6(1-\nu^2)}, \quad \frac{1}{3}<\beta<\frac{1}{2}.
\end{aligned}
\tag{3.3.13}
$$

In the case $q = 0$, $\beta = \frac{1}{2}$ we find that

$$
\Lambda_k = \Lambda_{k0} + \Lambda_{k1}\varepsilon + \cdots \quad (k = 1) \tag{3.3.14}
$$

where

$$
\begin{aligned}
\Lambda_{k0} &= \sqrt{1 + 3^{-1}(1-\nu^2)^{-1} z_0^4}, \\
\Lambda_{k1} &= -\tfrac{1}{4z_0^2\Lambda_{k0}}\left[3 - 8\nu^2 + 5\nu(1+2\nu)\Lambda_{k0}^2 + (1+\nu)\Lambda_{k0}^4\right].
\end{aligned}
\tag{3.3.15}
$$

Similarly, for $q = \beta$ in (3.3.8), we obtain

$$
\begin{aligned}
\Lambda_k &= \Lambda_{k0}\varepsilon^{-\beta} + \Lambda_{k2}\varepsilon^{\beta} + \cdots \quad (k = 2)\ 0<\beta\leq\tfrac{1}{2} \\
\Lambda_k &= \Lambda_{k0}\varepsilon^{-\beta} + \Lambda_{k2}\varepsilon^{2-3\beta} + \cdots \quad (k = 2)\ \tfrac{1}{2}<\beta<1
\end{aligned}
\tag{3.3.16}
$$

where

$$
\begin{aligned}
\Lambda_{k0} &= \tfrac{z_0}{\sqrt{1-\nu^2}}, \quad \Lambda_{k2} = \tfrac{4\nu^2+12\nu-1}{8Z_0\sqrt{1-\nu^2}}, \quad 0<\beta<\tfrac{1}{2} \\
\Lambda_{k2} &= \tfrac{4\nu^2+12\nu-1}{8Z_0\sqrt{1-\nu^2}} - \tfrac{\left(\nu^5+3\nu^4+2\nu^3+6\nu^2-35\nu+21\right)}{12(1-\nu^2)^2(1+\nu)^2\sqrt{1-\nu^2}} Z_0^3 \quad \beta = \tfrac{1}{2} \\
\Lambda_{k2} &= -\tfrac{\left(\nu^5+3\nu^4+2\nu^3+6\nu^2-35\nu+21\right)}{12(1-\nu^2)^2(1+\nu)^2\sqrt{1-\nu^2}} z_0^3 \\
&\tfrac{1}{2}<\beta<1.
\end{aligned}
\tag{3.3.17}
$$

In the case, $q \neq 0$, $q < \beta$ substituting (3.3.11) into (3.3.8) and keeping only the leading order terms we obtain the following limiting equation

$$\begin{aligned} D(z, \lambda, \varepsilon) = & \frac{64}{3\pi^2}\left(\frac{1+\nu}{1-\nu}\right)^2 \Lambda^4 \varepsilon^2 \langle \left[12\left(1-\nu^2\right)\Lambda_{k0}^2 z_0^2 \right. \\ & \left. + O\left(\varepsilon^{2\beta-2q}\right)\right]\varepsilon^{2\beta-2q} + \left\{4z_0^6 \right. \\ & \left. + O\left[\max\left(\varepsilon^{2\beta-2q}, \varepsilon^{2-2\beta}\right)\right]\right\}\varepsilon^{2-6\beta} = 0 \end{aligned} \tag{3.3.18}$$

Equation (3.3.18) imposes that $q = 2\beta - 1$. Since $q > 0$ we find that $\beta > \frac{1}{2}$. Thus, $\frac{1}{2} < \beta < 1$. Now we seek Λ_k in the form

$$\begin{aligned} \Lambda_k &= \Lambda_{k0}\varepsilon^{1-2\beta} + \Lambda_{k2}\varepsilon^{2\beta-1} + \cdots, \quad \frac{1}{2} < \beta < \frac{2}{3} \\ \Lambda_k &= \Lambda_{k0}\varepsilon^{1-2\beta} + \Lambda_{k2}\varepsilon^{3-4\beta} + \cdots, \quad \frac{2}{3} \leq \beta < 1 \end{aligned} \tag{3.3.19}$$

Substituting (3.3.19) into (3.3.8) we get

$$\Lambda_{k0} = \frac{z_0^2}{\sqrt{3(1-\nu^2)}}, \quad \begin{aligned} \Lambda_{k2} &= \frac{\sqrt{3(1-\nu^2)}}{2z_0^2}, \quad \frac{1}{2} < \beta < \frac{2}{3} \\ \Lambda_{k2} &= \frac{\sqrt{3(1-\nu^2)}}{2z_0^2} - \frac{z_0^4}{12(1-\nu)\sqrt{3(1-\nu^2)}}, \quad \beta = \frac{2}{3} \\ \Lambda_{k2} &= -\frac{z_0^4}{12(1-\nu)\sqrt{3(1-\nu^2)}}, \quad \frac{2}{3} < \beta < 1 \end{aligned} \tag{3.3.20}$$

Now let us consider the case when $z\varepsilon \to 0$ as $\varepsilon \to 0$, $\lambda\varepsilon \to const$ in the limit $\varepsilon \to 0$. We seek λ_n $(n = k-2, k = 3, 4, \ldots)$ in the following form:

$$\lambda_n = \frac{\delta_n}{\varepsilon} + O(\varepsilon), \quad z = z_0\varepsilon^{-\beta}, \quad 0 < \beta < 1. \tag{3.3.21}$$

Substituting (3.3.21) into (3.3.5) and resorting to the asymptotic expansions of functions $J_z(x)$, $Y_z(x)$ for large values of the argument δ_n and the index n we obtain the following equations

$$\sin\sqrt{\frac{2(1-2\nu)}{1-\nu}}\delta_n = 0, \tag{3.3.22}$$

$$\sin 2\delta_n = 0. \tag{3.3.23}$$

As noted in [11], the frequencies determined from these equations correspond to simple forms of transverse vibrations of an infinite plate (tension-compression and shear).

To construct the asymptotics of the zeros of the second group we set $\lambda \sim z$, $z\varepsilon \to const$, $\lambda\varepsilon \to const$ as $\varepsilon \to 0$, giving $z = z_0\varepsilon^{-1}$. We seek λ_p $(p = 1, 2, \ldots, p \neq k)$ in the following form:

$$\lambda_p = \frac{\gamma p}{\varepsilon} + O(\varepsilon). \tag{3.3.24}$$

As above, on substituting (3.3.24) into (3.3.5) and employing the asymptotic expansions of, $J_z(x)$, $Y_z(x)$ at large Z and x for γ_p we obtain the following equation

$$\begin{gathered}\left[\left(\gamma_p^2 + \chi_p^2\right)^2 sh\alpha_p ch\chi_p - 4\alpha_p\chi_p\gamma_p^2 ch\alpha_\rho sh\chi_\rho\right] \\ \times \left[\left(\gamma_p^2 + \chi_p^2\right)^2 ch\alpha_p sh\chi_p - 4\alpha_p\chi_p\gamma_p^2 sh\alpha_p ch\chi_p\right] = 0\end{gathered} \tag{3.3.25}$$

where

$$\alpha_p^2 = z_0^2 - \frac{1-20}{2(1-v)}\gamma_p^2, \quad \chi_p^2 = z_0^2 - \gamma_p^2$$

For a given z Eq. (3.3.25) defines a countable set of γ_p. Equation (3.3.25) actually coincides with Rayleigh-Lamb's frequency equation for an elastic layer, and it remains valid in the case of $z = z_0\varepsilon^{-\beta}$, $\lambda = \lambda_0\varepsilon^{-\beta}$, $\beta > 1$ as well.

We would also like to mention that, in principle, there is a particular case which is described by the relations

$$z = z_0\varepsilon^{-1}, \quad \lambda = \lambda_0\varepsilon^{-\beta} \text{ for } 0 < \beta < 1 \text{ at } \varepsilon \to 0. \tag{3.3.26}$$

Now, if we substitute (3.3.26) into (3.3.5) and transform it by means of the asymptotic expansions of Bessel functions in the leading term of the asymptotics, we obtain

$$D(z, \lambda, \varepsilon) = sh^2 2z_0 - 4z_0^2 + O(\varepsilon). \tag{3.3.27}$$

The real parameter z_0 cannot be a solution to the equation $sh^2 2z_0 - 4z_0^2 = 0$ as this equation has only complex zeros. Therefore, in this case, we get that $D(z, \lambda, \varepsilon) \neq 0$ which implies that there is no vibration in this case.

As noted above, the case $z^2 - \frac{1}{4} = 0$ $(n = 0)$ is particular. In this case,

$$U_r = U_0(\rho)e^{i\omega t}, \quad U_\theta \equiv 0, \quad \tau_{r\theta} \equiv 0 \tag{3.3.28}$$

Substituting (3.3.28) into Lame's equation and the boundary conditions (3.3.2), we obtain

$$U_0'' + \frac{2}{\rho}U_0' - \frac{2}{\rho^2}U_0 + \alpha^2 U_0 = 0 \tag{3.3.29}$$

$$\left[(1-\nu)U_0' + \frac{2\nu}{\rho}U_0\right]_{\rho=\rho_s} = 0 \quad (s = 1, 2) \tag{3.3.30}$$

The general solution of Eq. (3.3.29) has the form

$$U_0 = \frac{1}{\rho}\left[\left(\alpha\cos\alpha\rho - \frac{1}{\rho}\sin\alpha\rho\right)C_1 + \left(\alpha\sin\alpha\rho + \frac{1}{\rho}\cos\alpha\rho\right)C_2\right] \tag{3.3.31}$$

Inserting U_0 into the boundary conditions (3.3.30) we obtain the frequency equation

$$\begin{aligned} D_0(\lambda,\varepsilon) = &\left[\lambda^4 - \frac{8\nu}{1-\nu}\lambda^2 + 16\right.\\ &\left. - 2\lambda^2\left(\lambda^2 + \frac{4\nu}{1-\nu}\right)\varepsilon^2 + \lambda^4\varepsilon^4\right]\sin 2\alpha\varepsilon \\ &- 8\alpha\left(4 + \lambda^2 - \lambda^4\varepsilon^2\right)\varepsilon\cos 2\alpha\varepsilon = 0. \end{aligned} \tag{3.3.32}$$

Equation (3.3.32) in the limit $\varepsilon \to 0$ has a bounded zero with the following asymptotic property:

$$\lambda_k = \lambda_{k0} + \varepsilon^2\lambda_{k2} + \cdots. \tag{3.3.33}$$

Employing (3.3.32) in the relation (3.3.33) we obtain

$$\lambda_{k0}^2 = \frac{4(1+\nu)}{1-\nu}, \quad \lambda_{k2} = -\frac{4}{3(1-\nu)3\Lambda_{k0}}\left(3\nu^3 + 14\nu^2 - 35\nu + 14\right).$$

Equation (3.3.32) as $\varepsilon \to 0$ has a countable set of zeros, each of which tends to infinity.

Let us seek λ_n in terms of the expansion

$$\lambda_n = \frac{S_n}{\varepsilon} + O(\varepsilon) \quad (n = 1, 2, \ldots). \tag{3.3.34}$$

Substituting (3.3.34) into (3.3.32) we obtain

$$\sin 2\sqrt{\frac{1-2\nu}{2(1-\nu)}}S_n = 0. \tag{3.3.35}$$

In the first term of the asymptotic behaviour these frequencies coincide with the frequencies determined from the formulas (3.3.22) and correspond to purely radial vibrations of the sphere.

3. Now for comparison, we present the frequency equations obtained in Kirchhoff-Love theory. The frequency equation, based on A. I. Lurye's ratios of elasticity, has the form

$$\begin{aligned}&(1-\nu^2)\left[-4(1-\nu^2)\Lambda^4+4\Lambda^2z^2-4z^2+3(4\nu+1)\Lambda^2+9\right]\\&\quad+\frac{1}{3}\left\{-4z^6+\left[19+(1-\nu^2)\Lambda^2\right]z^4\right.\\&\quad-\left[4\nu^2-8\nu+\frac{51}{4}+2(1-\nu)(3-2\nu)\Lambda^2\right]z^2\\&\quad\left.+\frac{33}{16}-2\nu+\nu^2+\frac{1}{4}(1-\nu^2)(5-4\nu)\Lambda^2\right\}\varepsilon^2\\&\quad+\frac{1}{9}\left(4z^6-3z^4+\frac{3}{4}z^2-\frac{1}{16}\right)\varepsilon^4=0\end{aligned} \tag{3.3.36}$$

Here we will not dwell into all the cases discussed above. Let us conduct the asymptotic analysis of Kirchhoff-Love frequency equation for the most typical cases.

From (3.3.36) the following groups of zeros can be obtained

(a)
$$\Lambda_k=\Lambda_{k0}+\varepsilon^2\Lambda_{k2}+\cdots\quad(q=0,p=0) \tag{3.3.37}$$

$$\begin{aligned}\Lambda_{k0}^2&=2^{-3}(1-\nu^2)^{-1}\Big[4z^2+12\nu+3+(-1)^k\\&\qquad\times\sqrt{16z^4+(64\nu^2+96\nu-40)z^2+72\nu+153}\Big]\\\Lambda_{k2}&=6^{-1}(1-\nu^2)^{-1}\Lambda_{k0}^{-1}\left[8(1-\nu^2)\Lambda_{k0}^2\right.\\&\qquad\left.-4z^2-12\nu-3\right]^{-1}\left\{-4z^6+\left[19+(1-\nu^2)\Lambda_{k0}^2\right]z^4\right.\\&\qquad-\left(4\nu^2-8\nu+\frac{51}{4}+2(1-\nu^2)(3-2\nu)\Lambda_{k0}^2\right)z^2\\&\qquad\left.+\frac{33}{16}-2\nu+\nu^2+\frac{1}{4}(1-\nu^2)(5-4\nu)\Lambda_{k0}^2\right\}\end{aligned} \tag{3.3.38}$$

(b)
$$\Lambda_k=\Lambda_{k0}+\varepsilon\Lambda_{k1}+\cdots\quad\left(q=0,\ \beta=\frac{1}{2}\right) \tag{3.3.39}$$

$$\Lambda_{k0}^2 = 1 + \frac{z_0^4}{3(1 - \nu^2)}$$
$$\Lambda_{k1} = \frac{1}{8\Lambda_{k0}} z_0^{-2}\left[10 - \left(\nu^2 + 12\nu + 21\right)\Lambda_{k0}^2 + 3\left(1 - \nu^2\right)\Lambda_{k0}^4\right] \tag{3.3.40}$$

(c)
$$\Lambda_k = \Lambda_{k0}\varepsilon^{1-2\beta} + \varepsilon^{3-4\beta}\Lambda_{k2} + \cdots, \quad \frac{2}{3} \le \beta < 1 \tag{3.3.41}$$

$$\Lambda_{k0}^2 = \frac{z_0^4}{3(1-\nu^2)}, \quad \Lambda_{k2} = \frac{1}{2\Lambda_{k0}} + \frac{3\left(1-\nu^2\right)}{8\Lambda_{k0}z_0^2}\Lambda_{k0}^4 \quad \beta = \frac{2}{3}$$
$$\Lambda_{k2} = \frac{3\left(1-\nu^2\right)}{8Z_0^2}\Lambda_{k0}^3 \quad \frac{2}{3} < \beta < 1 \tag{3.3.42}$$

Comparing (3.3.37), (3.3.39), (3.3.41) with the exact expansions (3.3.9), (3.3.14), (3.3.19), we see that that the first terms of the expansions are the same but the subsequent terms differ substantially. As for the natural vibration frequencies defined by (3.3.21), (3.3.24) and (3.3.34), they are not available in the applied theory of shells.

Thus, for a sphere we get two frequencies in the leading term of the asymptotic expansions coinciding with the frequencies determined in the applied theory of shells and a countable set of frequencies which are not available in the applied theory of shells.

References

1. Goldenveizer, A.L., Lidskiy, V.B., Tovstik, P.E.: Free Vibrations of Thin Elastic Shells, 383p. Moscow: Nauka (1979)
2. Oniashvili, O.D.: Some Dynamic Problems of the Theory of Shells, 195p. Ed. USSR Academy of Sciences (1957)
3. Gontkevich, V.P.: Natural Vibrations of Plates and Shells, 102p. Kiev. Naukova Dumka (1964)
4. Birger, I.A., Panovko, Y.G.: Strength, Stability, Vibration, 567p. Directory, v.3. M.: Engineering (1968)
5. Achenbach, J.D.: An asymptotic method to analyze the vibrations of an elastic layer. Trans. ASME Ser. E J. Appl. Mech. **36**(1) (1969)
6. Moskalenko, V.: Natural vibrations of thick plates. News of Arm. SSR Acad. Sci. Mech. **21** (5), 57–64 (1968)
7. Aksentyan, O.C., Selezneva, T.N.: Determination of natural frequencies of circular plates. J. Appl. Math. Mech. **40**(1), 112–119 (1976)
8. Hrinchenko, V.T., Myaleshka, V.V.: Harmonic Vibrations and Waves in Elastic Bodies, 283p. Kiev. Naukova Dumka (1981)
9. Ustinov, Y.A.: On some peculiarities of the asymptotic method when applied to the study of vibrations of thin inhomogeneous elastic plates. In: Proceeding of I All-Union School on the Theory and Numerical Methods for the Calculation of Shells and Plates, pp. 395–403. Tbilisi (1975)

10. Shah, R., Datta: Study of the propagation of elastic waves in a hollow sphere by three-dimensional theory of elasticity and the theory of shells. Analytical research, p. 1. Proc. Amer. Society Eng.-Mech. Appl. Mech. **36**(3), 52–62 (1969)
11. Pao, Y.N., Mindlin, R.D.: Dispersion of flexural waves in an elastic, circular cylinder. J. Appl. Mech. **27**(3), 513–520 (1960)

Chapter 4
Asymptotic Analysis of Stress-Strain State of a Truncated Hollow Cone

Abstract It is well known that conical domains are the most difficult to study and the most interesting in relation to the research results among the other elastic domains bounded by coordinate surfaces of the four classical coordinate systems. In this chapter, we develop an asymptotic method of integration of axisymmetric equations of elasticity theory for a conical shell and a circular plate whose thickness varies linearly along the generatrix. An asymptotic theory is constructed for a general conical shell which remains true, in particular, for a cylindrical shell.

4.1 Construction of Homogeneous Solutions

We consider the axisymmetric problem of elastic equilibrium of a truncated hollow cone of variable thickness (Fig. 4.1). The position of the points of the cone in the space is defined by the spherical coordinates r, θ, φ, varying in the following intervals

$$r_1 \leq r \leq r_2, \theta_1 \leq \theta \leq \theta_2, 0 \leq \varphi \leq 2\pi$$

It is assumed that the conical part of the boundary is free from stress, i.e.

$$\sigma_\theta = 0, \tau_{r\theta} = 0 \quad \text{at } \theta = \theta_n \quad r_1 \leq r \leq r_2 \tag{4.1}$$

and the on the rest of the boundary the following boundary conditions are imposed:

$$\sigma_r = f_{1s}(\theta), \tau_{r\theta} = f_{2s}(\theta) \quad \text{at } r = r_s \quad (s = 1, 2) \tag{4.2}$$

The solutions of Lame's equations satisfying the homogeneous boundary conditions (4.1) are called homogeneous solutions for the cone. The considered problem was studied by the method of homogeneous solutions in [1]. The author of this work, however, builds a system of homogeneous solutions, as will be shown below, which is not complete and therefore does not satisfy arbitrary boundary conditions (4.2).

M. F. Mekhtiev, *Vibrations of Hollow Elastic Bodies*,
Advanced Structured Materials 88,
https://doi.org/10.1007/978-3-319-74354-7_4

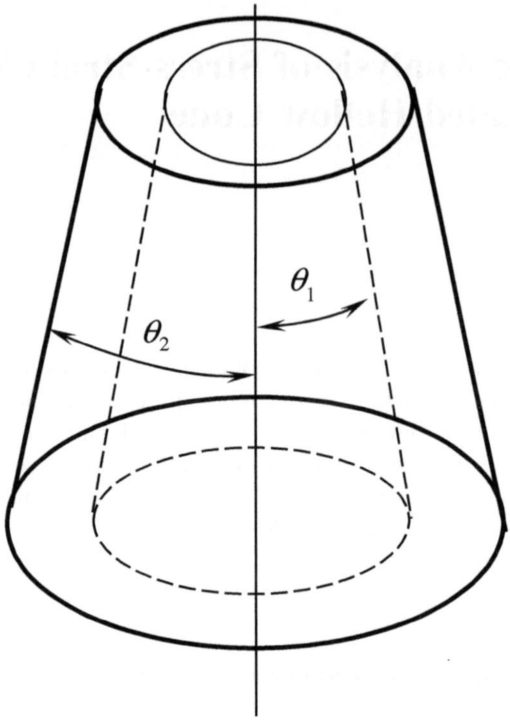

Fig. 4.1 Truncated hollow cone

Below, on the basis of asymptotic method, we will establish the existence of solutions of the homogeneous problem and present two different types of solutions except those which are found in [2]. To construct the homogeneous solutions for a cone we write equilibrium equations in the vector form as

$$\frac{1}{1-2v} grad\, div \bar{U} + \Delta \bar{U} = 0 \tag{4.1.1}$$

where $\bar{U}(U_r, U_\theta)$ is the displacement vector.

The notation here is the same as in Chap. 3.

Solutions of Eq. (4.1.1) will be sought in the form

$$U_r = r^\lambda a(\lambda, \theta), U_\theta = r^\lambda b(\lambda, \theta) \tag{4.1.2}$$

Substituting (4.1.2) into (4.1.1), and employing the separation of variables, we obtain the following system of ordinary differential equations with respect to a pair of functions (a, b):

$$\begin{aligned} L_1(\lambda)(a,b) &= a'' + ctg\theta a' + \frac{2(1-v)}{1-2v}(\lambda-1)(\lambda+2)a \\ &\quad + \left(\frac{\lambda-1}{1-2v} - 2\right)(b' + ctg\theta b) = 0 \\ L_2(\lambda)(a,b) &= \left(\frac{\lambda+2}{1-2v}\right)a' + \frac{2(1-v)}{1-2v} \\ &\quad \times \left(b'' + ctg\theta b' - \frac{1}{\sin^2\theta}b\right) + \lambda(\lambda+1)b = 0. \end{aligned} \tag{4.1.3}$$

Here primes denote the derivatives with respect to θ. Using the generalized Hooke's law and satisfying homogeneous boundary conditions on the conical surfaces (4.1) for the functions $a(\lambda,\theta)$, $b(\lambda,\theta)$, we obtain the boundary conditions

$$\begin{aligned} M_1(\lambda)(a,b)|_{\theta=\theta_n} &= [(v\lambda+1)a + (1-v)b' + vctg\theta b]_{\theta=\theta_n} = 0 \\ M_2(\lambda)(a,b)|_{\theta=\theta_n} &= [a' + (\lambda-1)b]_{\theta=\theta_n} = 0 \end{aligned} \tag{4.1.4}$$

Thus, the system of Eq. (4.1.3) together with boundary conditions (4.1.4) generates a spectral problem for a pair of (a,b).

Let us return to the study of the posed spectral problem. To construct the general solution of the system of Eq. (4.1.3), we search $a(\lambda,\theta)$ and $b(\lambda,\theta)$ in the form:

$$a(\lambda,\theta) = AT_\mu(\theta), b(\lambda,\theta) = B\frac{dT_\mu}{d\theta}(\theta) \tag{4.1.5}$$

where $T_\mu(\theta)$ satisfies the Legendre equation

$$T''_\mu + ctg\theta T'_\mu + \mu(\mu+1)T_\mu = 0 \tag{4.1.6}$$

Substituting (4.1.5) into (4.1.3) taking into account (4.1.6) with respect to μ we obtain the characteristic equation

$$\left(\chi^2 - \frac{1}{4}\right)^2 - 2(z^2 + 3/4)\left(\chi^2 - \frac{1}{4}\right) + \left(z^2 - \frac{1}{4}\right)(z^2 - 9/4) = 0 \tag{4.1.7}$$

where the notation

$$z = \lambda + 1/2, \chi = \mu + 1/2$$

is employed. It is a straightforward matter to show that Eq. (4.1.7) has the roots

$$\chi_1 = z + 1/2, \chi_2 = -z - 3/2, \chi_3 = z - 3/2, \chi_4 = -z + 1/2$$

It is known that the general solution of Eq. (4.1.6) contains two linearly independent functions $P_{\chi-1/2}(\cos\theta), Q_{\chi-1/2}(\cos\theta)$ where $P_{\chi-1/2}(\cos\theta), Q_{\chi-1/2}(\cos\theta)$ are Legendre functions of the first and second kinds respectively.

Further, we are given that

$$P_{-t-1}(\cos\theta) = P_t(\cos\theta),$$
$$Q_{-t-1}(\cos\theta) = Q_t(\cos\theta) - \pi ctg\,\pi t P_t(\cos\theta).$$

The general solution of (4.1.3) can be represented as:

$$\begin{aligned} a(z,\theta) &= (z-1/2)(z-7/2+4\nu)\psi_{z-3/2}(\theta) - (z+1/2)\psi_{z-1/2}(\theta) \\ b(z,\theta) &= (z+7/2-4\nu)\frac{d\psi_{z-3/2}(\theta)}{d\theta} - \frac{d\psi_{z+1/2}(\theta)}{d\theta} \\ \psi_z(\theta) &= A_z P_z(\cos\theta) + B_z Q_z(\cos\theta) \end{aligned} \tag{4.1.8}$$

Here A_z, B_z are arbitrary constants.

On using the boundary conditions (4.1.4), we obtain the following characteristic equation for the parameter z:

$$\begin{aligned} \Delta(z,\theta_1,\theta_2) = &-\left(z-\frac{1}{2}\right)^4 \varphi_1^2(z) D^{(0,0)}_{z-3/2}(\theta_1,\theta_2) D^{(1,1)}_{z+1/2}(\theta_1,\theta_2) \\ &-\left(z+\frac{1}{2}\right)^4 \varphi_1^2(-z) D^{(0,0)}_{z+1/2}(\theta_1,\theta_2) D^{(1,1)}_{z-3/2}(\theta_1,\theta_2) \\ &-4(1-\nu)z(z-1/2)^2 \varphi_1(z) ctg\theta_1 D^{(1,0)}_{z-3/2}(\theta_1,\theta_2) D^{(1,1)}_{z+1/2}(\theta_1,\theta_2) \\ &-4(1-\nu)z(z-1/2)^2 \varphi_1(z) ctg\theta_2 D^{(0,1)}_{z-3/2}(\theta_1,\theta_2) D^{(1,1)}_{z+1/2}(\theta_1,\theta_2) \\ &-16(1-\nu)^2 ctg\theta_1 ctg\theta_2 D^{(1,1)}_{z-3/2}(\theta_1,\theta_2) D^{(1,1)}_{z+1/2}(\theta_1,\theta_2) \\ &+4(1-\nu)z(z+1/2)^2 \varphi_1(-z) ctg\theta_1 D^{(1,0)}_{z+3/2}(\theta_1,\theta_2) D^{(1,1)}_{z-1/2}(\theta_1,\theta_2) \\ &+4(1-\nu)z(z+1/2)^2 \varphi_1(-z) ctg\theta_2 D^{(0,1)}_{z+1/2}(\theta_1,\theta_2) D^{(1,1)}_{z-3/2}(\theta_1,\theta_2) \\ &+\left(z^2-\frac{1}{2}\right)^2 \varphi_1(z)\varphi_1(-z)\Big[D^{(1,0)}_{z-3/2}(\theta_1,\theta_2) D^{(0,1)}_{z+1/2}(\theta_1,\theta_2) \\ &+D^{(0,1)}_{z-3/2}(\theta_1,\theta_2) D^{(1,0)}_{z+1/2}(\theta_1,\theta_2)\Big] + 2\left(z^2-\frac{1}{4}\right)^2 \\ &\times \varphi_1(z)\varphi_1(-z)\sin^{-1}\theta_1 \sin^{-1}\theta_2 = 0. \end{aligned} \tag{4.1.9}$$

Here, we us the notation

$$D^{(s,l)}_t(\varphi,\psi) = P^{(s)}_t(\cos\varphi) Q^{(l)}_t(\cos\psi) - P^{(l)}_t(\cos\psi) Q^{(s)}_t(\cos\varphi)$$
$$\varphi_1(z) = z^2 + z + 2\nu - 7/4,\ s,l = 0,1.$$

The transcendental equation (4.1.9) defines a countable set of roots z_k to which correspond the constants $A_{z_k-3/2}, B_{z_k-3/2}, A_{z_k+1/2}, B_{z_k+1/2}$ which are proportional to the cofactors of any row or column of the determinant of the system. Choosing the cofactors of the elements of the first row as the solutions of the system and summing over all the roots, we obtain the following homogeneous solutions:

$$
\begin{aligned}
U_r &= \frac{1}{\sqrt{r}} \sum_{k=0}^{\infty} C_k r^{z_k} U_{rk}(\theta) \\
U_\theta &= \frac{1}{\sqrt{r}} \sum_{k=0}^{\infty} C_k r^{z_k} U_{\theta k}(\theta) \\
\sigma_r &= \frac{2G}{r\sqrt{r}} \sum_{k=0}^{\infty} C_k r^{z_k} Q_{rk}(\theta) \\
\sigma_\varphi &= \frac{2G}{r\sqrt{r}} \sum_{k=0}^{\infty} C_k r^{z_k} Q_{\varphi k}(\theta) \\
\sigma_\theta &= \frac{2G}{r\sqrt{r}} \sum_{k=0}^{\infty} C_k r^{z_k} Q_{\theta k}(\theta) \\
\tau_{r\theta} &= \frac{G}{r\sqrt{r}} \sum_{k=0}^{\infty} C_k r^{z_k} T_k(\theta)
\end{aligned}
\tag{4.1.10}
$$

Here C_k are arbitrary constants and the functions entering the equations are defined by

$$
\begin{aligned}
U_{rk}(\theta) &= (z_k - 1/2)(z_k - 7/2 + 4v)F_1(z_k, \theta) - (z_k + 1/2)F_2(z_k, \theta) \\
U_{\theta k}(\theta) &= \varphi_2(z_k)\frac{dF_1(z_k, \theta)}{d\theta} - \frac{dF_2(z_k, \theta)}{d\theta} \\
Q_{rk}(\theta) &= (z_k - 1/2)\left(z_k^2 - 4z_k + 7/4 + 2v\right)F_1(z_k, \theta) - \left(z_k^2 - \frac{1}{4}\right)F_2(z_k, \theta) \\
Q_{\varphi k}(\theta) &= (z_k - 1/2)(z_k - 7/2 + 4v)F_1(z_k, \theta) + \varphi_2(z_k)ctg\theta \\
&\quad \times \frac{dF_1(z_k, \theta)}{d\theta} - (z_k + 1/2)^2 F_2(z_k, \theta) - ctg\theta \frac{dF_2(z_k, \theta)}{d\theta}, \\
Q_{\theta k}(\theta) &= -(z_k - 1/2)\varphi_1(z_k)F_1(z_k, \theta) - \varphi_2(z_k)ctg\theta \\
&\quad \times \frac{dF_1(z_k, \theta)}{d\theta} + (z_k + 1/2)^2 F_2(z_k, \theta) + ctg\theta \frac{dF_2(z_k, \theta)}{d\theta} \\
T_k(\theta) &= \varphi_1(-z_k)\frac{dF_1(z_k, \theta)}{d\theta} - (z_k - 1/2)\frac{dF_2(z_k, \theta)}{d\theta}
\end{aligned}
\tag{4.1.11}
$$

where

$$
\begin{aligned}
\varphi_2(z) &= z+7/2-4v \\
F_1(z_k,\theta) &= \left(z_k-\frac{1}{2}\right)(z_k^2-1/4)\varphi_1(-z_k)\left[\sin^{-1}\theta_2 D^{(0,1)}_{z_k-3/2}(\theta,\theta_1)\right. \\
&\quad \left. - D^{(1,0)}_{z_k+1/2}(\theta_1,\theta_2)D^{(0,1)}_{z_k-3/2}(\theta,\theta_2)\right] + 4(1-v)z_k\left(z_k-\frac{1}{2}\right)ctg\theta_2 \\
&\quad \times D^{(1,1)}_{z_k+1/2}(\theta_1,\theta_2)D^{(0,1)}_{z_k-3/2}(\theta,\theta_2)\Big] + \left(z_k-\frac{1}{2}\right)^3 \\
&\quad \times D^{(1,1)}_{z_k+1/2}(\theta_1,\theta_2)D^{(0,0)}_{z_k-3/2}(\theta,\theta_2) \\
F_2(z_k,\theta) &= -\left(z_k-\frac{1}{2}\right)^2\varphi_1(z_k)\varphi_1(-z_k)\left[\sin^{-1}\theta_2 D^{(0,1)}_{z_k+1/2}(\theta,\theta_1)\right. \\
&\quad \left. - D^{(1,0)}_{z_k-3/2}(\theta_1,\theta_2)D^{(0,1)}_{z_k+1/2}(\theta,\theta_2)\right] + 4(1-v)z_k\varphi_1(-z_k)ctg\theta_2 \\
&\quad \times D^{(1,1)}_{z_k-3/2}(\theta_1,\theta_2)D^{(0,1)}_{z_k+1/2}(\theta,\theta_2) + \left(z_k+\frac{1}{2}\right)^2 \\
&\quad \times \varphi_1^2(-z_k)D^{(1,1)}_{z_k-3/2}(\theta_1,\theta_2)D^{(0,0)}_{z_k+1/2}(\theta,\theta_2)
\end{aligned}
$$

4.2 Analysis of the Roots of the Characteristic Equation

We observe from Eq. (4.1.9) that the characteristic equation has a rather complex structure and is not subject to precise analysis. Therefore, as before, to effectively explore the set of the roots of Eq. (4.1.9) we propose

$$\theta_1 = \theta_0 - \varepsilon, \theta_2 = \theta_0 + \varepsilon \tag{4.2.1}$$

where θ_0 is the opening angle of the mid-surface of the shell, ε is a dimensionless parameter characterizing its thickness. We shall further assume that ε is a small parameter, and $0<\xi_1<\theta_0<\xi<\pi/2$ (ξ_1, ξ_2 are some constants).

$\theta_0 = \pi/2$ is a particular case and corresponds to a plate of variable thickness which will be examined later.

Substituting (4.2.1) into (4.1.9), we obtain

$$D(z,\varepsilon,\theta_0) = \Delta(z,\theta_1,\theta_2) = 0 \tag{4.2.2}$$

It can be shown that $D(z,\varepsilon,\theta_0)$ is an even function of z as well as ε.

We will prove the following statement with respect to the roots of the function $D(z,\varepsilon,\theta_0)$: the function has three sets of roots: (a) the first group consists of three multiple zeros not depending on a small parameter ε, $z_0 = 0, z_{1,2} = \pm 1/2$; (b) the second group consists of four roots that are order of $O(\varepsilon^{-1/2})$ in the limit $\varepsilon \to 0$ (c) the third group consists of a countable set of roots which are order of $O(\varepsilon^{-1})$ at $\varepsilon \to 0$.

To prove this statement, we represent here the function $D(z, \varepsilon, \theta_0)$ as follows:

$$
\begin{aligned}
D(z,\varepsilon,\theta_0) &= \left(z+\frac{1}{2}\right)^2 D_1(z,\varepsilon,\theta_0) + \left(z+\frac{1}{2}\right) D_2(z,\varepsilon,\theta_0) \\
&\quad \times \left(z+\frac{1}{2}\right)^{-1} D^{(1,1)}_{z+1/2}(z,\varepsilon,\theta_0) \\
D_1(z,\varepsilon,\theta_0) &= -\left(z+\frac{1}{2}\right)^2 \varphi_1^2(-z) D^{(0,0)}_{z+1/2}(z,\varepsilon,\theta_0) \\
&\quad \times D^{(1,1)}_{z-3/2}(z,\varepsilon,\theta_0) + 4(1-\nu) z\varphi_1(-z) \\
&\quad \times D^{(1,1)}_{z-3/2}(z,\varepsilon,\theta_0)\Big[ctg\theta_1 D^{(1,0)}_{z+1/2}(z,\varepsilon,\theta_0) \\
&\quad + ctg\theta_2 D^{(0,1)}_{z+1/2}(z,\varepsilon,\theta_0)\Big] + \left(z-\frac{1}{2}\right)^2 \varphi_1(z)\varphi_1(-z) \\
&\quad \times \Big[D^{(0,1)}_{z-3/2}(z,\varepsilon,\theta_0) D^{(1,0)}_{z+1/2}(z,\varepsilon,\theta_0) \\
&\quad + D^{(1,0)}_{z-3/2}(z,\varepsilon,\theta_0) D^{(0,1)}_{z+1/2}(z,\varepsilon,\theta_0)\Big] + 2\left(z-\frac{1}{2}\right)^2 \\
&\quad \times \varphi_1(z)\varphi_1(-z)\sin^{-1}\theta_1 \sin^{-2}\theta_2 \\
D_2(z,\varepsilon,\theta_0) &= -\left(z-\frac{1}{2}\right)^4 \varphi_1^2(z) D^{(0,0)}_{z-3/2}(z,\varepsilon,\theta_0) \\
&\quad - 4(1-\nu) z (z-1/2)^2 \varphi_1(z)\varphi_1(-z) \\
&\quad \times \Big[ctg\theta_1 D^{(1,0)}_{z-3/2}(z,\varepsilon,\theta_0) + ctg\theta_2 D^{(0,1)}_{z-3/2}(z,\varepsilon,\theta_0)\Big] \\
&\quad - 16(1-\nu)^2 ctg\theta_1 ctg\theta_2 D^{(1,1)}_{z-3/2}(z,\varepsilon,\theta_0)
\end{aligned}
\tag{4.2.3}
$$

After a great deal of algebra we can show that

$$
\begin{aligned}
&\lim_{z\to-\frac{1}{2}} D_1(z,\varepsilon,\theta_0) \neq, \ \lim_{z\to-\frac{1}{2}} (z+1/2)^{-1} D^{(1,1)}_{z+1/2}(z,\varepsilon,\theta_0) \neq 0, \\
&\lim_{z\to-\frac{1}{2}} D_2(z,\varepsilon,\theta_0) = 0.
\end{aligned}
\tag{4.2.4}
$$

This implies that $z_1 = -\frac{1}{2}$ is a double zero of the function $D(z, \varepsilon, \theta_0)$. Since $D(z, \varepsilon, \theta_0)$ is an even function of z it follows that $z_2 = \frac{1}{2}$ is also a double zero of the function $D(z, \varepsilon, \theta_0)$. Similarly, it is established that $z_0 = 0$ is a double zero as well. As in the first chapter, we can prove that all other zeros of $D(z, \varepsilon, \theta_0)$ increase indefinitely as $\varepsilon \to 0$. They can be divided into two groups according to their behaviour in the limit $\varepsilon \to 0$. The following limit relations are possible: (1) $\varepsilon z_k \to 0$ as $\varepsilon \to 0$; (2) $\varepsilon z_k \to const$ as $\varepsilon \to 0$. We firstly define those z_k that the product $\varepsilon z_k \to 0$ as $\varepsilon \to 0$. To this end, as before, we expand $D(z, \varepsilon, \theta_0)$ in a series in powers of ε:

$$\begin{aligned} D(z,\varepsilon,\theta_0) = {} & 16z^2\left(z^2-\frac{1}{4}\right)^2 \sin^{-2}\theta_0\varepsilon^2\langle 4(1-\nu^2)ctg^2\theta_0 \\ & +3^{-1}\{4z^4+2[8\nu-9-8(1-\nu^2)ctg^2\theta_0]z^2+12\nu^2-28\nu \\ & +65/4+24(1-\nu^2)ctg^2\theta_0+36(1-\nu^2)ctg^4\theta_0\}\varepsilon^2 \\ & +45^{-1}[-32z^6+\cdots]\varepsilon^4+\cdots\rangle = 0. \end{aligned} \tag{4.2.5}$$

Assume that the leading term of the asymptotic expansion of z_k is of the form

$$z_k = \gamma_0\varepsilon^{-\alpha}, \gamma_0 = O(1) \quad \text{if } \varepsilon \to 0, 0<\alpha<1. \tag{4.2.6}$$

Substituting (4.2.6) into (4.2.5) and keeping only the leading terms in it we obtain the following limit equation for γ_0:

$$\begin{aligned} & 4(1-\nu^2)ctg^2\theta_0+3^{-1}[4\gamma_0^4+O(\varepsilon^{2\alpha})]\varepsilon^{2-4\alpha} \\ & \quad +O[\max(\varepsilon^{4-6\alpha},\varepsilon^{2-2\alpha})] = 0. \end{aligned} \tag{4.2.7}$$

Now we consider three cases: (a) $0<\alpha<\frac{1}{2}$; (b) $\alpha=\frac{1}{2}$; (c) $\frac{1}{2}<\alpha<1$. In case a passing to limit in (4.2.7) as $\varepsilon \to 0$ we obtain $\gamma_0 = 0$ which contradicts the assumption (4.2.6). Similarly, in case b we obtain $\gamma_0 = 0$ and arrive at a contradiction. Finally, in case c we have

$$\gamma_0^4 + 3(1-\nu^2)ctg^2\theta_0 = 0 \tag{4.2.8}$$

Let us find, now, z_k in the form of an expansion

$$z_k = \alpha_k\varepsilon^{-1/2} + \alpha_k^{(0)} + \beta_k\varepsilon^{1/2} + \cdots (k = 3,4,5,6), \tag{4.2.9}$$

where $\alpha_k = \gamma_0, \alpha_k^{(0)} = 0$,

$$\beta_k = (40\alpha_k)^{-1}[24(1-\nu^2)ctg^2\theta_0 + 5(9-8\nu)]. \tag{4.2.10}$$

To construct the asymptotics of zeros of the third group we seek them in the form

$$z_n = \frac{\delta_n}{\varepsilon} + O \quad (\varepsilon)(n = k-6, k = 7,8,\ldots). \tag{4.2.11}$$

After substituting (4.2.11) into the characteristic Eq. (4.1.9) and transforming it by means of the asymptotic expansions of the functions $P_\nu(\cos\theta), Q_\nu(\cos\theta)$ [3]

$$\begin{aligned}P_\nu(\cos\theta) &= \frac{\Gamma(\nu+1)}{\Gamma(\nu+3/2)}\left(\frac{\pi}{2}\sin\theta\right)^{-1/2}\\ &\times\left\{\cos\left[(\nu+1/2)\theta-\frac{\pi}{4}\right]+O(\nu^{-1})\right\}\\ Q_\nu(\cos\theta) &= \frac{\Gamma(\nu+1)}{\Gamma(\nu+3/2)}\left(\frac{2}{\pi}\sin\theta\right)^{-1/2}\\ &\times\left\{\cos\left[(\nu+1/2)\theta+\frac{\pi}{4}\right]+O(\nu^{-1})\right\}\end{aligned} \tag{4.2.12}$$

we arrive at following equation for δ_n:

$$\sin^2 2\delta_n - 4\delta_n^2 = 0\,(n = 1, 2, \ldots). \tag{4.2.13}$$

It should be noted that Eq. (4.2.13) coincides with the equation determining the index of Saint-Venant's end effects in the theory of plates.

Note that when $\theta_0 \to 0$ the zeros determined from formulas (4.2.9) and (4.2.11) agree with the zeros obtained in [4] for a cylinder.

4.3 Analysis of the Stress-Strain State

Let us give the characteristics of stress-strain states defined by homogeneous solutions constructed in the previous section.

1. Substituting $z_0 = 0$ into expressions (4.1.10) we obtain that this root corresponds to the trivial solution.

   For $z_2 = \frac{1}{2}$ we get the following expressions

$$\begin{aligned}&U_r = C_2\cos\theta,\, U_\theta = -C_2\sin\theta\\ &\sigma_r = \sigma_\varphi = \sigma_\theta = \tau_{r\theta} = 0\end{aligned} \tag{4.3.1}$$

from which we deduce that this solution corresponds to the expression of the conical body as a solid.

In accordance with the three groups of the eigenvalues of the spectral problem (4.1.3), (4.1.4) the displacement vector and the stress tensor are transformed to the form

$$\begin{aligned}&\bar{U} = \bar{U}_1 + \bar{U}_2 + \bar{U}_3\\ &\underline{\underline{\sigma}} = \underline{\underline{\sigma}}^{(1)} + \underline{\underline{\sigma}}^{(2)} + \underline{\underline{\sigma}}^{(3)}\end{aligned} \tag{4.3.2}$$

The displacement vector $\bar{U}_1$ and the stress tensor $\underline{\underline{\sigma}}^{(1)}$ correspond to eigenvalues $z_1 = -\frac{1}{2}$ and their components are defined by the formulas

$$\begin{aligned} U_r^{(1)} &= -\frac{r_1}{\rho} C_1 [4(1-v)\cos\theta - (1-2v)l_1] \\ U_\theta^{(1)} &= \frac{r_1}{\rho} C_1 \left[(3-4v)\sin\theta - (1-2v)l_2 \sin^{-1}\theta \right. \\ &\quad \left. + (1-2v)l_1 ctg\theta\right] \end{aligned} \tag{4.3.3}$$

$$\begin{aligned} \sigma_r^{(1)} &= \frac{2GC_1}{\rho^2}[2(2-v)\cos\theta - (1-2v)l_1] \\ \sigma_\varphi^{(1)} &= -\frac{2G(1-2v)C_1}{\rho^2}\left[\cos\theta + l_2 ctg\theta \sin^{-1}\theta + l_1 \sin^2\theta\right] \\ \sigma_\theta^{(1)} &= \frac{-2G(1-2v)C_1}{\rho^2}\left[\cos\theta - l_2 ctg\theta \sin^{-1}\theta + l_1 ctg^2\theta\right] \\ \tau_{r\theta}^{(1)} &= \frac{-2G(1-2v)C_1}{\rho^2}\left[\sin\theta - l_2 \sin^{-1}\theta + l_1 ctg\theta\right] \end{aligned} \tag{4.3.4}$$

Here $\rho = rr_1^{-1}$ is a dimensionless coordinate, C_1 is an unknown constant, $l_1 = \cos\theta_1 + \cos\theta_2, l_2 = 1 + \cos\theta_1 \cos\theta_2$.

In [1] it is incorrectly stated that these zeros correspond to the trivial solution. Note that the solution (4.3.2), (4.3.3) coincides with the well-known Mitchell Neuber's solution for the cone [5].

The above equations are accurate ones on the basis of which we can easily obtain the approximate formulas suggesting that

$$\theta = \theta_0 + \varepsilon\eta, -1 \le \eta \le 1, \theta_1 = \theta_0 - \varepsilon, \theta_2 = \theta_0 + \varepsilon \tag{4.3.5}$$

2. Now we proceed to the study of homogeneous solutions corresponding to the second group of roots. As it follows from (4.2.9), four of the solutions correspond to this group of roots. Substituting expressions (4.3.5) into relationships (4.1.10) instead of $\theta, \theta_1, \theta_2$ and then expanding the formulas thus obtained in powers of ε, we obtain

$$\begin{aligned} U_r^{(2)} &= r_1 \left(\frac{\varepsilon}{\rho}\right)^{1/2} \sum_{k=3}^{6} C_k U_{rk} \\ U_\theta^{(2)} &= \frac{r_1}{\sqrt{\rho}} \sum_{k=3}^{6} C_k U_{\theta k} \end{aligned} \tag{4.3.6}$$

$$
\begin{aligned}
\sigma_r^{(2)} &= \frac{2G}{\rho\sqrt{\rho}} \sum_{k=3}^{6} C_k Q_{rk} \\
\sigma_\varphi^{(2)} &= \frac{2G}{\rho\sqrt{\rho}} \sum_{k=3}^{6} C_k Q_{\varphi k} \\
\sigma_\theta^{(2)} &= \frac{2G}{\rho\sqrt{\rho}} \varepsilon \sum_{k=3}^{6} C_k Q_{\theta k} \\
\tau_{r\theta}^{(2)} &= \frac{2G}{\rho} \left(\frac{\varepsilon}{\rho}\right)^{1/2} \sum_{k=3}^{6} C_k T_k
\end{aligned} \tag{4.3.7}
$$

with C_k being arbitrary constants and the functions appearing in the equations being defined by

$$
\begin{aligned}
U_{rk} &= 2(1-\nu)\big\{2\alpha_k\big(\alpha_k^2\eta + \nu ctg\theta_0\big) \\
&\quad + \sqrt{\varepsilon}\big[(\nu-2)ctg\theta_0 - 3\alpha_k^2\eta + \alpha_k\beta_k\big(\alpha_k^2\eta + \nu ctg\theta_0\big)\big] \\
&\quad + O(\varepsilon)\big\} \exp\left(\frac{\alpha_k}{\sqrt{\varepsilon}} \ln\, \rho\right) \\
U_{\theta k} &= -2(1-\nu)\alpha_k^2\big[2 + \beta_k \ln\, \rho\sqrt{\varepsilon} + O(\varepsilon)\big] \exp\left(\frac{\alpha_k}{\sqrt{\varepsilon}} \ln\, \rho\right) \\
Q_{rk} &= \big\{-12\big(1-\nu^2\big)\eta ctg^2\theta_0 + \big[16\alpha_k^3\beta_k\eta + 4(\nu-2)\alpha_k^3\eta \\
&\quad - 4\big(1-\nu^2\big)\alpha_k^2 ctg^2\theta_0 - 6\big(1-\nu^2\big)\eta\beta_k \ln\, \rho ctg^2\theta_0\big]\sqrt{\varepsilon} \\
&\quad + O(\varepsilon)\big\} \exp\left(\frac{\alpha_k}{\sqrt{\varepsilon}} \ln\, \rho\right) \\
Q_{\varphi k} &= \big\{-4\big(1-\nu^2\big)(3\nu\eta + 1)ctg^2\theta_0 + \big[4(1-2\nu)\eta\alpha_k^3 \\
&\quad - 2\big(1-\nu^2\big)(3\nu\eta + 1)\beta_k \ln\, \rho ctg^2\theta_0\big]\sqrt{\varepsilon} + O(\varepsilon)\big\} \exp\left(\frac{\alpha_k}{\sqrt{\varepsilon}} \ln\, \rho\right) \\
Q_{\theta k} &= \big(1-\eta^2\big)\big\{2\big(1-\nu^2\big)\big(3\alpha_k^2 - \nu\big)ctg^2\theta_0 + \big[\big(1-\nu^2\big)\big(3\alpha_k^2 - \nu\big) \\
&\quad \times \beta_k \ln\, \rho ctg^2\theta_0 - 2/3(1-2\nu)\alpha_k^3 ctg\theta_0\big]\sqrt{\varepsilon} + \cdots\big\} \exp\left(\frac{\alpha_k}{\sqrt{\varepsilon}} \ln\, \rho\right) \\
T_k &= 3\big(1-\nu^2\big)\big(\eta^2 - 1\big)ctg^2\theta_0\big[2\alpha_k + (\alpha_k\beta_k \ln\, \rho - 3)\sqrt{\varepsilon} + O(\varepsilon)\big] \\
&\quad \times \exp\left(\frac{\alpha_k}{\sqrt{\varepsilon}} \ln\, \rho\right)
\end{aligned} \tag{4.3.8}
$$

Note that this group of solutions was not found in [6].

3. For the third group of zeros, using the first term of the asymptotic of Legendre functions, we obtain two classes of solutions, the first of which corresponds to

the zeros of $\sin 2\delta_n + 2\delta_n$, and the second one to the zeros of $\sin 2\delta_n - 2\delta_n$. As noted in the first chapter, these solutions have the same structure:

$$
\begin{aligned}
U_r^{(3)} &= \frac{2r_1\varepsilon}{\sqrt{\rho}}\sum_{n=1}^{\infty} B_n\left[(1-\nu)\delta_n^{-1}F_n''(\eta) - \nu\delta_n F_n(\eta) + O(\varepsilon)\right]\exp\left(\frac{\delta_n}{\varepsilon}\ln\rho\right)\\
U_\theta^{(2)} &= \frac{-2r_1\varepsilon}{\sqrt{\rho}}\sum_{n=1}^{\infty} B_n\left[(2-\nu)F_n'(\eta) - (1-\nu)\delta_n^{-2}F_n'''(\eta) + O(\varepsilon)\right]\\
&\quad\times\exp\left(\frac{\delta_n}{\varepsilon}\ln\rho\right)\\
\sigma_r^{(3)} &= \frac{2G}{\sqrt{\rho}}\sum_{n=1}^{\infty} B_n\left[F_n''(\eta) + O(\varepsilon)\right]\exp\left(\frac{\delta_n}{\varepsilon}\ln\rho\right)\\
\sigma_\varphi^{(3)} &= \frac{2G}{\rho\sqrt{\rho}}\sum_{n=1}^{\infty} B_n\left[F_n''(\eta) + \delta_n^2F_n(\eta) + O(\varepsilon)\right]\exp\left(\frac{\delta_n}{\varepsilon}\ln\rho\right)\\
\sigma_\theta^{(3)} &= \frac{2G}{\rho\sqrt{\rho}}\sum_{n=1}^{\infty} B_n\left[\delta_n^2F_n(\eta) + O(\varepsilon)\right]\exp\left(\frac{\delta_n}{\varepsilon}\ln\rho\right)\\
\tau_{r\theta}^{(3)} &= -\frac{2G}{\sqrt{\rho}}\sum_{n=1}^{\infty} B_n\left[F_n'(\eta) + O(\varepsilon)\right]\exp\left(\frac{\delta_n}{\varepsilon}\ln\rho\right)
\end{aligned}
\tag{4.3.9}
$$

Here B_n are new unknown constants, $F_n(\eta)$ is P.F. Papkovich's function.

4. We investigate now the stress state picture described by homogeneous solutions (4.1.10).

Consider first the relationship of homogeneous solutions with the resultant vector of the stresses P acting in the cross section $r = const$.

We have:

$$
P = 2\pi r^2 \int_{\theta_1}^{\theta_2} (\sigma_r\cos\theta - \tau_{r\theta}\sin\theta)\sin\theta\, d\theta. \tag{4.3.10}
$$

Substituting now formulas (4.1.10) into the expression (4.3.10), we obtain

$$
P = C_1\gamma_0 + r^{1/2}\sum_{k=3}^{\infty} C_k r^{z_k}\gamma_k \tag{4.3.11}
$$

$$
\begin{aligned}
\gamma_0 &= -4\pi G(\cos\theta_2 - \cos\theta_1)\\
&\quad\times\left(\cos^2\theta_1 + 2\nu\cos\theta_1\cos\theta_2 + \cos^2\theta_2\right)
\end{aligned}
\tag{4.3.12}
$$

$$\gamma_k = 4\pi G \int\limits_{\theta_1}^{\theta_2} (Q_{rk} \cos\theta - T_k \sin\theta) \sin\theta d\theta \tag{4.3.13}$$

Let us prove that all γ_k $(k = 3, 4, \ldots)$ are equal to zero. For this purpose, we consider the boundary value problem

$$\begin{aligned} \sigma_r &= r_1^{z_k - 3/2} Q_{rs}, \tau_{r\theta} = r_1^{z_k - 3/2} T_s \quad (r = r_1) \\ \sigma_r &= r_2^{z_k - 3/2} Q_{rs}, \tau_{r\theta} = r_2^{z_k - 3/2} T_s \quad (r = r_2) \end{aligned} \tag{4.3.14}$$

It is easy to see that the solution of the problem (4.3.14) exists and is obtained from (4.1.10) if we suggest $C_k = \delta_{ks}$ in them where δ_{ks} is the Kronecker's symbol.

On the other hand, we know that a necessary condition for the solvability of the first problem of elasticity theory (stresses are given on the surface of the body) is the condition for the vanishing of the resultant vector and the principal moment of all external forces. In this case, the principal vector of external forces (4.3.14) in the projection on the symmetry axis $\theta = 0$ gives

$$P_s = \left(r_2^{z_k - 3/2} - r_1^{z_k - 3/2} \right) \gamma_s = 0. \tag{4.3.15}$$

As is easily seen, the last equality is satisfied only when $\gamma_s = 0$. For the resultant vector, we finally get

$$P = C_1 \gamma_0. \tag{4.3.16}$$

Thus, the stress state corresponding to the second and third groups is self-balanced in each section $r = const$.

Let us calculate the bending moment in the section $r = const$ and the shear force for the second and third groups of solutions. For simplicity, we set $\rho_1 = 0, \rho_2 = 1 (r = r_2 \rho)$ and calculate the shear force and the bending moment for each group of solutions.

$$\begin{aligned} M &= r_2^2 \int\limits_{\theta_1}^{\theta_2} \{\sigma_r \sin(\theta - \theta_0) - \tau_{r\theta}[1 - \cos(\theta - \theta_0)]\} \sin\theta d\theta \\ &= r_2^2 \sin\theta_0 \varepsilon^2 \int\limits_{-1}^{1} \eta \sigma_r d\eta + O(\varepsilon^3), \end{aligned} \tag{4.3.17}$$

$$Q = r_2 \int_{\theta_1}^{\theta_2} \{\sigma_r \sin(\theta - \theta_0) + \tau_{r\theta} \cos(\theta - \theta_0)\} \sin\theta d\theta$$
$$\approx r_2 \sin\theta_0 \varepsilon \int_{-1}^{1} \tau_{r\theta} d\eta + O(\varepsilon^2). \tag{4.3.18}$$

Substituting the expressions for the stresses for each group of solutions, respectively, we obtain

$$M_2 = -16(1 - \nu^2) G r_2^2 \cos\theta_0 ctg\theta_0 \frac{\varepsilon^2}{\rho\sqrt{\rho}} \times \sum_{k=3}^{6} C_k \exp\left(\frac{\alpha_k}{\sqrt{\varepsilon}} \ln \rho\right)$$
$$Q_2 = -16(1 - \nu^2) G r_2 \cos\theta_0 ctg\theta_0 \left(\frac{\varepsilon}{\rho}\right)^{3/2} \times \sum_{k=3}^{6} C_k \alpha_k \exp\left(\frac{\alpha_k}{\sqrt{\varepsilon}} \ln \rho\right) \tag{4.3.19}$$

$$M_3 = O(\varepsilon^3), Q_3 = O(\varepsilon^2) \tag{4.3.20}$$

Thus, the leading terms of the bending moment and the shear force determine the solution of the second group.

Based on the asymptotic analysis we will draw conclusions on the nature of the stress-strain state.

The solution (4.3.3), (4.3.4) defines the internal stress-strain state of the shell. The first terms of their asymptotic expansions in the thin-walledness parameter determines a momentless state of stress.

The solution (4.3.6) defines the boundary effects similar to the simple end effects in the application of the theory of shells. The first terms of the expansion of the solutions of (4.3.6) in the parameter ε together with (4.3.3) and (4.3.4) can be regarded as the solutions on the applied Kirchhoff-Love theory. The stress state, defined by (4.3.9), has the character of a boundary layer and is localized near the surface $r = const$. The first terms of its asymptotic expansion are completely equivalent to Saint-Venant's end effect in the theory of plates.

Thus, the analysis of the solutions shows that the stress state of a conical shell, as in the case of cylindrical and spherical shells, is composed of three types: internal stress state, the simple end effect and the boundary layer state.

4.4 Reduction to Infinite Systems

Consider the question of removing stress from the end surfaces of the shell.

Let the following stresses be given for $r = r_s \quad (s = 1, 2)$

$$\sigma_r = f_{1s}(\theta), \tau_{r\theta} = f_{2s}(\theta) \tag{4.4.1}$$

The functions $f_{js}(\theta)$ $(j = 1, 2)$ satisfy the conditions of equilibrium

$$\begin{aligned} &2\pi r_1^2 \int_{\theta_1}^{\theta_2} (f_{11} \cos\theta - f_{21} \sin\theta) \sin\theta \, d\theta \\ &= 2\pi r_2^2 \int_{\theta_1}^{\theta_2} (f_{12} \cos\theta - f_{22} \sin\theta) \sin\theta \, d\theta = P \end{aligned} \tag{4.4.2}$$

where P is the resultant vector of the forces acting in an arbitrary section $r = const.$

As shown above, non-self-balanced part of the stresses (4.4.1) can be removed using a penetrating solution (4.3.4), and the relationship of constant C_1 with the resultant vector is given by (4.3.16). Below, we assume that $p = 0$. We seek a solution in the form of (4.1.10) by virtue of the assumption $C_1 = 0$. To determine the arbitrary constants $C_k \quad (k = 3, 4, \ldots)$, the variation of which shall be regarded as independent, we use Lagrange's variational principle.

Since homogeneous solutions satisfy the equilibrium equation and the boundary conditions on the conical surface, the variational principle assumes the form

$$r_1 \sum_{s=1}^{2} \rho_s^2 \int_{\theta_1}^{\theta_2} [(\sigma_r - f_{1s})\delta U_r + (\tau_{r\theta} - f_{2s})\delta U_\theta]_{\rho=\rho_s} \sin\theta \, d\theta = 0. \tag{4.4.3}$$

Equating the coefficients at the independent variations to zero we obtain the following infinite system:

$$\sum_{k=1}^{\infty} m_{jk} C_k = N_j \quad (j = 1, 2, \ldots). \tag{4.4.4}$$

Here

$$\begin{aligned} m_{jk} = &\sum_{s=1}^{2} \exp(z_j + z_k) \\ &\times \ln \rho_s \int_{\theta_1}^{\theta_2} (Q_{rk} U_{rj} + T_k U_{\theta j}) \sin\theta \, d\theta \end{aligned} \tag{4.4.5}$$

$$N_j = \sum_{s=1}^{2} \exp(z_j + 3/2) \ln \rho_s \times \int_{\theta_1}^{\theta_2} (f_{1s} U_{rj} + f_{2s} U_{\theta j}) \sin\theta \, d\theta. \tag{4.4.6}$$

Using the smallness of the parameter ε of shell thin-walledness, we can construct the asymptotic solution of (4.4.4).

First of all, let us determine the assumptions about the external load more precisely.

Assume that $f_{1s} \sim 1$. Then if we consider that σ_r and $\tau_{r\theta}$, corresponding to the roots of the second group, have different orders ($\sigma_r^{(2)} \sim 1, \tau_{r\theta}^{(2)} \sim \sqrt{\varepsilon}$) then choosing the order f_{2s} we should be guided by the following considerations.

Using formulas (4.3.7), (4.3.9) and the fact that $F_k(\pm 1) = 0$ we obtain

$$\int_{-1}^{1} \tau_{r\theta} d\eta = -16G(1 - \nu^2) \frac{\sqrt{\varepsilon}}{\rho\sqrt{\rho}} ctg^2\theta_0 \times \sum_{k=3}^{6} C_k \alpha_k \exp\left(\frac{\alpha_k}{\sqrt{\varepsilon}} \ln \rho\right) \tag{4.4.7}$$

If now we present the given tangential stresses on the boundary in the form

$$f_{2s} = f_{2s}^{(1)} + f_{2s}^{(2)}, f_{2s}^{(1)} = \int_{\theta_1}^{\theta_2} f_{2s} d\eta \qquad f_{2s}^{(2)} = f_{2s} - f_{2s}^{(1)} \tag{4.4.8}$$

then, on the basis of the asymptotic formula (4.4.7), it is necessary to assume that $f_{2s}^{(1)}$ has the order $\sqrt{\varepsilon}$ and $f_{2s}^{(2)}$ may have the same order as f_{1s}, i.e. $f_{2s}^{(2)} \sim 1$.

Furthermore, using the formulas (4.3.7), (4.3.9), we will seek the unknown constants C_k, B_n in the form

$$\begin{aligned} C_k &= C_{k0} + C_{k1}\sqrt{\varepsilon} + \cdots, \\ B_n &= B_{n0} + B_{n1}\sqrt{\varepsilon} + \cdots. \end{aligned} \tag{4.4.9}$$

Taking into account the accepted order with respect to the expressions given on the boundary, on the basis of the variational principle we obtain the following system of equations with respect to the constants C_{k0}, B_{n0}.

$$\sum_{k=3}^{6} \prod_{jk} C_{k0} = E_j \quad (j = 3, 4, 5, 6) \tag{4.4.10}$$

$$\begin{aligned} &\sum_{n=1,3,\ldots}^{\infty} g_{tn} B_{n0} = H_t \quad (t = 1, 3, \ldots) \\ &\sum_{n=2,4,\ldots}^{\infty} g_{tn} B_{n0} = H_t \quad (t = 2, 4, \ldots) \end{aligned} \tag{4.4.11}$$

Here

$$\begin{aligned} \prod_{jk} &= 32(1 - \nu^2)(1 + \nu)G\alpha_k^2(\alpha_j - \alpha_k)ctg^2\theta_0 \\ &\times \sum_{s=1}^{2} \exp\left(\frac{\alpha_j + \alpha_k}{\sqrt{\varepsilon}} \ln \rho_s\right) \end{aligned} \tag{4.4.12}$$

$$\begin{aligned} E_j &= 4(1 - \nu)\alpha_j \int_{-1}^{1} (f_{1s}\alpha_j\eta - f_{2s})d\eta \\ &\times \sum_{s=1}^{2} \rho_s^{3/2} \exp\left(\frac{\alpha_j}{\sqrt{\varepsilon}} \ln \rho_s\right) \\ g_{tn} &= 2G\frac{\delta_t^2\delta_n^2(\sin^2\delta_t - \sin^2\delta_n)}{(\delta_t^2 - \delta_n^2)(\delta_t - \delta_n)}\left[\nu(\delta_t - \delta_n)^2 + 2\delta_t\delta_n\right] \\ &\sum_{s=1}^{2} \exp\left(\frac{\delta_t + \delta_n}{\varepsilon} \ln \rho_s\right)(t \neq n) \quad (n = 2, 4, \ldots) \\ g_{tt} &= 4G\delta_t^3\left(1 - \frac{2}{3}\sin^2\delta_t\right)\sum_{s=1}^{2} \exp\left(\frac{2\delta_t}{\varepsilon} \ln \rho_s\right) \\ H_t &= 2\sum_{s=1}^{2} \rho_s^{3/2} \exp\left(\frac{\delta_t}{\varepsilon} \ln \rho_s\right) \int_{-1}^{1} \{f_{1s}\left[(1 - \nu)\delta_t^{-1}F_n''(\eta)\right. \\ &\left. - \nu\delta_t F_t(\eta)\right] - f_{2s}\left[(1 - \nu)\delta_t^{-2}F_t'''(\eta) + (2 - \nu)F_n'(\eta)\right]\}d\eta \quad (t = 2, 4, \ldots) \\ H_t &= 2\sum_{s=1}^{2} \rho_s^{3/2} \exp\left(\frac{\delta_t}{\varepsilon} \ln \rho_s\right) \int_{-1}^{1} \{f_{1s}\left[(1 - \nu)\delta_t^{-1}F_n''(\eta) - \nu\delta_t F_t(\eta)\right] - \end{aligned}$$

(4.4.13)

For t, $n = 1, 3, \ldots$ the appropriate expressions g_{tn} are obtained by replacing $\cos\,\delta_n$ with $\sin\,\delta_n$ and $\sin\,\delta_n$ with $\cos\,\delta_n$ respectively in the aforementioned formulas.

From the structure of the resulting system one can conclude that the unknowns C_{k0} corresponding to the second group of zeros and unknowns B_{n0} corresponding to the third group of zeros are found independently.

The determination of C_{ki}, B_{ni} $(i = 1, 2, \ldots)$ is always reduced to the inversion of those matrices which coincide with the matrices (4.4.10), (4.4.11). The systems of Eqs. (4.4.10), (4.4.11) are considerably simplified if we investigate the condition of semi-infinite cone $(\rho_1 = 1, \rho_2 \to \infty)$ or a cone with vertex $(\rho_2 = 1, \rho_1 = 0)$.

In the first case, all the unknowns corresponding to zeros which have $\operatorname{Re}\alpha_k > 0, \operatorname{Re}\delta_n > 0$ should be set equal to zero, in the second case those unknowns for which $\operatorname{Re}\alpha_k < 0, \operatorname{Re}\delta_n < 0$ should be taken zero because of the boundedness of the solution at the top of the cone.

In both cases we obtain the same system written in the form

$$\begin{aligned} &\sum_{k=3}^{4} \prod\nolimits_{jk}^{(0)} C_{k0} = E_j^{(0)} (j = 3, 4), \\ &\sum_{n=1}^{\infty} g_{tn}^{(0)} B_{n0} = H_t^{(0)} (t = 1, 2, \ldots). \end{aligned} \tag{4.4.14}$$

The coefficients and right-hand sides of (4.4.14) can easily be obtained from the corresponding coefficients of the right-hand sides of (4.4.10), (4.4.11) if we set, in the first case $\rho_1 = 1, \rho_2 \to \infty$ and in the second $\rho_2 = 1, \rho_1 = 0$.

The general solution of the problem of determining the stress-strain state of a shell can be obtained by superposition of solutions corresponding to different groups of roots. We introduce the first term of the asymptotic expansions of a component of the vector of displacements and stresses:

$$\begin{aligned} U_r = {} & \frac{r_1}{\sqrt{\rho}} \left\{ \frac{-2}{\sqrt{\rho}} \cos\theta_0 C_{10} + 12(v^2 - 1) ctg\theta_0 \right. \\ & \times \sum_{k=3}^{6} \left(\eta\alpha_k + v\alpha_k^{-1} ctg\theta_0 \exp\left(\frac{\alpha_k}{\sqrt{\varepsilon}} \ln\rho \right) C_{k0}\sqrt{\varepsilon} \right. \\ & \left. + 2\varepsilon \sum_{n=1}^{\infty} \left[(1 - v)\delta_n^{-1} F_n''(\eta) - v\delta_n F_n(\eta) \right] \exp\left(\frac{\delta_n}{\varepsilon} \ln\rho \right) B_{n0} \right\} \\ U_\theta = {} & \frac{r_1}{\sqrt{\rho}} \left\{ \frac{2(v-1)}{\sqrt{\rho}} \sin\theta_0 C_{10} + 4(v-1) \sum_{k=3}^{6} \alpha_k^2 \right. \\ & \times \exp\left(\frac{\alpha_k}{\sqrt{\varepsilon}} \ln\rho \right) C_{k0} + 2\varepsilon \sum_{n=1}^{\infty} \left[(v-2) F_n'(\eta) + (v-1)\delta_n^{-2} F_n'''(\eta) \right] \\ & \left. \times \exp\left(\frac{\delta_n}{\varepsilon} \ln\rho \right) B_{n0} \right\} \end{aligned} \tag{4.4.15}$$

$$\sigma_r = \frac{2G}{\rho\sqrt{\rho}}\left\{\frac{2(1+\nu)}{\sqrt{\rho}}\cos\theta_0 C_{10} - 12(1-\nu^2)\eta ctg^2\theta_0 \right.$$
$$\left. \times \sum_{k=3}^{6} \alpha_k^2 \exp\left(\frac{\alpha_k}{\sqrt{\varepsilon}}\ln\rho\right)C_{k0} + \sum_{n=1}^{\infty}\left[F_n''(\eta)\right] + \exp\left(\frac{\delta_n}{\varepsilon}\ln\rho\right)B_{n0}\right\}$$
$$\sigma_\varphi = \frac{2G}{\rho\sqrt{\rho}}\left\{\frac{-4(1-2\nu)}{\sqrt{\rho}}\cos\theta_0\sin^{-2}\theta_0 C_{10} - 2(1-\nu^2)(3\nu\eta+1)ctg^2\theta_0 \right.$$
$$\left. \times \sum_{k=3}^{6} \exp\left(\frac{\alpha_k}{\sqrt{\varepsilon}}\ln\rho\right)C_{k0} + \sum_{n=1}^{\infty}\left[F_n''(\eta) + \delta_n^2 F_n(\eta)\right]\exp\left(\frac{\delta_n}{\varepsilon}\ln\rho\right)B_{n0}\right\}$$
$$\sigma_\theta = \frac{2G}{\rho\sqrt{\rho}}\left\{\frac{(1-2\nu)(\eta^2-1)}{\sqrt{\rho}}\cos\theta_0 C_{10}\varepsilon^2 + (1-\nu^2)(\eta^2-1)ctg^2\theta_0 \right.$$
$$\left. \times \sum_{k=3}^{6} (3\alpha_k^2 - \nu)\exp\left(\frac{\alpha_k}{\sqrt{\varepsilon}}\ln\rho\right)C_{k0}\varepsilon + \sum_{n=1}^{\infty}\delta_n^2 F_n(\eta)\exp\left(\frac{\delta_n}{\varepsilon}\ln\rho\right)B_{n0}\right\}$$
$$\tau_{r\theta} = \frac{G}{\rho\sqrt{\rho}}\left\{\frac{2(1-2\nu)(\eta^2-1)}{\sqrt{\rho}}\sin\theta_0 C_{10}\varepsilon^2 + 12(1-\nu^2)(\eta^2-1)ctg^2\theta_0 \right.$$
$$\left. \times \sum_{k=3}^{6} \alpha_k \exp\left(\frac{\alpha_k}{\sqrt{\varepsilon}}\ln\rho\right)\sqrt{\varepsilon}C_{k0} - 2\sum_{n=1}^{\infty}F_n'(\eta)\exp\left(\frac{\delta_n}{\varepsilon}\ln\rho\right)B_{n0}\right\}$$
(4.4.16)

In the formulas (4.4.15) and (4.4.16), the first and second terms of the right-hand sides correspond to the applied theory of shells, the remaining terms are corrections to the solution obtained by the Kirchhoff theory. On the boundary of the domain at $r = r_s$ the additional terms in the strains σ_r, σ_φ are of the same order as in the applied theory and the additional terms begin to play the main role in the strains $\sigma_\theta, \tau_{r\theta}$ as $\varepsilon \to 0$.

4.5 Construction of Refined Applied Theories for a Conical Shell

As in the first chapter, we will consider the construction of applied theories intended to relieve the existing stress from a conical part of the boundary. Suppose that the following conditions are given on the conical boundaries

$$\sigma_\theta = Q_n(r), \tau_{r\theta} = \tau_n(r) \quad \text{if } \theta = \theta_n \quad (n = 1, 2). \tag{4.5.1}$$

Carrying out the change of variable $t = \ln r$ the system of Lame's equations takes the form:

$$
\begin{aligned}
&\frac{2(1-v)}{1-2v}\left(\frac{\partial^2 U_r}{\partial t^2}+\frac{\partial U_r}{\partial t}-2U_r\right)+ctg\theta\frac{\partial U_r}{\partial \theta}+\frac{\partial^2 U_r}{\partial \theta^2}\\
&+\frac{1}{1-2v}\frac{\partial^2 U_\theta}{\partial t\partial \theta}-\frac{3-4v}{1-2v}\frac{\partial U_\theta}{\partial \theta}+\frac{1}{1-2v}ctg\theta\frac{\partial U_\theta}{\partial \theta}\\
&-\frac{3-4v}{1-2v}ctg\theta U_\theta=0\\
&\frac{1}{1-2v}\frac{\partial^2 U_r}{\partial t\partial \theta}+\frac{4(1-v)}{1-2v}\frac{\partial U_r}{\partial \theta}+\frac{\partial^2 U_\theta}{\partial t^2}+\frac{\partial U_\theta}{\partial t}\\
&+\frac{2(1-v)}{1-2v}\left(\frac{\partial^2 U_\theta}{\partial \theta^2}+ctg\theta\frac{\partial U_\theta}{\partial \theta}-\frac{U}{\sin^2\theta}\right)=0.
\end{aligned}
\tag{4.5.2}
$$

Assuming $\lambda=\frac{\partial}{\partial t}$ we obtain a system of ordinary differential equations with respect to U_r, U_θ:

$$
\begin{aligned}
&U_r''+ctg\theta\, U_r'+\frac{2(1-v)}{1-2v}(\lambda-1)(\lambda+2)U_r\\
&+\frac{1}{1-2v}(\lambda+4v-3)\left(U_\theta'+ctg\theta\, U_\theta\right)=0\\
&\frac{1}{1-2v}(\lambda+4+4v)U_r'+\frac{2(1-v)}{1-2v}\\
&\times\left(U_\theta''+ctg\theta\, U_\theta'-\frac{1}{\sin^2\theta}U_\theta\right)+\lambda(\lambda+1)U_\theta=0
\end{aligned}
\tag{4.5.3}
$$

The general solution of the system (4.5.3) is:

$$
\begin{aligned}
&U_r=(z-1/2)(z-7/2+4v)\psi_{z-3/2}(\theta)-(z+1/2)\psi_{z+1/2}(\theta)\\
&U_\theta=\varphi_2(z)\frac{d\psi_{z-3/2}}{d\theta}-\frac{d\psi_{z+1/2}}{d\theta}\\
&\psi_z=A_zP_z(\cos\theta)+B_zQ_z(\cos\theta),\, z=\lambda+1/2
\end{aligned}
\tag{4.5.4}
$$

Utilizing the boundary conditions on the conical boundaries (4.5.1) we obtain the inhomogeneous linear algebraic equations in the unknown constants $A_{z-3/2}, B_{z-3/2}, A_{z+1/2}, B_{z+1/2}$ whose solution may be given by Cramer's rule as

$$
\begin{aligned}
&A_{z-3/2}=\frac{\Delta_1}{\Delta}, B_{z-3/2}=\frac{\Delta_2}{\Delta}\\
&A_{z+1/2}=\frac{\Delta_3}{\Delta}, B_{z+1/2}=\frac{\Delta_4}{\Delta}.
\end{aligned}
\tag{4.5.5}
$$

Substituting (4.5.5) into (4.5.4) and expanding in the small parameter ε, we find

$$
\begin{aligned}
2G\Delta U_r = e^t\big[&-4(1-v)ctg\theta_0\{[2z^4+(4v-5)z^2\\
&+3(2v-3/2)z-3(2v-7/4)]\,(Q_2-Q_1)\\
&+ctg\theta_0\left[2z^2-4(1-2v)z+\frac{7}{2}-4v\right](\tau_2-\tau_1)\Big\}\varepsilon\\
&+\varepsilon^2\langle 4(1-v)ctg\theta_0\{2z^4-2(1-2v)z^3-12(1-v)z^2\\
&+3(v-1/2)z+\frac{63}{8}-9v-ctg^2\theta_0\left[2z^2-4(1-2v)z+\frac{7}{2}-4v\right]\Big\}\\
&\times(\tau_2+\tau_1)-2ctg\theta_0\big[2z^3+(4v-5)z^2+3(2v-3/2)z-3(2v-7/4)\big]\\
&\times(Q_2+Q_1)+\eta\big[2z^3+(4v-5)z^2-(2v-3/2)z+2v-7/4\big]\\
&\times\left[\left(4z^2+4v-5\right)(Q_2-Q_1)+4\left(vz+1-\frac{v}{2}\right)ctg\theta_0(\tau_2-\tau_1)\right]\Big\rangle+\cdots\Big]
\end{aligned}
\tag{4.5.6}
$$

$$
\begin{aligned}
2G\Delta U_\theta = e^t\big[&2z^2+2(1-2v)z+4v-7/2\big]\\
&\times\big[\left(4z^2+4v-5\right)(Q_2-Q_1)+(4vz+2v-4)ctg\theta_0(\tau_2-\tau_1)\big]\varepsilon\\
&+\varepsilon^2\langle\big[2z^2+2(1-2v)z+4v-7/2\big]\big[\left(5v-4-4z^2\right)\\
&\times\eta(Q-Q_1)+\left(4z^2-9\right)(1-v)(Q_2+Q_1)\big]ctg\theta_0\\
&+\big[8z^3-4\left(4v^2+2v-5\right)z^2-2\left(12v^2-16v+5\right)z\\
&+(7-8v)(1-2v)\big]\eta ctg^2\theta_0(\tau_2-\tau_1)\\
&+\{\,\big[8(1-v)(1-2v)z^2+24v(1-v)z+6(1-v)(7-8v)\big]\\
&\times ctg^2\theta_0-\left(z^2+2z-9/4\right)(z-4v+7/2)\\
&\times\left(4z^2+4v-5\right)\}(\tau_2+\tau_1)+\cdots\big]
\end{aligned}
$$

$$
\begin{aligned}
\Delta = 8z\left(z^2-\frac{1}{4}\right)\varepsilon^2\Big\langle&4\left(1-v^2\right)ctg\theta_0+\frac{1}{3}\{4z^2\\
&+2\big[8v-9-8\left(1-v^2\right)ctg^2\theta_0\big]z^2+12v^2-28v+65/4\\
&+24\left(1-v^2\right)ctg^2\theta_0+36\left(1-v^2\right)ctg^4\theta_0\}\varepsilon^2+\cdots\Big\rangle
\end{aligned}
\tag{4.5.7}
$$

As noted in Chap. 1, the relationships (4.5.6) can be used to construct approximate theories intended to relieve stress from a conical part of the boundary. Referring back to $t=\ln r$ and $\frac{\partial}{\partial t}=r\frac{\partial}{\partial r}=z-\frac{1}{2}$ we obtain the following system of ordinary differential equations:

$$
\begin{aligned}
GDU_r = {} & [-4(1-\nu)ctg\theta_0\{[2d_2d_1 + (4\nu - 5)d_2 \\
& + 3(2\nu - 3/2)d_1 - 3(2\nu - 7/4)](Q_2 - Q_1) \\
& + ctg\theta_0\left[2d_2 - 4(1-2\nu)d_1 + \frac{7}{2} - 4\nu\right](\tau_2 - \tau_1)\Big\}\varepsilon \\
& + \varepsilon^2\langle 4(1-\nu)ctg\theta_0\left[\{2d_2^2 - 2(1-2\nu)dd_1 - 12(1-\nu)d_2\right. \\
& + 3(\nu - 1/2)d_1 + \frac{63}{8} - 9\nu - ctg^2\theta_0[2d_2 - 4(1-2\nu)d_1 \\
& \left.\left. + \frac{7}{2} - 4\nu\right]\Big\}\right](\tau_2 + \tau_1) - 2ctg\theta_0[2d_2d_1 + (4\nu - 5)d_2 \\
& + 3(2\nu - 3/2)d_1 - 3(2\nu - 7/4)(Q_2 + Q_1) + \eta \\
& \times [2d_2d_1 + (4\nu - 5)d_2 - (2\nu - 3/2)d_1 + 2\nu - 7/4] \\
& \times \left[(4d_2 + 4\nu - 5)(Q_2 - Q_1) + 4\left(\nu d_1 + 1 - \frac{\nu}{2}\right)ctg\theta_0(\tau_2 - \tau_1)\right]\Big\rangle + \cdots\Big] \\
GDU_\theta = {} & [[2d_2 + 2(1-\nu)d_1 + 4\nu - 7/2][(4d_2 + 4\nu - 5) \\
& \times (Q_2 - Q_1) + (4\nu d_1 + 2\nu - 4)ctg\theta_0(\tau_2 - \tau_1)]\varepsilon + \varepsilon^2 \\
& \times \langle [2d_2 + 2(1-2\nu)d_1 + 4\nu + 7/2][(5\nu - 4 - L_1\nu d_2)\eta \\
& \times (Q_2 - Q_1) + (4d_2 - 9)(1-\nu)(Q_2 + Q_1)]ctg\theta_0 \\
& + \left[8d_2d_1 - 4\left(4\nu^2 + 2\nu - 5\right)d_2 - 2\left(12\nu^2 - 16\nu + 5\right)d_1\right. \\
& + (7 - 8\nu)(1-2\nu)]\eta ctg^2\theta_0(\tau_2 - \tau_1) + \{[8(1-\nu)(1-2\nu)d_2 \\
& + 24\nu(1-\nu)d_1 + 6(1-\nu)(7-8\nu)]ctg^2\theta_0 - (d_2 + 2d_1 - 9/4) \\
& \times (d_1 - 4\nu + 7/2)(4d_2 + 4\nu - 5)\}(\tau_2 + \tau_1)\rangle + \cdots]
\end{aligned}
\tag{4.5.8}
$$

where

$$
\begin{aligned}
D = {} & 16\varepsilon^2 d_1 d\langle 4\left(1-\nu^2\right)ctg^2\theta_0 + \frac{1}{3}\left\{4d^2 - 16(1-\nu)d\right. \\
& - 16\left(1-\nu^2\right)ctg^2\theta_0 d + 12(1-\nu)^2 + 20\left(1-\nu^2\right)ctg^2\theta_0 \\
& \left. + 36\left(1-\nu^2\right)ctg^4\theta_0\right\}\varepsilon^2 + \cdots\rangle \\
d_1 = {} & r\frac{d}{dr} + \frac{1}{2}, d_2 = r^2\frac{d^2}{dr^2} + 2r\frac{d}{dr} + \frac{1}{4}, d = d_2 - \frac{1}{4}
\end{aligned}
$$

In this way, it is possible to obtain applied theories with an arbitrary accuracy of the order of the smallness parameter ε.

In conclusion, we remark that:

(a) applied theories can be constructed by the proposed method only for smoothly varying loads (with a small index of variability);

(b) the application of the proposed method in conjunction with homogeneous solutions allow us to calculate the state of stress everywhere in the domain occupied by the shell with an accuracy of the order of the first discarded term in the asymptotic expansions;
(c) the relief of a stress state from the conical portion of the boundary can also be realized by means of solving an elasticity theory problem for an unbounded cone using Mellin's transform. This technique should be applied to non-smooth loads.

4.6 Axisymmetric Problem for a Plate of Variable Thickness

On the basis of homogeneous solutions the asymptotic behaviour of an axisymmetric stress-strain state of a plate of thickness $h = \varepsilon r, r$ being the distance from the centre of the plate together with the small parameter ε, is investigated. The com parison of the asymptotic solution with the Kirchhoff theory is conducted here.

As noted above, the case when $\theta_0 = \frac{\pi}{2}$ is a particular one and corresponds to a plate of variable thickness. Therefore, we will not be considering an arbitrary plate, but that particular form of a conical shell described in previous sections of this chapter which takes into account the degeneration of the mid-surface of the plane. Since this case of degeneration is a particular one, all the arguments of the previous sections have to be repeated anew. Thus, we consider an elastic body related to spherical coordinates r, θ, φ with the following range of parameters

$$r_1 \leq r \leq r_2, \frac{\pi}{2} - \varepsilon \leq \theta \leq \frac{\pi}{2} + \varepsilon, \; 0 \leq \varphi \leq 2\pi \tag{4.6.1}$$

Assume that the following homogeneous conditions on the conical boundaries $(\theta = \frac{\pi}{2} - \varepsilon, \theta = \frac{\pi}{2} + \varepsilon)$ are established:

$$\sigma_0 = 0, \tau_{r\theta} = 0. \tag{4.6.2}$$

The displacements and stresses can be represented as follows by the method of homogeneous solutions using the results for the cone:

$$U_r = \frac{1}{\sqrt{r}} \sum_{k=1}^{\infty} C_k r^{z_k} U_r(z_k, \theta)$$
$$U_\theta = \frac{1}{\sqrt{r}} \sum_{k=1}^{\infty} C_k r^{z_k} U_\theta U_r(z_k, \theta) \tag{4.6.3}$$

$$\begin{aligned}
\sigma_r &= \frac{2G}{r\sqrt{r}}\sum_{k=2}^{\infty} C_k r^{z_k} Q_r(z_k,\theta) \\
\sigma_\varphi &= \frac{2G}{r\sqrt{r}}\sum_{k=2}^{\infty} C_k r^{z_k} Q_\varphi(z_k,\theta) \\
\sigma_\theta &= \frac{2G}{r\sqrt{r}}\sum_{k=2}^{\infty} C_k r^{z_k} Q_\theta(z_k,\theta) \\
\tau_{r\theta} &= \frac{2G}{r\sqrt{r}}\sum_{k=2}^{\infty} C_k r^{z_k} T(z_k,\theta)
\end{aligned} \tag{4.6.4}$$

In the formulas (4.6.3) and (4.6.4) C_k are arbitrary constants, z_k are the zeros of the function $D(z,\varepsilon,\theta_0)$ at $\theta_0 = \frac{\pi}{2}$.

4.7 Analysis of the Characteristic Equation for a Plate of Variable Thickness

The asymptotic behaviour of homogeneous solutions was investigated above for $\theta_0 \neq \pi/2$ and $\varepsilon \to 0$. It was found that the zeros of the function $D(z,\varepsilon,\theta_0)$ can be divided into three groups by the nature of their asymptotic behaviour, each of which corresponds to its type of stress-strain state.

As we have mentioned above, the present case $\theta_0 = \frac{\pi}{2}$ is a particular one and requires separate consideration.

We start by investigating the zeros $D(z,\varepsilon,\theta_0)$ at $\theta_0 = \frac{\pi}{2}$. The zeros of the function $D(z,\varepsilon,\theta_0)$ at $\theta_0 = \frac{\pi}{2}$ can be divided into two groups by the nature of their asymptotic behaviour as $\varepsilon \to 0$. The first group consists of seven zeros and is characterized by the fact that all z_k $(k = \overline{0,6})$ of this group have a finite limit as $\varepsilon \to 0$, and three of them, namely $z_0 = 0, z_1 = \frac{1}{2}, z_2 = -\frac{1}{2}$ do not depend on the parameter ε. To find the remaining zeros of the first group we expand $D\left(z,\varepsilon,\frac{\pi}{2}\right)$ in a series in ε giving

$$\begin{aligned}
D\left(z,\varepsilon,\frac{\pi}{2}\right) &= -\frac{4}{3}z^2\left(z^2-\frac{1}{4}\right)^2\varepsilon^4\{16z^4 \\
&\quad + 8(8\nu-9)z^2 + (13-12\nu)(5-4\nu) \\
&\quad + \frac{1}{15}\left[-128z^6 + 16(57-48\nu)z^4 - 16(5-2\nu)\right. \\
&\quad \left.\times(31-30\nu)z^2 + 5\left(384\nu^2 - 832\nu + 449\right)\right]\varepsilon^2 + \cdots\} = 0
\end{aligned} \tag{4.7.1}$$

From the expansion (4.7.1) we obtain the following group of zeros:

$$z_k = (-1)^k \left[a_0 + \varepsilon^2 a_2 + \cdots\right] \quad (k = 3, 4, 5, 6) \tag{4.7.2}$$

where

$$a_0 = \frac{1}{2}\sqrt{5 - 4v}, a_2 = (3a_0)^{-1}\left(1 - v^2\right) \quad (k = 3, 4) \tag{4.7.3}$$

$$a_0 = \frac{1}{2}\sqrt{13 - 12v}, a_2 = 3(5a_0)^{-1}\left(1 - v^2\right) \quad (k = 5, 6) \tag{4.7.4}$$

For convenience, we assume $(-1)^k a_0 = \alpha_k, (-1)^k a_0 = \beta_k$.

The second group consists of a countable set of zeros and is characterized in that all $z_k \to \infty$ as $\varepsilon \to 0$ ($\varepsilon z_k \to const$ as $\varepsilon \to 0$). For this second group we obtain the following asymptotic representation in the same way used in Sect. 4.2:

$$z_k = \varepsilon^{-1}\delta_k + \gamma_k + \varepsilon\omega_k + \cdots \quad (k = 7, 8, \ldots) \tag{4.7.4}$$

where

$$\begin{aligned} &\sin 2\delta_k + 2\delta_k = 0, \gamma_k = 0 \\ &\omega_k = 2^{-3}\delta_k^{-1}\cos^{-2}\delta_k\left[16(1 - v) + (8v - 9)\cos^2\delta_k \right. \\ &\quad \left. + 8/3\delta_k^2\right] \quad (k = 8, 10, \ldots) \end{aligned} \tag{4.7.5}$$

$$\begin{aligned} &\sin 2\delta_k - 2\delta_k = 0, \gamma_k = 0 \\ &\omega_k = 2^{-3}\delta_k^{-1}\sin^{-2}\delta_k\left[16(1 - v) + (8v - 9)\sin^2\delta_k + 8/3\delta_k^2\right] \end{aligned} \tag{4.7.6}$$

It can be shown that the function $D\left(z, \varepsilon, \frac{\pi}{2}\right)$ has no other zeros other than the ones obtained.

4.8 Analysis of Stress-Strain State of a Plate

By substituting $z_0 = 0$ into the expression (4.6.3) we observe that this zero corresponds to the trivial solution.

The zeros $z_1 = \frac{1}{2}, z_2 = -\frac{1}{2}$ correspond to the following stress-strain state which is obtained from (4.3.3), (4.3.4) for $\theta_0 = \pi/2$:

$$\begin{aligned} U_r &= r_1\left[C_1 \sin \varepsilon\eta - \frac{4(1 - v)C_2}{\rho}\sin \varepsilon\eta\right] \\ U_\theta &= r_1\left\{C_1 \cos \varepsilon\eta + \frac{C_2}{\rho}[[(3 - 4v)\cos \varepsilon\eta \right. \\ &\quad \left. -(1 - 2v)\cos^2\varepsilon\cos^{-1}\varepsilon\eta]\right\} \end{aligned} \tag{4.8.1}$$

$$\begin{aligned} \sigma_r &= \frac{4G(v-2)}{\rho^2} C_2 \sin \varepsilon\eta \\ \sigma_\varphi &= \frac{2(1-2v)GC_2}{\rho^2} \left[\sin \varepsilon\eta + \cos^2 \varepsilon tg\varepsilon\eta \cos^{-1} \varepsilon\eta\right] \\ \sigma_\theta &= \frac{2(1-2v)GC_2}{\rho^2} \left[\sin \varepsilon\eta - \cos^2 \varepsilon tg\varepsilon\eta \cos^{-1} \varepsilon\eta\right] \\ \tau_{r\theta} &= \frac{2(1-2v)GC_2}{\rho^2} \left[\cos^2 \varepsilon \cos^{-1} \eta - \cos \varepsilon\eta\right] \end{aligned} \tag{4.8.2}$$

where C_1, C_2 are unknown constants.

Assuming $\theta = \frac{\pi}{2} + \varepsilon\eta$, $-1 \le \eta \le 1$ in (4.6.3), (4.6.4) and expanding into a series in the parameter ε, we obtain the following expressions for the remaining zeros of the first group:

$$U_r = \frac{r_1}{\sqrt{\rho}} \sum_{k=3}^{6} C_k U_{rk},\; U_\theta = \frac{r_1}{\sqrt{\rho}} \sum_{k=3}^{6} C_k U_{\theta k} \tag{4.8.3}$$

$$\begin{aligned} \sigma_r &= \frac{2G}{\rho\sqrt{\rho}} \sum_{k=3}^{6} C_k Q_{rk},\; \sigma_\varphi = \frac{2G}{\rho\sqrt{\rho}} \sum_{k=3}^{6} C_k Q_{\varphi k} \\ \sigma_\theta &= \frac{2G}{\rho\sqrt{\rho}} \sum_{k=3}^{6} C_k Q_{\theta k},\; \tau_{r\theta} = \frac{2G}{\rho\sqrt{\rho}} \sum_{k=3}^{6} C_k T_k. \end{aligned} \tag{4.8.4}$$

Here C_k are arbitrary constants and

$$\begin{aligned} U_{rk} &= \Big\{ 2(v-2) + 4v\alpha_k + \frac{1}{3}\varepsilon^2 \big[3\left(1-v^2\right)(3-2\alpha_k)\eta^2 \\ &\quad + (1-v)(1-5v) + 2\left(1-v^2\right)\alpha_k + 12v\beta_k + 12\beta_k \\ &\quad \times (v-2+2v\alpha_k) \ln \rho\big] + \cdots \Big\} \exp(\alpha_k \ln \rho) \\ U_{\theta k} &= \varepsilon\eta \Big\langle 4\left(1-v^2\right) + \frac{1}{3}\varepsilon^2 \big\{ \left[2(1-2v)\alpha_k - 2v^2 \right. \\ &\quad \left. + 4v - 1\right](1+v)\eta^2 + (1+v)[(1-2v)\alpha_k - 2(1-2v)] \\ &\quad - 24(1-v)\alpha_k\beta_k + 24\left(1-v^2\right)\beta_k \ln \rho \big\} + \cdots \Big\rangle \exp(\alpha_k \ln \rho) \end{aligned} \tag{4.8.5}$$

$$
\begin{aligned}
Q_{rk} &= \Big\langle 2(1+\nu)(1-2\alpha_k) + \frac{1}{3}\varepsilon^2\{3(1+\nu)[3\nu-4 \\
&\quad + 2(2-\nu)\alpha_k]\eta^2 + (1+\nu)(2-\nu-4\nu\alpha_k - 12\beta_k) \\
&\quad + 12(1+\nu)\beta_k(1-2\alpha_k)\ln\rho\} + \cdots\Big\rangle \exp(\alpha_k \ln\rho) \\
Q_{\varphi k} &= \Big\langle -4(1-\nu^2) + \frac{1}{3}\varepsilon^2\{3(1+\nu)\big[2\nu^2 - 2\nu - 1 \\
&\quad -2(1-2\nu)\alpha_k\big]\eta + 2\nu^2 - 2\nu + 1 + 2(1-2\nu)\alpha_k \\
&\quad -24\alpha_k\beta_k - 24(1-\nu^2)\beta_k \ln\rho\} + \cdots\Big\rangle \exp(\alpha_k\ln\rho) \\
Q_{\theta k} &= \varepsilon^2\left\{\frac{1}{3}(1-\nu^2)(1+3\eta^2) - 4\alpha_k\beta_k + O(\varepsilon^2)\right\} \\
&\quad \times \exp(\alpha_k \ln\rho) \\
T_k &= \eta(\eta^2-1)\varepsilon^3\{3^{-1}(1+\nu)(\nu-2+2\nu\alpha_k) + O(\varepsilon^2)\} \\
&\quad \times \exp(\alpha_k\ln\rho)
\end{aligned}
\tag{4.8.6}
$$

The coefficients α_k, β_k $(k = 3, 4)$ are defined by (4.7.3).

$$
\begin{aligned}
U_{rk} &= 2(1-\nu)\eta\Big\langle 3 - 2\alpha_k + \frac{1}{3}\varepsilon^2\{[3(7\nu-9) \\
&\quad -(3\nu-7)\alpha_k]\eta^2 + 9(\nu+2) - 6(1+\nu)\alpha_k \\
&\quad - 6\beta_k + 6\beta_k(3-2\alpha_k)\ln\rho\} + \cdots\Big\rangle \exp(\alpha_k\ln\rho) \\
U_{\theta k} &= \varepsilon^{-1}\langle 4(1-\nu) + \varepsilon^2\{\big[2(1-2\nu)\alpha_k - 6\nu^2 + 10\nu - 3\big] \\
&\quad \times \eta^2 - 12\nu^2 + 18\nu - 7 - 2(1-2\nu)\alpha_k + 8(1-\nu)\beta_k\ln\rho\} \\
&\quad + \cdots\rangle \exp(\alpha_k\ln\rho)
\end{aligned}
\tag{4.8.7}
$$

$$
\begin{aligned}
Q_{\varphi k} &= \eta\Big\langle 12\nu^2 - 16\nu + 2 - 4(1-2\nu)\alpha_k + \frac{1}{3}\varepsilon^2 \\
&\quad \times \big\{\big[18\nu^3 - 54\nu^2 + 56\nu - 22 + 4(1-2\nu)(2-3\nu)\alpha_k\big]\eta^2 \\
&\quad + 3\big(36\nu^3 - 36\nu^2 - 23\nu + 19\big) + 3(1-2\nu)(5-6\nu)\alpha_k \\
&\quad - 24\nu\alpha_k\beta_k - 12(1-2\nu)\beta_k + 12\beta_k \\
&\quad \times \big[6\nu^2 - 8\nu + 1 - 2(1-2\nu)\alpha_k\big]\ln\rho\big\} + \cdots\Big\rangle \exp(\alpha_k\ln\rho) \\
Q_{\varphi k} &= \eta(\eta^2-1)\varepsilon^2\big\{\big[2(1-2\nu)\alpha_k - 6\nu^2 + 8\nu - 1 + O(\varepsilon^2)\big\}\exp(\alpha_k\ln\rho) \\
T_k &= (\eta^2-1)\varepsilon\big[2(2-\nu)\alpha_k + 7\nu - 8 + O(\varepsilon^2)\big]\exp(\alpha_k\ln\rho)
\end{aligned}
\tag{4.8.8}
$$

where α_k, β_k are defined by (4.7.4).

The second group of zeros describes the stress-strain state as rapidly decaying away from the edge of the plate. Expanding the solutions of this group in powers of a small parameter ε, we find the following asymptotic expressions:

$$U_r = \frac{r_1}{\sqrt{\rho}}\varepsilon\sum_{n=1}^{\infty} B_n\langle 2\left[(1-v)\delta_n^{-1}F_n''(\eta) - v\delta_n F_n(\eta)\right] + \varepsilon\left\{\left(\frac{4-13v}{4} - 2v\delta_n\omega_n \ln\rho\right)F_n(\eta) + (1-v)\right. \times \left.\left[\frac{5(1-2v)}{4} + \delta_n^{-1}\omega_n\ln\rho\right]F_n''(\eta)\right\} + \cdots\rangle \exp\left(\varepsilon^{-1}\delta_n\ln\rho\right) \tag{4.8.9}$$

$$U_\theta = -\frac{r_1}{\sqrt{\rho}}\varepsilon\sum_{n=1}^{\infty} B_n\langle 2\left[(2-v)F_n'(\eta) + (1-v)\delta_n^{-2}F_n''(\eta)\right] + \varepsilon\left\{\left(\frac{26-19v}{4} - 2(2-v)\omega_n\ln\rho\right)F_n'(\eta) + (1-v)\delta_n^{-2}\right. \times \left.\left[11/4(1-2v)\delta_n^{-1} + \omega_n\ln\rho\right]F_n'''(\eta)\right\} + \cdots\rangle\exp\left(\varepsilon^{-1}\delta_n\rho\right)$$

$$\sigma_r = \frac{2G}{\rho\sqrt{\rho}}\sum_{n=1}^{\infty} B_n\left\{F_n''(\eta) + \varepsilon\left[(1+v)\delta_n F_n(\eta) + \left(\omega_n\ln\rho + \frac{3+4v}{4}\delta_n^{-1}\right)F_n''(\eta)\right] + \cdots\right\}\exp\left(\varepsilon^{-1}\delta_n\ln\rho\right)$$

$$\sigma_\varphi = \frac{2G}{\rho\sqrt{\rho}}\sum_{n=1}^{\infty} B_n\left\{F_n''(\eta) + \delta_n^2 F_n(\eta) + \varepsilon\left[\left(\frac{v}{4} + \delta_n\omega_n\right)\right.\right. \times \left.\left.\delta_n F_n(\eta) + \left(\frac{4+3v}{4}\delta_n^{-1} + \omega_n\ln\rho\right)F_n''(\eta)\right] + \cdots\right\}\exp\left(\varepsilon^{-1}\delta_n\ln\rho\right)$$

$$\sigma_\theta = \frac{2G}{\rho\sqrt{\rho}}\sum_{n=1}^{\infty} B_n\delta_n^2 F_n(\eta)\left[1 + \varepsilon\left(11/4\delta_n^{-1} + \omega_n\ln\rho + \cdots\right)\right]\exp\left(\varepsilon^{-1}\delta_n\ln\rho\right)$$

$$\tau_{r\theta} = -\frac{2G}{\rho\sqrt{\rho}}\sum_{n=1}^{\infty} B_n\delta_n F_n'(\eta)\left[1 + \varepsilon\left(5/4\delta_n^{-1} + \omega_n\ln\rho\right) + \cdots\right]\exp\left(\varepsilon^{-1}\delta_n\ln\rho\right) \tag{4.8.10}$$

Here B_n are new arbitrary constants, $F_n(\eta)$ are P.F. Papkovich's functions.

From formulas (4.8.9), (4.8.10) it can be seen that the first terms of the asymptotic expansions of stresses and displacements coincide with the solution of the boundary layer of the plate of constant thickness.

We shall analyse the stress-strain state corresponding to different groups of solutions.

The zero $z_1 = \frac{1}{2}$ corresponds to the rigid body motion of the plate. The stress state corresponding to the zero $z_2 = -\frac{1}{2}$ is equivalent to the resultant vector P of the loads directed along the axis of symmetry. Here the constant C_2 is related to P by the relation

$$
\begin{aligned}
P &= -2\pi r^2 \varepsilon \int_{-1}^{1} (\sigma_r \sin \varepsilon\eta + \tau_{r\theta} \cos \varepsilon\eta) \cos \varepsilon\eta d\eta \\
&= 16\pi G(1-\nu) \sin^2 \varepsilon C_2.
\end{aligned}
\tag{4.8.11}
$$

The resultant vector of stresses in the cross section $r = const$ for the rest of the homogeneous solutions is equal to zero. The stress state defined by (4.8.4), (4.8.6) is equivalent to the stresses T_r,T_φ referred to the midsurface of the plate given by

$$
\begin{aligned}
T_r &= \frac{E\varepsilon}{1-\nu^2} \frac{1}{\sqrt{\rho}} \sum_{k=3,4} C_k \rho^{\alpha_k} T_{rk} \\
T_\varphi &= \frac{E\varepsilon}{1-\nu^2} \frac{1}{\sqrt{\rho}} \sum_{k=3,4} C_k \rho^{\alpha_k} T_{\varphi k}
\end{aligned}
\tag{4.8.12}
$$

where

$$
\begin{aligned}
T_{rk} &= 8(1-\nu^2)(1-2\alpha_k) + \frac{4}{3}[2(1-\nu)(2-\nu)\alpha_k \\
&\quad + 24\nu\alpha_k\beta_k - 2(1-\nu^2)\beta_k - (1-\nu)(2-\nu)(1-2\nu) \\
&\quad + 12(1-\nu^2)(1-2\alpha_k)\beta_k \ln \rho]\varepsilon^2 + \cdots \\
T_{\varphi k} &= -16(1-\nu)(1-\nu^2) + \frac{4}{3}[2\nu(1-\nu)(2-\nu)\alpha_k \\
&\quad + 24\nu^2\alpha_k\beta_k + (1-\nu)(2-\nu)^2 - 12(1-\nu)(1-\nu^2)\beta_k \ln \rho\Big]\varepsilon^2 + \cdots
\end{aligned}
$$

From formulas (4.8.8) it can be seen that the stresses correspond to the bending moments of the form

$$
\begin{aligned}
M_r &= \frac{E\varepsilon^2}{6(1-\nu^2)} \sqrt{\rho} \sum_{k=5,6} C_k \rho^{\alpha_k} M_{rk} \\
M_\varphi &= \frac{E\varepsilon^2}{6(1-\nu^2)} \sqrt{\rho} \sum_{k=5,6} C_k \rho^{\alpha_k} M_{\varphi k}
\end{aligned}
\tag{4.8.13}
$$

where

$$
\begin{aligned}
M_{rk} &= 2(1-\nu)[2(\nu-2)\alpha_k + 8 - 7\nu] + \frac{1}{3}\{24(1-\nu)\alpha_k\beta_k \\
&+ 12(1-\nu)(\nu-2)\beta_k - \left(21\nu^3 - 60\nu^2 + 53\nu - 16\right)\alpha_k \\
&+ 123\nu^3 - 406\nu^2 + 225\nu - 70 + 12(1-\nu)[2(\nu-2)\alpha_k \\
&+ 8\nu - 7]\beta_k \ln \rho\}\varepsilon^2 + \cdots \\
M_{\varphi k} &= -2(1-\nu)\left(6\nu^2 - 8\nu + 1\right) + \frac{1}{3}\{12(1-\nu)[2\nu\alpha_k\beta_k \\
&+ (1-2\nu)\beta_k] - \left(6\nu^2 - 8\nu + 1\right)\left[2(1-2\nu)\alpha_k + 21\nu^2\right. \\
&\left.-32\nu + 12\right] - 12(1-\nu)\left(6\nu^2 - 8\nu + 1\right)\beta_k \ln \rho\}\varepsilon^2 + \cdots
\end{aligned}
$$

For comparison, we introduce here the expressions for the bending moments and forces produced by the Kirchhoff theory:

$$
\begin{aligned}
M_r^* = \frac{E\varepsilon^2}{6(1-\nu^2)}\sqrt{\rho}\sum_{k=5,6} 2(1-\nu) \\
\times [2(\nu-2)\alpha_k + 8 - 7\nu]C_k'\rho^{\alpha_k}
\end{aligned}
\tag{4.8.14}
$$

$$
\begin{aligned}
M_\varphi^* &= \frac{E\varepsilon^2}{6(1-\nu^2)}\sqrt{\rho}\sum_{k=5,6} -2(1-\nu)\left(6\nu^2 - 8\nu + 1\right)C_k'\rho^{\alpha_k} \\
T_r^* &= \frac{E\varepsilon}{1-\nu^2}\frac{1}{\sqrt{\rho}}\sum_{k=3,4} 8\left(1-\nu^2\right)(1-2\alpha_k)C_k'\rho^{\alpha_k} \\
T_\varphi^* &= \frac{E\varepsilon}{1-\nu^2}\frac{1}{\sqrt{\rho}}\sum_{k=3,4} -16(1-\nu)\left(1-\nu^2\right)C_k'\rho^{\alpha_k}
\end{aligned}
\tag{4.8.15}
$$

The comparison of (4.8.14) with (4.8.13) and the formulas (4.8.15) with (4.8.12) implies that the stress state in the Kirchhoff theory of plates of variable thickness coincides only with the first term of the asymptotics of the internal three-dimensional stress state. In conclusion, we note that in [7] it is erroneously stated that the solution (4.8.15) is not the first term of the asymptotic three-dimensional internal stress.

4.9 Reduction of a Boundary Value Problem for a Plate of Variable Thickness and Infinite Systems at Given Stresses

Let the following stresses be given at $r = r_s \quad (s = 1, 2)$

$$
\sigma_r = \sigma_s(\theta),\ \tau_{r\theta} = \tau_s(\theta). \tag{4.9.1}
$$

The functions $\sigma_s(\theta)$, $\tau_s(\theta)$ satisfy the conditions of equilibrium

$$\begin{gathered}2\pi r_1^2\varepsilon\int\limits_{-1}^{1}[\sigma_1(\theta)\sin\varepsilon\eta+\tau_1(\theta)\cos\varepsilon\eta]\cos\varepsilon\eta d\eta\\=2\pi r_2^2\varepsilon\int\limits_{-1}^{1}[\sigma_2(\theta)\sin\varepsilon\eta+\tau_2(\theta)\cos\varepsilon\eta]\cos\varepsilon\eta d\eta=P.\end{gathered}\tag{4.9.2}$$

Here P is the resultant vector of the forces acting in an arbitrary section $r=const$ As noted above, non-self-balanced part of the stresses (4.6.4) can be removed using a penetrating solution (4.8.2), and the relationship of constant C_2 with the resultant vector of forces P is given by (4.8.11). Therefore, below we will assume that $P=0$.

We seek a solution in the form of (4.6.3), (4.6.4) by virtue of the accepted assumption that $C_1=C_2=0$.

Then, based on the Lagrange variational principle we obtain the following infinite algebraic system of equations:

$$\sum_{k=3}^{\infty}m_{jk}C_k=N_j\quad(j=3,4,\ldots)\tag{4.9.3}$$

where

$$\begin{aligned}m_{jk}&=\sum_{s=1}^{2}\exp(z_j+z_k)\ln\rho_s\int\limits_{-1}^{1}(Q_{rk}U_{rj}+T_kU_{\theta j})\cos\varepsilon\eta d\eta\\N_j&=\sum_{s=1}^{2}\exp(z_j+3/2)\ln\rho_s\int\limits_{-1}^{1}(\sigma_sU_{rj}+T_sU_{\theta j})\cos\varepsilon\eta d\eta\end{aligned}\tag{4.9.4}$$

Using the smallness of the parameter ε of thin-walledness it is possible to construct the asymptotic solution of (4.9.3). It is only necessary to consider the fact that the magnitudes $\sigma_r,\tau_{r\theta}$ corresponding to the roots of the first group have different orders namely $\sigma_r\sim1,\tau_{r\theta}\sim\varepsilon^3$ (for $k=3,4$), $\sigma_r\sim1,\tau_{r\theta}\sim\varepsilon$ (for $k=5,6$).

Further, using formulas (4.8.3), (4.8.4), (4.8.9), (4.8.10), we seek the unknowns C_k,B_n in terms of the expansions

$$\begin{aligned}C_k&=C_{k0}+\varepsilon C_{k1}+\varepsilon^2C_{k2}+\cdots\\B_n&=B_{n0}+\varepsilon B_{n1}+\varepsilon^2B_{n2}+\cdots\end{aligned}\tag{4.9.5}$$

On the basis of the variational principle, we obtain the following system of equations with respect to C_{k0}, B_{n0}:

$$\sum_{k=3,4} \prod_{jk} C_{k0} = E_j \quad (j = 3, 4), \sum_{k=5,6} \prod_{jk} C_{k0} = E_j \quad (j = 5, 6) \tag{4.9.6}$$

$$\begin{aligned} &\sum_{n=1,3,\ldots}^{\infty} g_{tn} B_{n0} = H_t \quad (t = 1, 3, \ldots) \\ &\sum_{n=2,4,\ldots}^{\infty} g_{tn} B_{n0} = H_t \quad (t = 2, 4, \ldots) \end{aligned} \tag{4.9.7}$$

where

$$\begin{aligned} \prod_{jk} &= 16G(1+v)(v - 2 + 2v\alpha_j)(1 - 2\alpha_k) \\ &\times \sum_{s=1}^{2} \exp(\alpha_j + \alpha_k) \ln \rho_s \quad (j = 3, 4) \end{aligned} \tag{4.9.8}$$

$$\begin{aligned} \prod_{jk} &= \frac{16}{3} G(1 - v)(1 - 2\alpha_j)[7v - 8 + 2(2 - v)\alpha_k] \\ &\times \sum_{s=1}^{2} (\alpha_j + \alpha_k) \ln \rho_s \quad (j = 5, 6) \end{aligned}$$

$$\begin{aligned} E_j &= 2\sum_{s=1}^{2} \rho_s^{3/2} \exp(\alpha_j \ln \rho_s) \int_{-1}^{1} [(v - 2 + 2v\alpha_j)\sigma_s(\theta) \\ &+ 2(1 - v^2)\eta\varepsilon\tau_s(\theta)] d\eta \quad (j = 3, 4) \\ E_j &= 2\sum_{s=1}^{2} \rho_s^{3/2} \exp(\alpha_j \ln \rho_s) \int_{-1}^{1} [(3 - 2\alpha_j)\eta\sigma_s(\theta) \\ &+ 2\tau_s(\theta)\varepsilon^{-1}] d\eta \quad (j = 5, 6) \end{aligned} \tag{4.9.9}$$

$$H_t = 2\sum_{s=1}^{2} \rho_s^{3/2} \exp\left(\frac{\delta_t}{\varepsilon}\ln \rho_s\right) \int_{-1}^{1} \{\sigma_s[(1-\nu)\delta_t^{-1}F_t''(\eta) - \nu\delta_t F_t(\eta)] - \tau_s\left[(1-\nu)\delta_t^{-2}F_t'''(\eta) + (2-\nu)F_t'(\eta)\right]\}d\eta + 16G\nu(1+\nu)(1-2\alpha_k)\frac{\sin^2\delta_t}{\delta_t}C_{k0}\sum_{s=1}^{2}\exp\left(\frac{\delta_t}{\varepsilon}\ln \rho_s\right) \quad (t = 2, 4, \ldots)$$

$$H_t = 2\sum_{s=1}^{2} \rho_s^{3/2} \exp\left(\frac{\delta_t}{\varepsilon}\ln \rho_s\right) \int_{-1}^{1} \{\sigma_s[(1-\nu)\delta_t^{-1}F_t''(\eta) - \nu\delta_t F_t(\eta)] - \tau_s\left[(1-\nu)\delta_t^{-2}F_t'''(\eta) + (2-\nu)F_t'(\eta)\right]\}d\eta - 16G\nu[7\nu - 8 + 2(2-\nu)\alpha_k]\frac{\sin^2\delta_t}{\delta_t}C_{k0}\sum_{s=1}^{2}\exp\left(\frac{\delta_t}{\varepsilon}\ln \rho_s\right) \quad (t = 1, 3, \ldots)$$

The coefficients of the system (4.9.7) are defined by (4.4.13) as above.

4.10 Construction of Applied Theories for the Plates of Variable Thickness

Unloading the conical part of the boundary can be accomplished by means of applied theories constructed for a cone (4.5.8) where we should assume that $\theta_0 = \pi/2$ for the plate. However, below, modifying the method of constructing the solution, we divide the problem into two independent problems: the problem of plate tension-compression and the problem of plate bending.

Therefore, let the conditions be set on conic boundaries as:

$$\sigma_\theta = Q_n(r),\ \tau_{r\theta} = \tau_n(r) \quad \text{at } \theta = \frac{\pi}{2} + (-1)^n\varepsilon \quad (n = 1, 2) \tag{4.10.1}$$

The character of the boundary conditions on the spherical boundaries will not be specified here.

Using the results of Sect. 4.5, the general solution of this problem can be represented in the form:

$$U_r = \left[\left(z - \frac{1}{2}\right)(z + 4\nu - 7/2)\psi_{z-3/2} - (z + 1/2)\psi_{z+1/2}\right]$$
$$U_\theta = \left[\varphi_2(z)\frac{d\psi_{z-3/2}}{d\theta} - \frac{d\psi_{z+1/2}}{d\theta}\right] \tag{4.10.2}$$

where

$$\psi_z(\theta) = A_z T_z(\theta) + B_z F_z(\theta),$$
$$T_z(\theta) = P_z(\cos\theta) + P_z(-\cos\theta),$$
$$F_z(\theta) = P_z(\cos\theta) - P_z(-\cos\theta).$$

Here, instead of employing the traditional linearly dependent solutions $P_z(\cos\,\theta)$ and $Q_z(\cos\,\theta)$ of the Legendre equation, we introduce, for the sake of convenience, solutions that are even and odd with respect to $\theta = \pi/2$ and are denoted respectively by $T_z(\theta)$ and $F_z(\theta)$. This form of the solution allows us to split the general problem for a plate into two independent problems.

We introduce the magnitudes

$$\begin{aligned} q_1 &= \frac{1}{2}(Q_1 + Q_2), q_2 = \frac{1}{2}(Q_2 - Q_1) \\ s_1 &= \frac{1}{2}(\tau_1 + \tau_2), s_2 = \frac{1}{2}(\tau_2 - \tau_1) \end{aligned} \tag{4.10.3}$$

by means of which we divide the general boundary value problem into two parts:

Problem A

$$\sigma_\theta = q_1,\ \tau_{r\theta} = (-1)^n s_1 \quad \text{for } \theta = \frac{\pi}{2} + (-1)^n \varepsilon \tag{4.10.4}$$

Problem B

$$\sigma_\theta = (-1)^n q_2, \tau_{r\theta} = s_2 \quad \text{for } \theta = \frac{\pi}{2} + (-1)^n \varepsilon \tag{4.10.5}$$

By the nature of symmetry with respect to the plane $\theta = \pi/2$ problem A may be called a plate tension-compression problem, whereas problem B may be called as a plate bending problem.

The arbitrary constants $A_{z-3/2}, A_{z+1/2}, B_{z-3/2}, B_{z+1/2}$ are determined from the boundary conditions (4.10.4), (4.10.5)

$$A_{z-3/2} = \Delta_1^{-1}\Delta_{11}, A_{z+1/2} = \Delta_1^{-1}\Delta_{12} \tag{4.10.6}$$

$$B_{z-3/2} = \Delta_2^{-1}\Delta_{21}, B_{z+1/2} = \Delta_2^{-1}\Delta_{22} \tag{4.10.7}$$

Substituting (4.10.6), (4.10.7) into (4.10.2), respectively, we obtain:for problem A

$$\begin{aligned} \Delta_1 U_r &= [(z - 1/2)(z + 4\nu - 7/2)T_{z-3/2}(\theta)\Delta_{11} - (z + 1/2)T_{z+1/2}(\theta)\Delta_{12}] \\ \Delta_1 U_\theta &= \left[\varphi_2(z)\frac{dT_{z-3/2}(\theta)}{d\theta}\Delta_{11} - \frac{dT_{z+1/2}(\theta)}{d\theta}\Delta_{12}\right] \end{aligned} \tag{4.10.8}$$

for problem B

$$\Delta_2 U_r = [(z-1/2)(z+4v-7/2)F_{z-3/2}(\theta)\Delta_{21} - (z+1/2)F_{z+1/2}(\theta)\Delta_{22}]$$
$$\Delta_2 U_\theta = \left[\varphi_2(z)\frac{dF_{z-3/2}}{d\theta}(\theta)\Delta_{21} - \frac{dF_{z+1/2}}{d\theta}(\theta)\Delta_{22}\right] \tag{4.10.9}$$

where

$$\begin{aligned}
\Delta_1 &= (z+1/2)\varphi_1(-z)T_{z+1/2}(\theta_1)\frac{dT_{z-3/2}}{d\theta}(\theta_1) \\
&\quad - (z-1/2)^2\varphi_1(z)T_{z-3/2}(\theta_1)\frac{dT_{z+1/2}}{d\theta}(\theta_1) \\
&\quad - 4(1-v)ztg\varepsilon\frac{dT_{z-3/2}(\theta_1)}{d\theta}\frac{dT_{z+1/2}}{d\theta}(\theta_1) \\
\Delta_{11} &= (z-1/2)\frac{dT_{z+1/2}}{d\theta}(\theta_1)q_1(t) + \left[(z+1/2)^2\right. \\
&\quad \left.\times T_{z+1/2}(\theta_1) + tg\varepsilon\frac{dT_{z+1/2}}{d\theta}(\theta_1)\right]s_1(t) \\
\Delta_{12} &= \varphi_1(-z)\frac{dT_{z-3/2}}{d\theta}(\theta_1)q_1(t) + [(z+1/2)\varphi_1(z) \\
&\quad \left.\times T_{z-3/2}(\theta_1) + \varphi_2(z)tg\varepsilon\frac{dT_{z-3/2}}{d\theta}(\theta_1)\right]s_1(t) \\
\theta_1 &= \frac{\pi}{2} - \varepsilon.
\end{aligned} \tag{4.10.10}$$

The magnitudes $\Delta_2, \Delta_{21}, \Delta_{22}$ have the same structure and only in (4.10.10) the function T_z should be replaced by F_z and $q_1(t), s_1(t)$ should be replaced by $q_2(t)$ and $s_2(t)$, respectively.

Expanding in the parameter ε we obtain:

for the problem A

$$\begin{aligned}
2G\Delta_1 U_r &= 4(1-v)s_1(t) + 2(2-v-2vz)q_1(t)\varepsilon + \cdots \\
2G\Delta_1 U_\theta &= 2\varepsilon\eta[(v-2-2vz)s_1(t) + 2(1-v) \\
&\quad \times (z^2 - 9/4)q_1(t)\varepsilon + \cdots]
\end{aligned} \tag{4.10.11}$$

for the problem B

$$\begin{aligned}
2G\Delta_2 U_r &= 2\varepsilon\eta\{(1-v)(2z-3)q_2(t) \\
&\quad + \left[4v - 3 - 2(1-v)\left(z^2 - \frac{1}{4}\right)\right]s_2(t)\varepsilon + \cdots \\
2G\Delta_2 U_\theta &= -4(1-v)q_2(t) + (1-vz)(2z-1)s_2(t)\varepsilon + \cdots
\end{aligned} \tag{4.10.12}$$

where

$$\Delta_1 = \varepsilon\left\{-4z^2 + 5 - 4\nu + \frac{1}{3}\left[4z^4 + 2(6\nu - 7)z^2 + \frac{77}{4} - 19\nu\right]\varepsilon^2 + \cdots\right\}$$

$$\Delta_2 = 4\left(z^2 - \frac{1}{4}\right)\varepsilon^3\left\{-4z^2 + 13 - 12\nu + \frac{1}{10}\right.$$
$$\left.\times\left[28z^4 + 2(50\nu - 71)z^2 + \frac{855}{4} - 205\right]\varepsilon^2 + \cdots\right\}$$

As noted in Sect. 4.5, the relationships (4.10.11) and (4.10.12) can be used to construct approximate theories intended to relieve stress from a conical part of the plate boundary. Since $t = \ln r$ and $z = r\frac{d}{dr} + \frac{1}{2}$ we obtain the following system of ordinary differential equations:

for the problem A

$$\begin{aligned} 2G\varepsilon D_1 U_r &= 4(1-\nu)s_1(r) + 4\left(1 - \nu - \nu_r\frac{d}{dr}\right)q_1(r)\varepsilon + \cdots \\ 2G\varepsilon D_1 U_\theta &= \varepsilon\eta\left\{-4\left(1 + \nu_r\frac{d}{dr}\right)s_1(r) + 4(1-\nu)\right. \\ &\quad \left.\times\left[\frac{1}{r^2}\frac{d}{dr}\left(r^2\frac{d}{dr}\right) - 2\right]q_1(r)\varepsilon^2 + \cdots\right\} \end{aligned} \tag{4.10.14}$$

for the problem B

$$\begin{aligned} 8G\varepsilon^3 D_2 U_r &= 2\varepsilon\eta\left\{2(1-\nu)\left(r\frac{d}{dr} - 1\right)\tau_2(r)\right. \\ &\quad \left. + \left[4\nu - 3 - 2(1-\nu)\frac{1}{r^2}\frac{d}{dr}\left(r^2\frac{d}{dt}\right)\right]s_2(r)\varepsilon + \cdots\right\} \\ 8G\varepsilon^3 D_r U_\theta &= 4(1-\nu)q_2(r) + 2(1-2\nu)r\frac{ds_2(r)}{dr}\varepsilon + \cdots \end{aligned} \tag{4.10.15}$$

where

$$D_1 = 4 - 4\nu - \frac{4}{r^2}\frac{d}{dr}\left(r^2\frac{d}{dr}\right) + \frac{1}{3}\varepsilon^2\left\{4\left[\frac{1}{r^2}\frac{d}{dr}\left(r^2\frac{d}{dr}\right) + 1/4\right]^2\right.$$
$$\left. + 2(6\nu - 7)\frac{1}{r^2}\frac{d}{dr}\left(r^2\frac{d}{dr}\right) + \frac{63}{4} - 16\nu\right\} + \cdots$$

$$D_2 = \frac{1}{r^2}\frac{d}{dr}\left(r^2\frac{d}{dr}\right)\left\langle 12 - 12\nu - \frac{4}{r^2}\frac{d}{dr}\left(r^2\frac{d}{dr}\right) + \frac{1}{10}\right.$$
$$\times\left\{28\left[\frac{1}{r^2}\frac{d}{dr}\left(r^2\frac{d}{dr}\right) + 1/4\right]^2 + 2(50\nu - 71)\frac{1}{r^2}\frac{d}{dr}\right.$$
$$\left.\left.\times\left(r^2\frac{d}{dr}\right) + \frac{773}{4} - 180\nu\right\}\varepsilon^2 + \cdots\right\rangle$$

Regarding the use of this system one can make the same remarks as in Sect. 4.5.

4.11 Investigation of Elastic Equilibrium of a Hollow Cone with a Fixed Side Surface and Mixed Boundary Conditions on the Side Surface

1. Suppose that the side surface of a cone is firmly sealed, i.e.

$$U_r = 0,\ U_\theta = 0 \text{ for } \theta = \theta_n \tag{4.11.1}$$

The nature of the boundary conditions at the ends of the cone will not be specified. However, we assume them to be of such a kind that the shell is in equilibrium.

Using the results of Sect. 4.1 and satisfying homogeneous conditions (4.11.1), we obtain the characteristic equation

$$\begin{aligned}\Delta(z,\theta_1,\theta_2) &= -2\left(z^2-\frac{1}{4}\right)\varphi_0(z)\varphi_0(-z)\sin^{-1}\theta_1\\ &\times \sin^{-1}\theta_2 - \left(z-\frac{1}{2}\right)^2\varphi_0^2(z)D^{(0,0)}_{z-3/2}(\theta_1,\theta_2)D^{(1,1)}_{z+1/2}(\theta_1,\theta_2)\\ &-\left(z+\frac{1}{2}\right)^2\varphi_0^2(-z)D^{(0,0)}_{z+1/2}(\theta_1,\theta_2)D^{(1,1)}_{z-3/2}(\theta_1,\theta_2)\\ &-\left(z^2-\frac{1}{4}\right)\varphi_0(-z)\Big[D^{(0,1)}_{z-3/2}(\theta_1,\theta_2)D^{(1,0)}_{z-1/2}(\theta_1,\theta_2)\\ &+D^{(1,0)}_{z-3/2}(\theta_1,\theta_2)D^{(0,1)}_{z+1/2}(\theta_1,\theta_2)\Big]=0\\ \varphi_0(z) &= z-7/2+4\nu\end{aligned} \tag{4.11.2}$$

Let us refer to the study of the behaviour of the roots of Eq. (4.11.2) for small values of the parameter ε.

It can be seen from (4.11.2) that $z_{1,2} = \pm 1/2$ are the roots of the characteristic equation. The second group of the roots consists of a countable set of zeros; they all tend to infinity as $\varepsilon \to 0$ and $\varepsilon z_k \to const$ when $\varepsilon \to 0$. To construct the asymptotics of the zeros of the second group we seek them in the form:

$$z_n = \frac{\delta_n}{\varepsilon} + O(\varepsilon) \quad (n = 3, 4, \ldots). \tag{4.11.3}$$

On substituting (4.11.3) into the characteristic equation and rewriting it by means of asymptotic expansions for the Legendre functions, we obtain

$$(3-4\nu)^2\sin^2 2\delta_n - 4\delta_n^2 = 0. \tag{4.11.4}$$

This equation has already occurred in the first chapter and, as noted there, coincides with the characteristic equation for the elastic layer.

Direct verification establishes that the roots $z = \pm\frac{1}{2}$ corresponds to the trivial solution.

We now present the first terms of the asymptotic expansions corresponding to zeros of the second group.

For displacements and stresses in the first approximation we obtain two classes of solutions the first of which corresponds to the zeros $(3-4\nu)\sin 2\delta + 2\delta$ and the second one to the zeros of function $(3-4\nu)\sin 2\delta - 2\delta$:

$$\begin{aligned} U_r^{(1)} &= \frac{r_1\varepsilon}{\sqrt{\rho}}\sum_{n=1}^{\infty} C_n\{[(3-4\nu)\cos\delta_n + \delta_n\sin\delta_n]\sin\delta_n\eta \\ &\quad + \eta\delta_n\cos\delta_n\cos\delta_n\eta + O(\varepsilon)\}\exp\left(\frac{\delta_n}{\varepsilon}\ln\rho\right) \\ U_\theta^{(1)} &= \frac{r_1\varepsilon}{\sqrt{\rho}}\sum_{n=1}^{\infty}\delta_n C_n[\sin\delta_n\cos\delta_n\eta - \eta\cos\delta_n\sin\delta_n\eta + O(\varepsilon)]\times\exp\left(\frac{\delta_n}{\varepsilon}\ln\rho\right) \end{aligned} \tag{4.11.5}$$

$$\begin{aligned} \sigma_r^{(1)} &= \frac{2G}{\rho\sqrt{\rho}}\sum_{n=1}^{\infty}\delta_n C_n\{[(3-2\nu)\cos\delta_n + \delta_n\sin\delta_n]\sin\delta_n\eta \\ &\quad + \eta\delta_n\cos\delta_n\eta + O(\varepsilon)\}\exp\left(\frac{\delta_n}{\varepsilon}\ln\rho\right) \\ \sigma_\varphi^{(1)} &= \frac{4\nu G}{\rho\sqrt{\rho}}\sum_{n=1}^{\infty}\delta_n C_n[\cos\delta_n\sin\delta_n\eta + O(\varepsilon)]\exp\left(\frac{\delta_n}{\varepsilon}\ln\rho\right) \\ \sigma_\theta^{(1)} &= \frac{-2G}{\rho\sqrt{\rho}}\sum_{n=1}^{\infty}\delta_n C_n\{[(1-2\nu)\cos\delta_n + \delta_n\sin\delta_n]\sin\delta_n\eta \\ &\quad + \eta\delta_n\cos\delta_n\cos\delta_n\eta + O(\varepsilon)\}\exp\left(\frac{\delta_n}{\varepsilon}\ln\rho\right) \end{aligned} \tag{4.11.6}$$

$$\begin{aligned} \tau_{r\theta}^{(1)} &= \frac{4G}{\rho\sqrt{\rho}}\sum_{n=1}^{\infty}\delta_n C_n\{[2(1-\nu)\cos\delta_n + \delta_n\sin\delta_n]\cos\delta_n\eta \\ &\quad - \eta\delta_n\cos\delta_n\sin\delta_n\eta + O(\varepsilon)\}\exp\left(\frac{\delta_n}{\varepsilon}\ln\rho\right) \\ U_r^{(2)} &= \frac{r_1\varepsilon}{\sqrt{\rho}}\sum_{n=1}^{\infty} B_n\{[(3-4\nu)\sin\delta_n + \delta_n\cos\delta_n]\cos\delta_n\eta \\ &\quad - \eta\delta_n\sin\delta_n\sin\delta_n\eta + O(\varepsilon)\}\exp\left(\frac{\delta_n}{\varepsilon}\ln\rho\right) \end{aligned} \tag{4.11.7}$$

$$
\begin{aligned}
U_\theta^{(2)} &= \frac{r_1\varepsilon}{\sqrt{\rho}}\sum_{n=1}^{\infty} B_n\delta_n[\cos\delta_n\sin\delta_n\eta - \eta\sin\delta_n\cos\delta_n\eta + O(\varepsilon)] \times \exp\left(\frac{\delta_n}{\varepsilon}\ln\rho\right)\\
\sigma_r^{(2)} &= \frac{2G}{\rho\sqrt{\rho}}\sum_{n=1}^{\infty}\delta_n B_n\{[(3-2\nu)\sin\delta_n + \delta_n\cos\delta_n]\cos\delta_n\eta\\
&\quad - \eta\delta_n\sin\delta_n\sin\delta_n\eta + O(\varepsilon)\}\exp\left(\frac{\delta_n}{\varepsilon}\ln\rho\right)\\
\sigma_\varphi^{(2)} &= \frac{4\nu G}{\rho\sqrt{\rho}}\sum_{n=1}^{\infty}\delta_n B_n[\sin\delta_n\cos\delta_n\eta + O(\varepsilon)]\exp\left(\frac{\delta_n}{\varepsilon}\ln\rho\right)\\
\sigma_\theta^{(2)} &= \frac{-2G}{\rho\sqrt{\rho}}\sum_{n=1}^{\infty}\delta_n B_n\{[(1-2\nu)\sin\delta_n - \delta_n\cos\delta_n]\cos\delta_n\eta\\
&\quad - \eta\delta_n\sin\delta_n\sin\delta_n\eta + O(\varepsilon)\}\exp\left(\frac{\delta_n}{\varepsilon}\ln\rho\right)\\
\tau_{r\theta}^{(2)} &= \frac{4G}{\rho\sqrt{\rho}}\sum_{n=1}^{\infty}\delta_n B_n\{[2(1-\nu)\sin\delta_n - \delta_n\cos\delta_n]\sin\delta_n\eta\\
&\quad + \eta\delta_n\sin\delta_n\cos\delta_n\eta + O(\varepsilon)\}\exp\left(\frac{\delta_n}{\varepsilon}\ln\rho\right).
\end{aligned}
\tag{4.11.8}
$$

Here C_n, B_n, are arbitrary constants. It may be observed from formulas (4.11.4), (4.11.5) and (4.11.6), (4.11.7) that the first group of solutions describes the shell bending problem and the second one the shell stretching-compression problem. Since the solutions (4.11.4), (4.11.5) and (4.11.6), (4.11.7) do not depend on θ_0, they are valid at the value $\theta_0 = \pi/2$ which corresponds to a plate of variable thickness.

In the general case of loading the arbitrary constants C_n, B_n, given as above, may be determined using the Lagrange variational principle. Under special conditions of the shell edge bearing they are determined exactly by the generalized orthogonality condition for a cone that will be established later.

2. Now consider the case when one of the boundary conditions holds true on the side surface of the cone

$$U_\theta = 0, \tau_{r\theta} = 0 \text{ at } \theta = \theta_n \tag{4.11.9}$$

$$U_r = 0, \sigma_\theta = 0 \text{ at } \theta = \theta_n \tag{4.11.10}$$

We investigate the case (4.11.9); the case (4.11.10) is investigated similarly. As above, using the results of Sect. 4.1 on employing the homogeneous boundary conditions (4.11.9), we obtain the characteristic equation

$$\Delta(z,\theta_1,\theta_2) = -16(1-\nu)^2 z^2 D^{(1,1)}_{z-3/2}(\theta_1,\theta_2) D^{(1,1)}_{z+1/2}(\theta_1,\theta_2) = 0 \qquad (4.11.11)$$

Equation (4.11.11), besides the root $z=0$, has also the roots $z_{1,2}=\pm 1/2, z_{3,4}=\pm 3/2$ independent of the small parameter ε. It is not difficult to show that these roots correspond to trivial solutions. The remaining roots of Eq. (4.11.11) tend to infinity when $\varepsilon \to 0$. Finding $z_k = \frac{\delta_k}{\varepsilon} + O(\varepsilon)$ and using the first term of the asymptotics of the Legendre functions, for δ_k we obtain

$$\sin^2 \delta_k = 0 \qquad (4.11.12)$$

In the case of displacements and stresses, in the first terms of their asymptotic expansions we obtain two classes of solutions, the first of which corresponds to the zeros of the function $\cos\delta$ and the second one corresponds to the zeros of $\sin\delta$ which are presented as

$$\begin{aligned}
U_r &= r_1\varepsilon \sum_{k=1}^{\infty} C_k[\sin \delta_k\eta + O(\varepsilon)] \exp\left(\frac{\delta_k}{\varepsilon}\ln \rho\right)\\
U_\theta &= r_1\varepsilon \sum_{k=1}^{\infty} C_k[\cos \delta_k\eta + O(\varepsilon)] \exp\left(\frac{\delta_k}{\varepsilon}\ln \rho\right)\\
\sigma_r &= 2G \sum_{k=1}^{\infty} C_k[\delta_k \sin \delta_k\eta + O(\varepsilon)] \exp\left(\frac{\delta_k}{\varepsilon}\ln \rho\right)\\
\sigma_\varphi &= O(\varepsilon)\\
\sigma_\theta &= -2G \sum_{k=1}^{\infty} C_k[\delta_k \sin \delta_k\eta + O(\varepsilon)] \exp\left(\frac{\delta_k}{\varepsilon}\ln \rho\right)\\
\tau_{r\theta} &= 2G \sum_{k=1}^{\infty} C_k[\delta_k \cos \delta_k\eta + O(\varepsilon)] \exp\left(\frac{\delta_k}{\varepsilon}\ln \rho\right)
\end{aligned} \qquad (4.11.13)$$

$$\begin{aligned}
U_r &= r_1\varepsilon \sum_{k=1}^{\infty} B_k[-\cos \delta_k\eta + O(\varepsilon)] \exp\left(\frac{\delta_k}{\varepsilon}\ln \rho\right)\\
U_\theta &= r_1\varepsilon \sum_{k=1}^{\infty} B_k[\sin \delta_k\eta + O(\varepsilon)] \exp\left(\frac{\delta_k}{\varepsilon}\ln \rho\right)\\
\sigma_r &= 2G \sum_{k=1}^{\infty} B_k[-\delta_k \cos \delta_k\eta + O(\varepsilon)] \exp\left(\frac{\delta_k}{\varepsilon}\ln \rho\right)\\
\sigma_\varphi &= O(\varepsilon)\\
\sigma_\theta &= 2G \sum_{k=1}^{\infty} B_k[\delta_k \cos \delta_k\eta + O(\varepsilon)] \exp\left(\frac{\delta_k}{\varepsilon}\ln \rho\right)\\
\tau_{r\theta} &= 2G \sum_{k=1}^{\infty} B_k[\delta_k \sin \delta_k\eta + O(\varepsilon)] \exp\left(\frac{\delta_k}{\varepsilon}\ln \rho\right)
\end{aligned} \qquad (4.11.14)$$

The arbitrary constants C_k, B_k are determined from the state of the boundary conditions being valid at the ends of the cone.

We now prove the generalized orthogonality condition for the cone.

Let one of the boundary conditions (4.11.1), (4.11.9), (4.11.10) hold true on the side surface. We seek the solution as a sum of elementary solutions

$$\begin{aligned} U_r &= \sum_{k=1}^{\infty} r^{\lambda_k} a_k(\theta), U_\theta = \sum_{k=1}^{\infty} r^{\lambda_k} b_k(\theta) \\ \sigma_r &= \frac{2G}{r} \sum_{k=1}^{\infty} r^{\lambda_k} \{[(1-v)\lambda_k + 2v] a_k \\ &\quad + v\left(b_k' + ctg\theta b_k\right)\} \\ \tau_{r\theta} &= \frac{G}{r} \sum_{k=1}^{\infty} r^{\lambda_k} \left[a_k' + (\lambda_k - 1) b_k\right] \end{aligned} \tag{4.11.15}$$

Let a_k, b_k, a_p, b_p be the solutions corresponding to the eigenvalues λ_k and λ_p, respectively. Then, according to Betty theorem, for any r the following equality holds true

$$\begin{aligned} &\int_{\theta_1}^{\theta_2} \left(\sigma_{rk} U_{rp} + \tau_{rk} U_{\theta p}\right) r^2 \sin\theta d\theta \\ &= \int_{\theta_1}^{\theta_2} \left(\sigma_{rp} U_{rk} + \tau_{rp} U_{\theta k}\right) r^2 \sin\theta d\theta. \end{aligned} \tag{4.11.16}$$

Substituting (4.11.15) into (4.11.16) and integrating by parts we obtain:

$$\begin{aligned} r^{\lambda_k + \lambda_p + 1} \Bigg\{ &\left(\lambda_k - \lambda_p\right) \left[\frac{2(1-v)}{1-2v} \left(a_k, a_p\right) + \left(a_k, a_p\right)\right] \\ &+ \frac{1}{1-2v} \left[\left(a_k', b_p\right) - \left(a_p', b_k\right)\right] + \frac{2v}{1-2v} \sin\theta \left(a_p b_k - a_k b_p\right)_{\theta_1}^{\theta_2} \Bigg\} = 0 \end{aligned}$$

Hence, we obtain for any r

$$\begin{aligned} &\left(\lambda_k - \lambda_p\right) \left[\frac{2(1-v)}{1-2v} \left(a_k, a_p\right) + \left(b_k, b_p\right)\right] + \frac{1}{1-2v} \left[\left(a_k', b_p\right) - \left(a_p', b_k\right)\right] \\ &+ \frac{2v}{1-2v} \sin\theta \left(b_k a_p - a_k b_p\right)_{\theta_1}^{\theta_2} = 0. \end{aligned} \tag{4.11.17}$$

In (4.11.17) the last term, for any of the boundary conditions (4.11.1), (4.11.9) and (4.11.10), is equal to zero. Therefore, we finally obtain

$$(\lambda_k - \lambda_p)\left[\frac{2(1-\nu)}{1-2\nu}(a_k, a_p) + (b_k, b_p)\right] + \frac{1}{1-2\nu}\left[(a'_k, b_p) - \left(a'_p, b_k\right)\right] = 0$$
$$(k \neq p)$$

where $(f, g) = \int_{\theta_1}^{\theta_2} fg \sin\theta \, d\theta$.

4.12 Asymptotic Analysis of the Solutions of Some Axisymmetric Problems for Plates of Variable Thickness

A stress-strain state of a plate with linearly varying thickness subjected to the action of non-axisymmetric loads applied to a spherical surface is studied here. The behaviour of the solutions is investigated as the parameter ε of thin-walledness tends to zero.

Following a similar analysis as above, we consider the plate with linearly varying thickness $h = \varepsilon r$:

$$r_1 \le r \le r_2, \pi/2 - \varepsilon \le \theta \le \pi/2 + \varepsilon, 0 \le \varphi \le 2\pi. \quad (4.12.1)$$

It is assumed that the conical boundary surfaces are free of stress, i.e. the following conditions hold true:

$$\sigma_\theta = 0, \tau_{r\theta} = 0, \tau_{\theta\varphi} = 0 \text{ at } \theta = \pi/2 \mp \varepsilon \quad (4.12.2)$$

The boundary conditions on the remaining part of the boundary are given as

$$\sigma_r = f_{1s}(\theta, \varphi), \tau_{r\theta} = f_{2s}(\theta, \varphi), \tau_{r\varphi} = f_{3s}(\theta, \varphi) \quad \text{at } r = r_s, (s = 1, 2) \quad (4.12.3)$$

To construct the general solution of the equilibrium equations satisfying conditions (4.12.2) we start from Papkovich-Neuber's [5] general solution which in spherical coordinates has the form

$$\begin{aligned}
U_r &= 4(1-\nu)B_r - \frac{\partial}{\partial r}(rB_r + B_0), \\
U_\theta &= 4(1-\nu)B_\theta - \frac{1}{r}\frac{\partial}{\partial \theta}(rB_r + B_0), \\
U_\varphi &= 4(1-\nu)B_\varphi - \frac{1}{r \sin\theta}\frac{\partial}{\partial \varphi}(rB_r + B_0).
\end{aligned} \quad (4.12.4)$$

The functions $B_r, B_\theta, B_\varphi, B_0$ satisfy the following differential equations

$$\begin{aligned}
&\Delta B_r - \frac{2}{r^2}B_r - \frac{2}{r^2}\frac{\partial B_\theta}{\partial\theta} - \frac{2ctg\theta}{r^2}B_\theta - \frac{2}{r^2\sin\theta}\frac{\partial B_\varphi}{\partial\varphi} = 0\\
&\Delta B_\theta + \frac{2}{r^2}\frac{\partial B_r}{\partial\theta} - \frac{1}{r^2\sin\theta}B_\theta - \frac{2ctg\theta}{r^2\sin\theta}\frac{\partial B_\varphi}{\partial\varphi} = 0\\
&\Delta B_\varphi + \frac{2}{r^2\sin\theta}\frac{\partial B_r}{\partial\varphi} + \frac{2ctg\theta}{r^2\sin\theta}\frac{\partial B_\theta}{\partial\varphi} - \frac{1}{r^2\sin\theta}B_\varphi = 0\\
&\Delta B_0 = 0
\end{aligned} \tag{4.12.5}$$

where

$$\Delta = \frac{\partial^2}{\partial r^2} + \frac{2}{r}\frac{\partial}{\partial r} + \frac{ctg\theta}{r^2}\frac{\partial}{\partial\theta} + \frac{1}{r^2}\frac{\partial^2}{\partial\theta^2} + \frac{1}{r^2\sin^2\theta}\frac{\partial^2}{\partial\varphi^2}.$$

Seeking the solutions of (4.12.5) in the form

$$B_t = r^\lambda b_{tm}(\theta)e^{im\varphi}, (t = r,\theta), B_\varphi = ir^\lambda b_{\varphi m}(\theta)e^{im\varphi}, B_0 = r^{\lambda+1}b_{0m}(\theta)e^{im\varphi} \tag{4.12.6}$$

and using the arbitrariness contained in Papkovich-Neuber's solution, we get the following expressions for $b_{tm}, b_{\varphi m}, b_{0m}$

$$\begin{aligned}
b_{rm}(\theta) &= -\lambda r^\lambda\psi_{\lambda-1}(\theta), b_{\theta m}(\theta) = \psi'_{\lambda-1}(\theta)\\
&\quad + m\sin^{-1}\theta\psi_\lambda(\theta), b_{\phi m}(\theta) = -\psi'_\lambda(\theta)\\
&\quad - m\sin^{-1}\theta\psi_{\lambda-1}(\theta), b_{0m}(\theta) = \psi_{\lambda+1}(\theta)\\
\psi_\lambda(\theta) &= A_\lambda T_\lambda(\theta) + B_\lambda F_\lambda(\theta),\\
T_\lambda(\theta) &= P^m_\lambda(\cos\theta) + P^m_\lambda(-\cos\theta)\\
F_\lambda(\theta) &= P^m_\lambda(\cos\theta) - P^m_\lambda(-\cos\theta),
\end{aligned} \tag{4.12.7}$$

(A_λ, B_λ are arbitrary constants) where $P^m_\lambda(\cos\theta)$ are the associated Legendre functions; the functions $T_\lambda(\theta), F_\lambda(\theta)$ constitute two linearly independent solutions of Legendre equation and introduced for the sake of convenience, since the first one is even and the second one is odd with respect to variable $\xi = \theta - \pi/2$.

Substituting (4.12.7) into (4.12.4) and using the generalized Hooke's law we obtain the following equations for displacements and stresses (hereinafter the amplitude values of functions by ϕ are provided)

$$
\begin{aligned}
U_r &= r^\lambda\left[\lambda(\lambda+4\nu-3)\psi_{\lambda-1}(\theta)+(\lambda+1)\psi_{\lambda+1}(\theta)\right]\\
U_\theta &= r^\lambda\left[(\lambda+4-4\nu)\psi'_{\lambda-1}(\theta)+m\sin^{-1}\theta\psi_\lambda(\theta)+\psi'_{\lambda+1}(\theta)\right]\\
U_\varphi &= -r^\lambda\left[(\lambda+4-4\nu)m\sin^{-1}\theta\psi_{\lambda-1}(\theta)+\psi'_\lambda(\theta)+m\sin^{-1}\theta\psi_{\lambda+1}(\theta)\right]\\
\sigma_r &= 2Gr^{\lambda-1}\left[\lambda(\lambda^2-3\lambda-2\nu)\psi_{\lambda-1}(\theta)+\lambda(\lambda+1)\psi_{\lambda+1}(\theta)\right]\\
\sigma_\varphi &= 2Gr^{\lambda-1}\left\{\left[\lambda(\lambda-3+2\nu-4\nu\lambda)-(\lambda+4-4\nu)m^2\sin^{-2}\theta\right]\psi_{\lambda-1}(\theta)\right.\\
&\quad+(\lambda+4-4\nu)ctg\theta\psi'_{\lambda-1}(\theta)+m\sin^{-1}\theta\left[ctg\theta\psi_\lambda(\theta)-\psi'_\lambda(\theta)\right]\\
&\quad\left.+\left(\lambda+1-m^2\sin^{-2}\theta\right)\psi_{\lambda+1}(\theta)+ctg\theta\psi'_{\lambda+1}(\theta)\right\}\\
\sigma_\theta &= 2Gr^{\lambda-1}\left\{\left[(\lambda+4-4\nu)m^2\sin^{-2}\theta-\lambda\left(\lambda^2+2\lambda-1+2\nu\right)\right]\psi_{\lambda-1}(\theta)-\right.\\
&\quad+(\lambda+4-4\nu)ctg\theta\psi'_{\lambda-1}(\theta)+m\sin^{-1}\theta\left[\psi'_\lambda(\theta)-ctg\theta\psi_\lambda(\theta)\right]\\
&\quad\left.+\left[m^2\sin^{-2}\theta-(\lambda+1)^2\right]\psi_{\lambda+1}(\theta)-ctg\theta\psi'_{\lambda+1}(\theta)\right\}\\
\tau_{r\varphi} &= -Gr^{\lambda-1}\left[\left(\lambda^2-2+2\nu\right)m\sin^{-1}\theta\psi_{\lambda-1}(\theta)+1/2(\lambda-1)\psi'_\lambda(\theta)+\lambda m\sin^{-1}\theta\psi_{\lambda+1}(\theta)\right]\\
\tau_{r\theta} &= Gr^{\lambda-1}\left[\left(\lambda^2-2+2\nu\right)\psi'_{\lambda-1}(\theta)+2^{-1}(\lambda-1)m\sin^{-1}\theta\psi_\lambda(\theta)+\lambda\psi'_{\lambda+1}(\theta)\right]\\
\tau_{\theta\varphi} &= Gr^{\lambda-1}\left\{(\lambda+4-4\nu)m\sin^{-1}\theta\left[ctg\theta\psi_{\lambda-1}(\theta)-\psi'_{\lambda-1}(\theta)\right]\right.\\
&\quad+\left[1/2\lambda(\lambda+1)-m^2\sin^{-2}\theta\right]\psi_\lambda(\theta)+ctg\theta\psi'_\lambda(\theta)+m\sin^{-1}\theta\\
&\quad\left.\times\left[ctg\theta\psi_{\lambda+1}(\theta)-\psi'_{\lambda+1}(\theta)\right]\right\}.
\end{aligned}
\tag{4.12.8}
$$

The chosen form of solutions allow us to divide the general problem for a plate into two independent ones: problem I: plate tension-compression problem; problem II: plate bending problem. In the first case, in formulas (4.12.7), (4.12.8) we should assume $A_\lambda = B_{\lambda-1} = B_{\lambda+1} = 0$, in the second case we should assume $B_\lambda = A_{\lambda-1} = A_{\lambda+1} = 0$.

Constants $A_{\lambda-1}, B_{\lambda-1}, A_\lambda, B_\lambda, A_{\lambda+1}, B_{\lambda+1}$ are determined by satisfying homogeneous boundary conditions (4.12.2).

Substitution of the expressions for stresses into each of the conditions (4.12.2) leads to a system of six linear homogeneous equations for the constants $A_{\lambda-1}, \ldots B_{\lambda+1}$. Obviously, this system is divided into two independent systems of third order, one of which corresponds to problem I and comprises only unknowns $A_{\lambda-1}, B_\lambda, A_{\lambda+1}$ and the second one corresponds to problem II and contains only unknowns $B_{\lambda-1}, A_\lambda, B_{\lambda+1}$.

The conditions for the existence of nontrivial solutions of the systems corresponding to problems I and II lead, respectively, to the following characteristic equations

$$D_1(z, \varepsilon) = 0 \tag{4.12.9}$$

$$D_2(z, \varepsilon) = 0 \tag{4.12.10}$$

where

$$
\begin{aligned}
D_1(z,\varepsilon) = \Big\{ & 4(1-\nu)z tg\varepsilon T'_{z-3/2}(\theta_1)T'_{z+1/2}(\theta_1) \\
& + \Big[(z-1/2)^2\varphi_1(z) - (z-1/2)\varphi_2(z)m^2\cos^{-2}\varepsilon\Big] \\
& \times T_{z+3/2}(\theta_1)T'_{z+1/2}(\theta_1) - \varphi_1(-z)\big(z^2+z+1/4-m^2\cos^{-2}\varepsilon\big) \\
& \times T_{z+1/2}(\theta_1)T'_{z-3/2}(\theta_1)\Big\}\big[tg\varepsilon F'_{z-1/2}(\theta_1) + 1/2(z^2-1/4 \\
& -2m^2\cos^{-2}\varepsilon)F_{z-1/2}(\theta_1)\big] + m^2\cos^{-2}\varepsilon\big[tg\varepsilon F_{z-1/2}(\theta_1) \\
& - F'_{z-1/2}(\theta_1)\big]\big[4(1-\nu)zT'_{z-3/2}(\theta_1)T'_{z+1/2}(\theta_1) + \varphi_1(-z) \\
& \times tg\varepsilon T_{z+1/2}(\theta_1)T'_{z-3/2}(\theta_1) - (z-1/2)\varphi_2(z)tg\varepsilon T_{z-3/2}(\theta_1) \\
& \times T'_{z+1/2}(\theta_1)\Big] + 1/2(z-3/2)m^2\cos^{-2}\varepsilon F_{z-1/2}(\theta_1) \\
& \times \big\{2(1-\nu)z(2z+3)tg\varepsilon T_{z-3/2}(\theta_1)T_{z+1/2}(\theta_1) - \varphi_2(z) \\
& \times \big(z^2+z+1/4+tg^2\varepsilon - m^2\cos^{-2}\varepsilon\big)T_{z+1/2}(\theta_1)T'_{z-3/2}(\theta_1) \\
& + \big[(z-1/2)\varphi_1(z) - \varphi_2(z)m^2\cos^{-2}\varepsilon + \varphi_2(z)tg^2\varepsilon\big] \\
& \times T_{z-3/2}(\theta_1)T'_{z+1/2}(\theta_1)\big\}.
\end{aligned}
\tag{4.12.11}
$$

Here,

$$z = \lambda + 1/2, \varphi_1(z) = z^2 + z + 2\nu - 7/4, \varphi_2(z) = z + 7/2 - 4\nu, \theta_1 = \pi/2 - \varepsilon.$$

To obtain expressions for $D_2(z,\varepsilon)$ in (4.12.11) the functions $T_{z-3/2}(\theta)$, $T_{z+1/2}(\theta)$ should be replaced with functions $F_{z-3/2}(\theta)$, $F_{z+1/2}(\theta)$ and the functions $F_{z-1/2}(\theta)$ with functions $T_{z-1/2}(\theta)$ respectively. From the structure $D_1(z,\varepsilon)$ and $D_2(z,\varepsilon)$ it can be concluded that $D_1(z,\varepsilon)$ is an even function of ε and $D_2(z,\varepsilon)$ is an odd function of ε.

In addition, we can prove that $D_1(z,\varepsilon)$ and $D_2(z,\varepsilon)$ are odd functions of z.

Each root z_k apparently corresponds to a homogeneous solution. Assuming the completeness of these solutions the general solution of the equilibrium equations satisfying the boundary conditions (4.12.2) can be represented as

$$
\begin{aligned}
U_r &= r_1\rho^{-1/2}\sum_{k=0}^{\infty} C_k\rho^{z_k}U(z_k,\theta) \\
U_\theta &= r_1\rho^{-1/2}\sum_{k=0}^{\infty} C_k\rho^{z_k}W(z_k,\theta) \\
U_\varphi &= r_1\rho^{-1/2}\sum_{k=0}^{\infty} C_k\rho^{z_k}V(z_k,\theta)
\end{aligned}
\tag{4.12.12}
$$

$$\sigma_r = 2G(1-2\nu)^{-1}\rho^{-3/2}\sum_{k=0}^{\infty} C_k \rho^{z_k} Q_r(z_k,\theta)$$
$$\sigma_\varphi = 2G(1-2\nu)^{-1}\rho^{-3/2}\sum_{k=0}^{\infty} C_k \rho^{z_k} Q_\varphi(z_k,\theta)$$
$$\sigma_\theta = 2G(1-2\nu)^{-1}\rho^{-3/2}\sum_{k=0}^{\infty} C_k \rho^{z_k} Q_\theta(z_k,\theta)$$
$$\tau_{r\varphi} = Gi\rho^{-3/2}\sum_{k=0}^{\infty} C_k \rho^{z_k} T_1(z_k,\theta) \tag{4.12.13}$$
$$\tau_{r\theta} = G\rho^{-3/2}\sum_{k=0}^{\infty} C_k \rho^{z_k} T_2(z_k,\theta)$$
$$\tau_{r\varphi} = Gi\rho^{-3/2}\sum_{k=0}^{\infty} C_k \rho^{z_k} T_3(z_k,\theta)$$

where, for problem I we have

$$\begin{aligned}
U(z_k,\theta) = {} & \left[(z_k^2 - 1/4)F_{z_k-1/2}(\theta_1) + 2tg\varepsilon F'_{z_k-1/2}(\theta_1)\right] \\
& \times \Big[(z_k+1/2)\varphi_1(z_k)T_{z_k+1/2}(\theta)T_{z_k-3/2}(\theta_1) + (z_k-1/2)^2\varphi_2(-z_k) \\
& \times T_{z_k-3/2}(\theta)T_{z_k+1/2}(\theta_1)\Big] - \{(z_k-3/2)[(z_k-1/2)\varphi_2(-z_k) \\
& \times T_{z_k-3/2}(\theta)T_{z_k+1/2}(\theta_1) + (z_k+1/2)\varphi_2(z_k)T_{z_k+1/2}(\theta)T_{z_k-3/2}(\theta_1)] \\
& \times tg\varepsilon\cos^{-2}\varepsilon + (z_k^2-1/4)\varphi_2(-z_k)\left[T_{z_k-3/2}(\theta)T_{z_k+1/2}(\theta_1)\right. \\
& \left.-T_{z_k+1/2}(\theta)T_{z_k-3/2}(\theta_1)\right]\cos^{-2}\varepsilon\}m^2F_{z_k-1/2}(\theta_1) \\
W(z_k,\theta) = {} & \left[(z_k^2 - 1/4)F_{z_k-1/2}(\theta_1) + 2tg\varepsilon F'_{z_k-1/2}(\theta_1)\right] \\
& \times \Big[\varphi_1(-z_k)T'_{z_k+1/2}(\theta)T'_{z_k-3/2}(\theta_1) - (z_k-1/2) \\
& \times \varphi_2(z_k)T'_{z_k-3/2}(\theta)T'_{z_k+1/2}(\theta_1)\Big] + m^2F_{z_k-1/2}(\theta_1) \\
& \times \{(z_k+1/2)\varphi_2(z_k)T_{z_k-3/2}(\theta)T'_{z_k+1/2}(\theta_1) + (z_k-1/2) \\
& \times \varphi_2(-z_k)T'_{z_k+1/2}(\theta)T'_{z_k-3/2}(\theta_1)\Big]\cos^{-2}\varepsilon + (z_k-3/2) \\
& \times \varphi_2(z_k)\Big[T_{z_k+1/2}(\theta_1)T'_{z_k-3/2}(\theta) - T_{z_k-3/2}(\theta_1)T_{z_k+1/2}(\theta)\Big]tg\varepsilon\cos^2\varepsilon\Big\} \\
& - 2m^2\cos^{-1}\varepsilon\times\cos^{-1}\varepsilon\eta F_{z_k-1/2}(\theta_1)\{4(1-\nu)z_kT'_{z_k-3/2}(\theta_1)T'_{z_k+1/2}(\theta_1) \\
& + tg\varepsilon[\varphi_1(-z_k)T_{z_k+1/2}(\theta_1)T'_{z_k-3/2}(\theta_1) - (z_k-1/2) \\
& \times \varphi_2(z_k)T_{z_k-3/2}(\theta_1)T_{z_k+1/2}(\theta_1)]\}
\end{aligned} \tag{4.12.14}$$

$$
\begin{aligned}
V(z_k,\theta) &= \cos^{-1}\varepsilon\eta[\varphi_1(-z_k)T_{z_k+1/2}(\theta)T'_{z_k-3/2}(\theta_1)\\
&\quad-(z_k-1/2)\varphi_2(z_k)T_{z_k-3/2}(\theta)T_{z_k+1/2}(\theta_1)]\big[(z_k^2-1/4)\\
&\quad\times F_{z_k-1/2}(\theta_1)+2tg\varepsilon F'_{z_k-1/2}(\theta_1)]+m^2\cos^{-1}\varepsilon\eta F_{z_k-1/2}(\theta_1)\\
&\quad\times\{(z_k+1/2)\varphi_2(z_k)T_{z_k-3/2}(\theta)T'_{z_k+1/2}(\theta_1)+(z_k-1/2)\\
&\quad\times\varphi_2(-z_k)T_{z_k+1/2}(\theta)T'_{z_k-3/2}(\theta_1)\big]\cos^{-2}\varepsilon+(z_k-3/2)\\
&\quad\times\varphi_2(z_k)\big[T_{z_k-3/2}(\theta)T_{z_k+1/2}(\theta_1)-T_{z_k+1/2}(\theta)T_{z_k-3/2}(\theta_1)\big]\\
&\quad\times tg\varepsilon\cos^{-2}\varepsilon\}-2\cos^{-1}\varepsilon F'_{z_k-1/2}(\theta)\{4(1-\nu)z_kT'_{z_k-3/2}(\theta_1)\\
&\quad\times T'_{z_k+1/2}(\theta_1)+tg\varepsilon\big[\varphi_1(-z_k)T_{z_k+1/2}(\theta_1)T'_{z_k-3/2}(\theta_1)\\
&\quad-(z_k-1/2)\varphi_2(z_k)T_{z_k-3/2}(\theta_1)T'_{z_k+1/2}(\theta_1)\big]\big\}\\
&\quad-(z_k-1/2)\varphi_2(z_k)T_{z_k-3/2}(\theta_1)T'_{z_k+1/2}(\theta_1)\big]\big\}\\
Q_r(z_k,\theta) &= [(1-\nu)z_k+1/2(5\nu-1)]U(z_k,\theta)\\
&\quad+\nu\big[W'(z_k,\theta)+ctg\theta W(z_k,\theta)-m\sin^{-1}\theta V(z_k,\theta)\big]\\
Q_\varphi(z_k,\theta) &= [\nu z_k+1/2(2-\nu)]U(z_k,\theta)\\
&\quad+\nu W'(z_k,\theta)+(1-\nu)\big[ctg\theta W(z_k,\theta)-m\sin^{-1}\theta V(z_k,\theta)\big]\\
Q_\theta(z_k,\theta) &= [\nu z_k+1/2(2-\nu)]U(z_k,\theta)\\
&\quad+(1-\nu)W'(z_k,\theta)+\nu\big[ctg\theta W(z_k,\theta)-m\sin^{-1}\theta V(z_k,\theta)\big]\\
T_1(z_k,\theta) &= m\sin^{-1}\theta U(z_k,\theta)+(z_k-3/2)V(z_k,\theta)\\
T_2(z_k,\theta) &= U'(z_k,\theta)+(z_k-3/2)V(z_k,\theta)\\
T_3(z_k,\theta) &= m\sin^{-1}\theta W(z_k,\theta)+V'(z_k,\theta)-ctg\theta V(z_k,\theta)
\end{aligned}
\tag{4.12.15}
$$

Here C_k are arbitrary constants, $\theta=\pi/2+\varepsilon\eta$, $-1\le\eta\le 1$. Expressions for displacements and stresses for problem II are obtained from (4.12.14), (4.12.15) by interchanging T and F.

4.13 Asymptotic Analysis of the Characteristic Equation

We now investigate the behaviour of the roots of Eqs. (4.12.9), (4.12.10) depending on the parameter ε. As in the axially symmetric case the zeros of functions $D_1(z,\varepsilon)$ and $D_2(z,\varepsilon)$ can be divided into two groups.

The first group of each equation consists of seven zeros and is characterized in that all z_k of this group have a finite limit as $\varepsilon\to 0$ and three of them, namely, $z_0^{(1)}=z^{(2)}=0; z_0^{(3)}=z_0^{(4)}=-1/2, z_0^{(5)}=z_0^{(6)}=1/2$ are independent of ε. Further, the zeros of the function $D_1(z,\varepsilon)$ will be denoted by x_k $(k=1,2,\ldots)$ and the zeros of $D_2(z,\varepsilon)$ by y_k $(k=1,2,\ldots)$.

For the remaining zeros of the first group we expand $D_1(z,\varepsilon) = D_1(x,\varepsilon)$ and $D_2(z,\varepsilon) = D_2(y,\varepsilon)$ in a series in the small parameter ε. On doing so we obtain

$$\begin{aligned} D_1(x,\varepsilon) = x(x^2 - 1/4)\varepsilon^2\langle 4x^4 + 2(2v - 7 - 4m^2)x^2 + 9/4(5 - 4v) \\ - 2(4v - 5)m^2 + 4m^4 - \frac{1}{3!}\{12x^6 + (28v - 81 - 36m^2)x^4 \\ + [745/4 - 122v + 2(53 - 2v)m^2 + 36m^4]x^2 + 1/16(2124v - 2331) \\ + 1/4(431 - 316v)m^2 - (24v + 25)m^4 - 12m^6\}\varepsilon^2 + \cdots\rangle = 0 \end{aligned} \tag{4.13.1}$$

$$\begin{aligned} D_2(y,\varepsilon) = y(y^2 - 1/4)\varepsilon^3\langle -16y^4 + 8(7 - 6v + 4m^2)y^2 \\ + 12v - 13 + 8(12v - 11)m^2 - 16m^4 + \frac{1}{10}\{112y^6 \\ + 4(100v - 209 - 84m^2)y^4 + [1837 - 1640v \\ + 40(33 - 14v)m^2 + 336m^4]y^2 + 1/4(1540 - 1635) \\ + (2380v - 2197)m^2 + 4(40v - 121)m^4 - 112m^6\}\varepsilon^2 + \cdots\rangle = 0. \end{aligned} \tag{4.13.2}$$

From the expansions of (4.13.1) and (4.13.2) we obtain the following group of roots

$$x_k = \alpha_k + \beta_k\varepsilon^2 + \cdots (k = 1,2,3,4) \tag{4.13.3}$$

$$y_k = \gamma_k + \delta_k\varepsilon^2 + \cdots (k = 1,2,3,4) \tag{4.13.4}$$

where

$$\begin{aligned} \alpha_k = \pm 1/2\left[7 - 2v + 2\sqrt{(1+v)^2 + 12(2-v)m^2}\right]^{1/2} (k = 1,2) \\ \alpha_k = \pm 1/2\left[7 - 2v - 2\sqrt{(1+v)^2 + 12(2-v)m^2}\right]^{1/2} (k = 3,4) \end{aligned} \tag{4.13.5}$$

$$\begin{aligned} \beta_k = 2(3\alpha_k)^{-1}(1+v)\left(4\alpha_k^2 + 2v - 7 - 4m^2\right)^{-1}[4(1 - v + m^2)\alpha_k^2 \\ - 9(1 - v) - (8v - 11)m^2 - 4m^4](k = 1,2,3,4) \\ \gamma_k = \pm 1/2\left[7 - 6v + 4m^2 + 2\sqrt{9(1 - v^2) + 4(3v - 2)m^2}\right]^{1/2} (k = 1,2) \\ \gamma_k = \pm 1/2\left[7 - 6v + 4m^2 - 2\sqrt{9(1 - v^2) + 4(3v - 2)m^2}\right]^{1/2} (k = 3,4) \end{aligned} \tag{4.13.6}$$

It can be shown that all the other zeros of the functions $D_1(x,\varepsilon)$ and $D_2(y,\varepsilon)$ tend to infinity as $\varepsilon \to 0$ and $\varepsilon y_k, \varepsilon x_k$ have a finite limit as $\varepsilon \to 0$. For the roots in question the following asymptotic representations are obtained:

$$x_n = \frac{\omega_{n1}}{\varepsilon} + O(\varepsilon), y_n = \frac{\omega_{n2}}{\varepsilon} + O(\varepsilon), (n = k - 4, k \geq 5) \tag{4.13.7}$$

where

$$\sin \omega_{n1}(\sin \omega_{n1} + 2\omega_{n1}) = 0 \ (n = 1,2,\ldots) \tag{4.13.8}$$

$$\cos \omega_{n2}(\sin \omega_{n1} + 2\omega_{n1}) = 0 \ (n = 1,2,\ldots) \tag{4.13.9}$$

Each of the expressions (4.13.8) (4.13.9) generates two groups of zeros, one of which is real, and the other complex. Here we obtain once again, in the first term of the asymptotic behaviour, the complete analogy with the theory of homogeneous plates, where the first group corresponds to the vortex solution, and the second one to the potential solution.

In what follows, the roots of equations $\sin \omega_{n1} = 0, \cos \omega_{n2} = 0$ will be denoted by $l_{p1}, l_{p2} (p = 1,2,\ldots)$.

4.14 Construction of Asymptotic Formulas for Displacements and Stresses

The roots $z_0^{(i)}, (i = \overline{1,6})$ correspond to the trivial solution.

To construct solutions corresponding to roots $x_k, y_k \quad (k = 1,2,3,4)$ we transform (4.12.12), (4.12.13) taking into account the smallness of ε and formulas (4.13.3)–(4.13.6). Assuming $\theta = \pi/2 + \varepsilon\eta, -1 \leq \eta \leq 1$ and expanding the solution of the first group in the small parameter ε, we find the following asymptotic expressions for them:

$$\begin{aligned} U_{r1}^{(i)} &= r_1 \rho^{-1/2} \sum_{k=1}^{4} A_k^{(i)} U_k^{(i)} \\ U_{\theta 1}^{(i)} &= r_1 \rho^{-1/2} \sum_{k=1}^{4} A_k^{(i)} W_k^{(i)} \\ U_{\varphi 1}^{(i)} &= r_1 \rho^{-1/2} m \sum_{k=1}^{4} A_k^{(i)} V_k^{(i)} \end{aligned} \tag{4.14.1}$$

$$\begin{aligned}
\sigma_{r1}^{(i)} &= 2G\rho^{-3/2}\sum_{k=1}^{4} A_k^{(i)} Q_{rk}^{(i)} \\
\sigma_{\varphi 1}^{(i)} &= 2G\rho^{-3/2}\sum_{k=1}^{4} A_k^{(i)} Q_{\varphi k}^{(i)} \\
\sigma_{\theta 1}^{(i)} &= G\rho^{-3/2}\sum_{k=1}^{4} A_k^{(i)} Q_{\theta k}^{(i)} \\
\tau_{r\varphi 1}^{(i)} &= G\rho^{-3/2} m\sum_{k=1}^{4} A_k^{(i)} T_{1k}^{(i)} \\
\tau_{r\theta 1}^{(i)} &= G\rho^{-3/2}\sum_{k=1}^{4} A_k^{(i)} T_{2k}^{(i)} \\
\tau_{\theta\varphi 1}^{(i)} &= G\rho^{-3/2} m\sum_{k=1}^{4} A_k^{(i)} T_{3k}^{(i)}
\end{aligned} \tag{4.14.2}$$

Here $A_k^{(i)}$ $(i = 1, 2)$ are arbitrary constants corresponding to problem I for $i = 1$, and corresponding to problem II for $i = 2$.

$$\begin{aligned}
U_k^{(1)} = {} & \big[-4v\alpha_k^3 + 2(2-v)\alpha_k^2 + v\big(9+4m^2\big)\alpha_k + 9/2(v-2) \\
& + 2(5v-4)m^2 + \frac{1}{3!}\langle 4(1+2v)\alpha_k^2 + 6(2v-3)\alpha_k^4 - 2[16v+9 \\
& + 4(1+2v)m^2\big]\alpha_k^3 - 72v\alpha_k^2\beta_k + \big[77 - 52v + 48(1-v)m^2\big]\alpha_k^2 \\
& + 24(2-v)\alpha_k\beta_k + \big[9/4(14v+9) + 2(28v+3)m^2 + 4(1+2v)m^4\big] \\
& \times \alpha_k + 6v\big(9+m^4\big)\beta_k + 3\eta^2\{-4(1-v)\alpha_k^5 + 6(1-v)\alpha_k^4 + 2(1-v) \\
& \times \big(9+4m^2\big)\alpha_k^3 - 3(1-v)\big(9+4m^2\big)\alpha_k^2 + \big[2(1+v)m^2 - 81/4(1-v) \\
& - 4(1-v)m^4\big]\alpha_k + 243/8(1-v) - 3(1+v)m^2 + 6(1-v)m^4\}\rangle\varepsilon^2 + \cdots\big] \\
& \times \exp(x_k \ln\rho) \\
W_k^{(1)} = {} & \varepsilon\eta\big[4(1-v)\alpha_k^4 - 2(1-v)\big(9+4m^2\big)\alpha_k^2 + 81/4(1-v) \\
& - 2(1+v)m^2 + 4(1-v)m^4 + \frac{1}{3!}\langle 4(2v-3)\alpha_k^6 - 4(1-2v)\alpha_k^5 \\
& + \big[87 - 70v + 12(3-2v)m^2\big]\alpha_k^4 + 96(1-v)\alpha_k^3\beta_k + 2(1-2v) \\
& \times \big(9+4m^2\big)\alpha_k^3 + [1/4(774v - 837) + (132v - 130)m^2 \\
& + 12(2v-3)m^4\big]\alpha_k^2 - 24(1-v)\big(9+4m^2\big)\alpha_k\beta_k + (1-2v) \\
& \times \big(6m^2 - 4m^4 - 81/4\big)\alpha_k + 81/6(33-34v) + 1/4(271 - 350v)m^2 \\
& + (43 - 62v)m^4 + 4(3-2v)m^6 + \eta^2\big\{4v\alpha_k^6 + 4(1-2v)\alpha_k^5 \\
& - \big(2 + 15v + 12vm^2\big)\alpha_k^4 - -2(1-2v)\big(9+4m^2\big)\alpha_k^3 \\
& + \big[1/4(27v+36) + (4+6v)m^2 + 12vm^4\big]\alpha_k^2 + (1-2v) \\
& \times \big(81/4 - 6m^2 + 4m^4\big)\alpha_k + 81/6(3v-2) + 1/4(29v-52)m^2 \\
& + (9v-2)m^4 - 4vm^6\big\}\rangle\varepsilon^2 + \cdots\big] \exp(x_k \ln\rho)
\end{aligned} \tag{4.14.3}$$

$$
\begin{aligned}
V_k^{(1)} = {} & \left[4\nu\alpha_k^2 + 4(1-2\nu)\alpha_k + 14 - 13\nu - 4\nu m^2 + \frac{1}{3!}\langle -4(1+2\nu)\alpha_k^4\right.\\
& - 12(1-2\nu)\alpha_k^3 + \left[68\nu - 40 + 8(1+2\nu)m^2\right]\alpha_k^2 + 48\nu\alpha_k\beta_k + (1-2\nu)\\
& \times \left(31 + 12m^2\right)\alpha_k + 24(1-2\nu)\beta_k + 1/4(497 - 466\nu) + 4(13 - 23\nu)m^2\\
& - 4(1+2\nu)m^4 + 3\eta^2\left\{4(1-\nu)\alpha_k^4 + 2\left[7\nu - 9 - 4(1-\nu)m^2\right]\alpha_k^2\right.\\
& \left.\left. - 4(1-2\nu)\alpha_k + 1/12(75 - 87\nu) - 2(1-\nu)m^2 + 4(1-\nu)m^4\right\}\rangle\varepsilon^2 + \cdots\right]\\
& \times \exp(x_k \ln\,\rho)\\
Q_{rk}^{(1)} = {} & \left[4(1+\nu)\alpha_k^3 - 2\left(1 + \nu + 2\nu m^2\right)\alpha_k^2 - [9(1+\nu) + 4(2-\nu)m^2]\right.\\
& \times \alpha_k + 9/2(1+\nu) + (4\nu - 5)m^2 + 4\nu m^4 + \frac{1}{3!}\langle 4\alpha_k^6 - 4(5+2\nu)\alpha_k^5\\
& + [4\nu - 9 - 8(1-\nu)m^2]\alpha_k^4 + \left[2(16\nu + 43) + 4(13 - 2\nu)m^2\right]\alpha_k^3\\
& + 72(1+\nu)\alpha_k^2\beta_k - 24\left(1 + \nu + 2\nu m^2\right)\alpha_k\beta_k - [1/4(64\nu + 73)\\
& + 6(2\nu + 3)m^2 + 4(4\nu - 1)m^4]\alpha_k^2 - [9/4(14\nu + 41) + 3(39 - 10\nu)m^2\\
& + 16(2-\nu)m^4]\alpha_k - 6\left[9(1+\nu) - 4(2-\nu)m^2\right]\beta_k + 9/16(28\nu + 73)\\
& + 3/2(38 - 37\nu)m^2 + (32\nu + 15)m^4 + 8\nu m^6 + 3\eta^2\left\{-4\alpha_k^6 + 4(2-\nu)\alpha_k^5\right.\\
& + \left[12\nu + 15 + 4(\nu+2)m^2\right]\alpha_k^4 + 2(\nu - 2)\left(9 + 4m^2\right)\alpha_k^3 - [9/4(4\nu + 3)\\
& + 2(9\nu - 4)m^2 + 4(1+2\nu)m^4]\alpha_k^2 + [81/4(2-\nu) - 2(2+5\nu)m^2\\
& + 4(2-\nu)m^4]\alpha_k + 81/16(2\nu - 3) + 3/4(2 + 11\nu)m^2 - 3m^4\\
& \left.\left. + 4\nu m^6\right\}\rangle\varepsilon^2 + \cdots\right]\exp(x_k \ln\,\rho),
\end{aligned}
$$

$$
\begin{aligned}
Q_{\varphi k}^{(1)} = {} & \left[4\nu\alpha_k^4 + \left(4 - 14\nu - 4\nu m^2\right)\alpha_k^2 + 4(1-2\nu)m^2\alpha_k + 9/4\right.\\
& \times (5\nu - 4) + 3(2-\nu)m^2 + \frac{1}{3!}\langle -8\nu\alpha_k^6 + 4(1-2\nu)\alpha_k^5 + [18(3\nu - 1)\\
& + 4(4\nu - 1)m^2]\alpha_k^4 + 96\nu\alpha_k^3\beta_k - 2(1-2\nu)\left(9 + 10m^2\right)\alpha_k^3 + [1/2\\
& \times (154 - 251\nu) + 8(1 - 9\nu)m^2 + 8(1-\nu)m^4]\alpha_k^2 + 12\left(4 - 14\nu - 4\nu m^2\right)\\
& \times \alpha_k\beta_k + 24(1-2\nu)m^2\beta_k + (1-2\nu)\left(81/4 + 37m^2 + 16m^4\right)\alpha_k\\
& + 9/8(89\nu - 73) + 1/4(41 + 68\nu)m^2 + (22 - 6\nu)m^4 - 4m^6 + 3\eta^2\\
& \times \left\{-4\nu\alpha_k^6 - 4(1-2\nu)\alpha_k^5 + \left[19\nu - 2 + 4(1+2\nu)m^2\right]\alpha_k^4 + 2(1-2\nu)\right.\\
& \times \left(9 + 4m^2\right)\alpha_k^3 + \left[9/4(4 - 11\nu) - 4m^2 - 4(2+\nu)m^2\right]\alpha_k^2 - (1-2\nu)\\
& \times \left(81/4 - 2m^2 + 4m^4\right)\alpha_k + 81/16(\nu - 2) + 1/4(85 - 50\nu)m^2\\
& \left.\left. - (3\nu + 4)m^4 + 4m^6\right\}\rangle\varepsilon^2 + \cdots\right]\exp(x_k \ln\,\rho)\\
Q_{\theta k}^{(1)} = {} & \frac{1}{3!}\varepsilon^2\left[\langle -12\alpha_k^6 + \left(69 - 16\nu + 36m^2\right)\alpha_k^4 + \left[68\nu - \frac{529}{4}\right.\right.\\
& \left. - (8\nu + 82)m^2 - 36m^4\right]\alpha_k^2 + 96\alpha_k^3\beta_k + 24\left(2\nu - 7 - 4m^2\right)\\
& \times \alpha_k\beta_k + 12(1-2\nu)m^2\alpha_k + \frac{1359}{16} - 72\nu + (46\nu - 239/4)m^2\\
& + (13 + 24\nu)2m^4 + 12m^6 + 3\eta^2\left\{4(1-\nu)\alpha_k^4 - [18(1-\nu) + 4(2-\nu)\right.\\
& \left. \times m^2]\alpha_k^2 - 4(1-2\nu)m^2\alpha_k + 81/4(1-\nu) + (11\nu - 16)m^2 + 4m^4\right\}\rangle\\
& \left. + \cdots\right]\exp(x_k \ln\,\rho)
\end{aligned}
$$

$$
\begin{aligned}
T_{1k}^{(1)} &= \big[4v\alpha_k^3 - 6v\alpha_k^2 + \big(4 - 5v - 4m^2\big)\alpha_k + 3/2(5v-4) + 2(2-v) \\
&\quad \times m^2 + \frac{1}{3!}\big\langle -4(1+2v)\alpha_k^5 + 6(1+2v)\alpha_k^4 + \big[32v - 2 + 8(1+2v)m^2\big] \\
&\quad \times \alpha_k^3 + 72v\alpha_k^2\beta_k - 72v\alpha_k\beta_k + \big(7 - 56v - 24m^2\big)\alpha_k^2 \\
&\quad + \big[1/4(115 - 110v) + (14 - 56v)m^2 - 4(1+2v)m^4\big]\alpha_k \\
&\quad + 6\big(4 - 5v - 4m^2\big)\beta_k + 3/8(158v - 139) + 6(v+3)m^2 \\
&\quad + 6(3-2v)m^4 + 3\eta^2\big\{4(1-v)\alpha_k^5 - 6(1-v)\alpha_k^4 - (1-v) \\
&\quad \times \big(9 + 8m^2\big)\alpha_k^3 + \big[23 - 19v + 12(1-v)m^2\big]\alpha_k^2 + [1/4(65 - 97v) \\
&\quad - 2(1+v)m^2 + 4(1-v)m^4\big]\alpha_k + 1/8(147v - 123) + (7 - 5v)m^2 \\
&\quad - 6(1-v)m^4\big\}\big\rangle\varepsilon^2 + \cdots\big] \exp(x_k \ln \rho) \\
T_{2k}^{(1)} &= \eta\big(\eta^2 - 1\big)\varepsilon^3\big[\big\langle 4\alpha_k^7 - 6\alpha_k^6 + \big(8v - 23 - 12m^2\big)\alpha_k^5 \\
&\quad + \big[1/2(61 - 8v) + (8v - 22)m^2\big]\alpha_k^4 + [1/4(171 - 114v) \\
&\quad + (24v - 26)m^2 + 12m^4\big]\alpha_k^3 + [1/8(144v - 369) + (45 - 48v) \\
&\quad \times m^2 + 2(8v - 13)m^4\big]\alpha_k^2 + [81/16(8v - 5) + 1/4(187 - 120v)m^2 \\
&\quad + (16v - 3)m^4 - 4m^6\big]\alpha_k + 81/32(7 - 8v) + 1/2(252v - 507)m^2 \\
&\quad + 1/2(8v - 19)m^4 + 2(5 - 4v)m^6\big\rangle + \cdots\big] \exp(x_k \ln \rho)
\end{aligned}
\tag{4.14.4}
$$

$$
\begin{aligned}
U_k^{(2)} &= \eta[2(1-v)(3 - 2\gamma_k) + \frac{1}{3!}\langle 6(1-2v)(2\gamma_k - 1)\gamma_k^2 \\
&\quad + 3\big[22v - 21 - 4(1-2v)m^2\big]\gamma_k - 24(1-v)\delta_k + 3/2(61 - 62v) \\
&\quad - 6(1-2v)m^2 + \eta^2\big\{4(2-v)\gamma_k^3 + 2(7v - 8)\gamma_k^2 + [2 - 3v \\
&\quad + 4(v-2)m^2\big]\gamma_k + 1/2(5v - 4) + 2(10 - 11v)m^2\big\}\big\rangle\varepsilon^2 + \cdots\big] \\
&\quad \times \exp(y_k \ln \rho) \\
W_k^{(2)} &= \varepsilon^{-1}[4(1-v) + \frac{1}{2!}\big\{4(2v - 3)\gamma_k^2 - 4(1-2v)\gamma_k \\
&\quad + 25 - 26v + 4(3-2v)m^2 + \eta^2\big[4v\gamma_k^2 + 4(1-2v)\gamma_k \\
&\quad + 7v - 6 - 4vm^2\big]\big\}\varepsilon^2 + \cdots\big] \exp(y_k \ln \rho)
\end{aligned}
\tag{4.14.5}
$$

$$
\begin{aligned}
Q_{rk}^{(2)} &= \eta\langle -4\gamma_k^2 + 4(2-\nu)\gamma_k + 2\nu - 3 + 4\nu m^2 + \frac{1}{3!}\left[12\gamma_k^4\right. \\
&- 12(1-2\nu)\gamma_k^3 - 12\left[\nu + 5 + (1+2\nu)m^2\right]\gamma_k - 48\gamma_k\delta_k + 24(2-\nu) \\
&\times \delta_k + (41 - 22\nu)\gamma_k + 3/4(44\nu - 61) + 3(1+4\nu)m^2 + 24\nu m^4 + \eta^2 \\
&\times \{8\gamma_k^4 + 4(\nu-2)\gamma_k^3 + 2\left[5 - \nu - 2(\nu+2)m^2\right]\gamma_k^2 + 3[\nu - 1 - 4(\nu-2) \\
&\times m^2]\gamma_k + 1/2(2-3\nu) + (11\nu - 10)m^2 + 4\nu m^4\}]\varepsilon^2 + \cdots\rangle \exp(y_k \ln\rho) \\
&\times \delta_k - 9(1-2\nu)\gamma_k + 1/2(33 - 57\nu) + (78\nu - 33)m^2 + 12m^4 + \eta^2 \\
&\times \{4\nu\gamma_k^4 + 8(1-2\nu)\gamma_k^3 + \left[10\nu - 16 - 4(\nu+2)m^2\right]\gamma_k^2 - 2(1-2\nu) \\
&(5 + 6m^2)\gamma_k + 1/4(32 - 43\nu) + (44 - 35\nu)m^2 + 8m^4\}]\varepsilon^2 + \cdots\rangle \exp(y_k \ln\rho) \\
Q_{\theta k}^{(2)} &= \frac{1}{3!}\eta\varepsilon^2\langle\{12\gamma_k^4 + 6(4\nu - 7 - 4m^2)\gamma_k^2 - 12(1-2\nu)\gamma_k \\
&+ 9/4(7 - 8\nu) + 6(13 - 12\nu)m^2 + 12m^4 + \eta^2\left[-4\gamma_k^4 + (14 + 8m^2)\right. \\
&\times \gamma_k^2 + 12(1-2\nu)\gamma_k + 1/4(48\nu - 37) + (24\nu - 34)m^2 - 4m^4]\} \\
&+ \cdots\rangle \exp(y_k \ln\rho) \\
T_{1k}^{(2)} &= \eta\langle 12(1-\nu)(2\gamma_k - 3) + \frac{1}{3!}[-12(1-2\nu)\gamma_k^3 + 8(1-2\nu)\gamma_k^2 \\
&+ \left[9(3-2\nu) + 12(1-2\nu)m^2\right]\gamma_k + 144(1-\nu)\delta_k + 1/2(102\nu - 105) \\
&- 6(1-2\nu)m^2 + \eta^2\{4(\nu-2)\gamma_k^3 - 6(\nu-2)\gamma_k^2 + \left[14 - 17\nu + 4(2-\nu)m^2\right] \\
&\times \gamma_k + 1/2(35\nu - 34) + 2(7\nu - 8)m^2\}]\varepsilon^2 + \cdots\rangle \exp(y_k \ln\rho) \\
T_{3k}^{(2)} &= 6(\eta^2 - 1)\varepsilon\{\left[-2\gamma_k^2 - 2(1-2\nu)\gamma_k + 1/2(11 - 12\nu)\right. \\
&+ 2m^2] + \cdots\} \exp(y_k \ln\rho)
\end{aligned}
\tag{4.14.6}
$$

We now present the first terms of the asymptotic expansions for the displacements and stresses corresponding to the zeros of the functions $\sin\omega \pm \omega, \sin\omega, \cos\omega$. After carrying out the necessary transformations we obtain:

$$
\begin{aligned}
U_{r2}^{(i)} &= r_1\rho^{-1/2}\varepsilon\sum_{n=1}^{\infty} B_{ni}\{2\left[(1-\nu)\omega_{n1}^{-1}\psi_{ni}''(\eta) - \nu\omega_{ni}\psi_{ni}(\eta)\right] \\
&+ O(\varepsilon)\}\exp\left(\frac{\omega_{ni}}{\varepsilon}\ln\rho\right) \\
U_{\theta 2}^{(i)} &= -r_1\rho^{-1/2}\varepsilon\sum_{n=1}^{\infty} B_{ni}\{2\left[(2-\nu)\psi_{ni}'(\eta) + (1-\nu)\omega_{ni}^{-2}\psi_{ni}'''(\eta)\right] \\
&+ O(\varepsilon)\}\exp\left(\frac{\omega_{ni}}{\varepsilon}\ln\rho\right)
\end{aligned}
\tag{4.14.7}
$$

$$U_{\varphi 2}^{(i)} = -r_1 m\rho^{-1/2}\varepsilon^2 \sum_{n=1}^{\infty} B_{ni}\{2[(1-\nu)\omega_{ni}^{-1}\psi_{ni}''(\eta) - \nu\omega_{ni}\psi_{ni}(\eta)] + O(\varepsilon)\} \exp\left(\frac{\omega_{ni}}{\varepsilon}\ln\rho\right) \quad (4.14.8)$$

$$\sigma_{r2}^{(i)} = 2G\rho^{-3/2}\sum_{n=1}^{\infty} B_{ni}[\psi_{ni}''(\eta) + O(\varepsilon)]\exp\left(\frac{\omega_{ni}}{\varepsilon}\ln\rho\right)$$

$$\sigma_{\varphi 2}^{(i)} = 2G\rho^{-3/2}\sum_{n=1}^{\infty} B_{ni}[\psi_{ni}''(\eta) + \omega_{ni}^2\psi_{ni}(\eta) + O(\varepsilon)]\exp\left(\frac{\omega_{ni}}{\varepsilon}\ln\rho\right)$$

$$\sigma_{\theta 2}^{(i)} = 2G\rho^{-3/2}\sum_{n=1}^{\infty} B_{ni}[\omega_{ni}^2\psi_{ni}(\eta) + O(\varepsilon)]\exp\left(\frac{\omega_{ni}}{\varepsilon}\ln\rho\right)$$

$$\tau_{r\varphi 2}^{(i)} = 2Gm\rho^{-3/2}\varepsilon\sum_{n=1}^{\infty} B_{ni}\{2\omega_{ni}[(1-\nu)\omega_{ni}^{-1}\psi_{ni}''(\eta) - \nu\omega_{ni}\psi_{ni}(\eta)] + O(\varepsilon)\}\exp\left(\frac{\omega_{ni}}{\varepsilon}\ln\rho\right)$$

$$\tau_{r\theta 2}^{(i)} = -2G\rho^{-3/2}\sum_{n=1}^{\infty} B_{ni}[\omega_{ni}\psi_{ni}'(\eta) + O(\varepsilon)]\exp\left(\frac{\omega_{ni}}{\varepsilon}\ln\rho\right) \quad (4.14.9)$$

$$\tau_{\theta\varphi 2}^{(i)} = 2Gm\varepsilon\rho^{-3/2}\sum_{n=1}^{\infty} B_{ni}[\omega_{ni}\psi_{ni}'(\eta) + O(\varepsilon)]\exp\left(\frac{\omega_{ni}}{\varepsilon}\ln\rho\right)$$

$$U_{r3}^{(i)} = U_{\theta 3}^{(i)} \approx 0;\ \sigma_{r3}^{(i)} = \sigma_{\phi 3}^{(i)} = \sigma_{\theta 3}^{(i)} = \tau_{r\theta 3}^{(i)} \approx 0$$

$$U_{\varphi 3}^{(i)} = r_1\rho^{-1/2}\varepsilon\sum_{p=1}^{\infty} D_{pi}L_{pi}(\eta)\exp\left(\frac{l_{pi}}{\varepsilon}\ln\rho\right)$$

$$\tau_{r\varphi 3}^{(i)} = 2G\rho^{-3/2}\sum_{p=1}^{\infty} D_{pi}L_{pi}(\eta)l_{pi}\exp\left(\frac{l_{pi}}{\varepsilon}\ln\rho\right)$$

$$\tau_{\theta\varphi 3}^{(i)} = 2G\rho^{-3/2}\sum_{p=1}^{\infty} D_{pi}L_{pi}'(\eta)l_{pi}\exp\left(\frac{l_{pi}}{\varepsilon}\ln\rho\right) \quad (4.14.10)$$

In formulas (4.14.7)–(4.14.10), $B_{ni}(i = 1,2; n = 1,2,3,\ldots), D_{pi}, (i = 1,2;\ p = 1,2,3,\ldots)$ are arbitrary constants, $\psi_{ni}(\eta)$ is Papkovich's function given by

$$\psi_{n1}(\eta) = (\omega_{n1}^{-1}\sin\omega_{n1} + \cos\omega_{n1})\cos\omega_{n1}\eta + \eta\sin\omega_{n1}\sin\omega_{n1}\eta(n = 1,2,3,\ldots)$$
$$\psi_{n2}(\eta) = (\sin\omega_{n2} - \omega_{n2}^{-1}\cos\omega_{n2})\sin\omega_{n2}\eta + \eta\cos\omega_{n2}\cos\omega_{n2}\eta(n = 1,2,3,\ldots)$$
$$L_{p1}(\eta) = \cos l_{p1}\eta(p = 1,2,\ldots)$$
$$L_{p2}(\eta) = \sin l_{p2}\eta(p = 1,2,\ldots)$$

The arbitrary constants A_{ki}, B_{ni}, D_{pi} are determined from an infinite system of linear algebraic equations.

From the comparison of solutions of the first and second groups we may conclude that the first group defines the basic strain-stress state of the plate, and the second one defines the boundary effect similar to Saint-Venant's end effect of the theory of homogeneous plates.

4.15 Kirsch Problem for Plates of Variable Thickness

Here we consider a plate of thickness $h = \varepsilon r$.

Let the conical part of the boundary be stress-free and on the spherical part of the boundary the following boundary conditions hold true:

$$\begin{aligned} &\sigma_r = 0, \ \ \tau_{r\varphi} = 0, \ \ \tau_{r\theta} = 0, \ \text{при} \ \ r = r_1 \\ &\sigma_r = P(1 + \cos 2\varphi), \ \tau_{r\varphi} = 0, \ \tau_{r\theta} = 0, \ \text{при} \ \ r = r_2 \end{aligned} \tag{4.15.1}$$

Using $\sigma_r = \sigma_r^{(0)} + \sigma_r^{(2)} \cos 2\varphi$ and the results of Sect. 4.12 the stresses can be written as $(m = 0, \ m = 2)$

$$\begin{aligned} &\sigma_r = \sigma_r^{(0)} + \sigma_r^{(2)} \cos 2\varphi, \sigma_\varphi = \sigma_\varphi^{(0)} + \sigma_\varphi^{(2)} \cos 2\varphi, \\ &\sigma_\theta = \sigma_\theta^{(0)} + \sigma_\theta^{(2)} \cos 2\varphi, \tau_{r\varphi} = \tau_{r\varphi}^{(2)} \sin 2\varphi, \\ &\tau_{r\theta} = \tau_{r\theta}^{(0)} + \tau_{r\theta}^{(2)} \cos 2\varphi, \tau_{\theta\varphi} = \tau_{\theta\varphi}^{(2)} \sin 2\varphi. \end{aligned} \tag{4.15.2}$$

As shown in Sects. 4.8 and 4.12 the stress-strain state in such a plate is composed of primary stress state and the state of boundary layer type.

The strain-stress states of boundary layer type decay exponentially as we move away from the edge of the plate to the interior. Thus, far from the edge, the stress state is identical with the main stress state. We show that the conditions (4.15.1) can determine the main stress state in an elementary way, without solving the infinite linear algebraic equations. To this end, taking the average of the boundary conditions (4.15.1) and stress (4.15.2) by the thickness respectively, we obtain:

$$\begin{aligned} &\bar{\sigma}_r = 0, \bar{\tau}_{r\varphi} = 0, \bar{\tau}_{r\theta} = 0, \ \ \text{at } r = r_1 \\ &\bar{\sigma}_r = P(1 + \cos 2\varphi), \bar{\tau}_{r\varphi} = 0, \bar{\tau}_{r\theta} = 0, \ \ \text{at } r = r_2 \end{aligned} \tag{4.15.3}$$

$$\begin{aligned} &\bar{\sigma}_r = \bar{\sigma}_r^{(0)} + \bar{\sigma}_r^{(2)} \cos 2\varphi, \bar{\sigma}_\varphi = \bar{\sigma}_\varphi^{(0)} + \bar{\sigma}_\varphi^{(2)} \cos 2\varphi \\ &\bar{\sigma}_\theta = \bar{\sigma}_\theta^{(0)} + \bar{\sigma}_\theta^{(2)} \cos 2\varphi, \bar{\tau}_{r\varphi} = \bar{\tau}_{r\varphi}^{(2)} \sin 2\varphi \\ &\bar{\tau}_{r\theta} = 0, \bar{\tau}_{\theta\varphi} = 0 \end{aligned} \tag{4.15.4}$$

where

$$\bar{\sigma}_r^{(0)} = \rho^{-3/2}\sum_{k=1}^{2} C_k\rho^{x_k}\bar{Q}_{r0}(x_k,\varepsilon)$$
$$\bar{\sigma}_\varphi^{(0)} = \rho^{-3/2}\sum_{k=1}^{2} C_k\rho^{x_k}\bar{Q}_{\varphi 0}(x_k,\varepsilon) \tag{4.15.5}$$

$$\bar{\sigma}_\theta^{(0)} = \rho^{-3/2}\sum_{k=1}^{2} C_k\rho^{x_k}\bar{Q}_{\theta 0}(x_k,\varepsilon)$$
$$\bar{\sigma}_r^{(2)} = \rho^{-3/2}\sum_{k=1}^{6} C_k\rho^{x_k}\bar{Q}_{r2}(x_k,\varepsilon)$$
$$\bar{\sigma}_\varphi^{(2)} = \rho^{-3/2}\sum_{k=1}^{6} C_k\rho^{x_k}\bar{Q}_{\varphi 2}(x_k,\varepsilon) \tag{4.15.6}$$
$$\bar{\sigma}_\theta^{(2)} = \rho^{-3/2}\sum_{k=1}^{6} C_k\rho^{x_k}\bar{Q}_{\theta 2}(x_k,\varepsilon)$$
$$\bar{\tau}_{r\varphi}^{(2)} = \rho^{-3/2}\sum_{k=1}^{6} C_k\rho^{x_k}\bar{T}_{r2}(x_k,\varepsilon)$$

In the expansions (4.14.4) after averaging all values with respect to thickness and assuming $m=0, m=2$ we obtain the quantities $\bar{Q}_{r0}(x_k,\varepsilon), \bar{Q}_{\varphi 0}(x_k,\varepsilon)$, $\bar{Q}_{\theta 0}(x_k,\varepsilon)$ and $\bar{Q}_{r2}(x_k,\varepsilon),\ldots\bar{T}_{r2}(x_k,\varepsilon)$ respectively.

Here C_k are arbitrary constants, ε is a small parameter,x_k are zeros of $D_1(x,\varepsilon)$ (4.12.9) for $m=0(k=1,2)$ and $m=2(k=3,4,5,6)$ with the following asymptotic expansions

$$x_k = \alpha_k + \beta_k\varepsilon^2 + \cdots (k=1,2,3,4,5,6) \tag{4.15.7}$$

where the constants α_k are given by

$$\alpha_k = (-1)^{k+1}1/2\sqrt{5-4\nu},\ \beta_k = (3\alpha_k)^{-1}(1-\nu^2)(k=1,2) \tag{4.15.8}$$

$$\alpha_k = 1/2(-1)^{k+1}\left(23-2\nu+2\sqrt{\nu^2-46\nu+97}\right)^{1/2}(k=3,4) \tag{4.15.9}$$

$$\alpha_k = 1/2(-1)^{k+1}\left(23-2\nu-2\sqrt{\nu^2-46\nu+97}\right)^{1/2}(k=5,6) \tag{4.15.10}$$

Substituting (4.15.5), (4.15.6) into (4.15.3) we obtain the following system of equations with respect to C_k

$$\left.\begin{aligned} &\sum_{k=1}^{2} \bar{Q}_{r0}(x_k, \varepsilon) C_k = 0 \\ &\rho_2^{-3/2} \sum_{k=1}^{2} \rho_2^{x_k} \bar{Q}_{r0}(x_k, \varepsilon) C_k = P \end{aligned}\right\} \tag{4.15.11}$$

$$\left.\begin{aligned} &\sum_{k=3}^{6} \bar{Q}_{r2}(x_k, \varepsilon) C_k = 0 \\ &\sum_{k=1}^{6} \bar{T}_{r2}(x_k, \varepsilon) C_k = 0 \\ &\rho_2^{-3/2} \sum_{k=3}^{6} \rho_2^{x_k} \bar{Q}_{r2}(x_k, \varepsilon) C_k = P \\ &\sum_{k=3}^{6} \rho_2^{x_k} \bar{T}_{r2}(x_k, \varepsilon) C_k = 0 \end{aligned}\right\} \tag{4.15.12}$$

Taking into account that $x_{2k} = -x_{2k-1}$ from the system (4.15.11) and (4.15.12) we get

$$\begin{aligned} C_1 &= \bar{Q}_{r0}^{-1}(x_1) \rho_2^{3/2} \left(\rho_2^{x_1} - \rho_2^{-x_1}\right)^{-1} P \\ C_2 &= \bar{Q}_{r0}^{-1}(-x_1) \rho_2^{3/2} \left(\rho_2^{x_1} - \rho_2^{-x_1}\right)^{-1} P \end{aligned} \tag{4.15.13}$$

$$\begin{aligned} C_3 &= \rho_2^{3/2} \Delta^{-1} \Delta_1 P, C_4 = \rho_2^{3/2} \Delta^{-1} \Delta_2 P \\ C_5 &= \rho_2^{3/2} \Delta^{-1} \Delta_3 P, C_6 = \rho_2^{3/2} \Delta^{-1} \Delta_4 P \end{aligned} \tag{4.15.14}$$

where

$$\begin{aligned} &\Delta = \Phi(x_3, x_5, \varepsilon) - \Phi(-x_3, x_5, \varepsilon) - 2\gamma(x_3, -x_3, \varepsilon)\gamma(x_5, -x_5, \varepsilon) \\ &\Phi(x, y, \varepsilon) = \gamma(x, y, \varepsilon)\gamma(-x, -y, \varepsilon) ch[(x+y) \ln \rho_2] \\ &\gamma(x, y, \varepsilon) = \bar{Q}_{r2}(x, \varepsilon)\bar{T}_{r2}(y, \varepsilon) - \bar{Q}_{r2}(y, \varepsilon)\bar{T}_{r2}(x, \varepsilon) \\ &\Delta_1 = \bar{T}_{r2}(x_5, \varepsilon)\gamma(-x_3, -x_5, \varepsilon) \exp(x_5 \ln \rho_2) \\ &\quad - \bar{T}_{r2}(-x_5, \varepsilon)\gamma(-x_3, x_5, \varepsilon) \exp(-x_5 \ln \rho_2) \\ &\quad - \bar{T}_{r2}(-x_3, \varepsilon)\gamma(x_5, -x_5, \varepsilon) \exp(-x_3 \ln \rho_2) \\ &\Delta_3 = \bar{T}_{r2}(-x_3, \varepsilon)\gamma(x_3, -x_5, \varepsilon) \exp(-x_3 \ln \rho_2) \\ &\quad - \bar{T}_{r2}(x_3, \varepsilon)\gamma(-x_3, -x_5, \varepsilon) \exp(x_3 \ln \rho_2) \\ &\quad - \bar{T}_{r2}(-x_5, \varepsilon)\gamma(x_3, -x_3, \varepsilon) \exp(-x_5 \ln \rho_2) \\ &\Delta_2(x_3, x_5, \varepsilon) = -\Delta_1(-x_3, x_5, \varepsilon) \\ &\Delta_4(x_3, x_5, \varepsilon) = -\Delta_3(x_3, -x_5, \varepsilon) \end{aligned}$$

Substituting (4.15.3), (4.15.4) into (4.15.2) we obtain

$$\bar{\sigma}_r^{(0)} = \left(\frac{\rho_2}{\rho}\right)^{3/2} (\rho_2^{x_1} - \rho_2^{-x_1})^{-1} (\rho^{x_1} - \rho^{-x_1}) P$$
$$\bar{\sigma}_\varphi^{(0)} = \left(\frac{\rho_2}{\rho}\right)^{3/2} (\rho_2^{x_1} - \rho_2^{-x_1})^{-1} P \times \left[\frac{\bar{Q}_{\varphi 0}(x_1, \varepsilon)}{\bar{Q}_{r0}(x_1, \varepsilon)} \rho^{x_1} - \frac{\bar{Q}_{\varphi 0}(-x_1, \varepsilon)}{\bar{Q}_{r0}(-x_1, \varepsilon)} \rho^{-x_1}\right] \tag{4.15.15}$$
$$\bar{\sigma}_\theta^{(0)} = \left(\frac{\rho_2}{\rho}\right)^{3/2} (\rho_2^{x_1} - \rho_2^{-x_1})^{-1} P \left[\frac{\bar{Q}_{\theta 0}(x_1, \varepsilon)}{\bar{Q}_{r0}(x_1, \varepsilon)} \rho^{x_1} - \frac{\bar{Q}_{\theta 0}(-x_1, \varepsilon)}{\bar{Q}_{r0}(-x_1, \varepsilon)} \rho^{-x_1}\right]$$

$$\bar{\sigma}_r^{(2)} = \Delta^{-1} P \left(\frac{\rho_2}{\rho}\right)^{3/2} [\bar{Q}_{r2}(x_3, \varepsilon)\Delta_1(x_3, x_5, \varepsilon)\rho^{x_3} - \bar{Q}_{r2}(-x_3, \varepsilon)\Delta_1(-x_3, x_5, \varepsilon)\rho^{-x_3} + \bar{Q}_{r2}(x_5, \varepsilon)\Delta_3(x_3, x_5)\rho^{x_5} - \bar{Q}_{r2}(-x_5, \varepsilon)\Delta_3(x_3, -x_5, \varepsilon)\rho^{-x_5}]$$
$$\bar{\sigma}_r^{(2)} = \Delta^{-1} P \left(\frac{\rho_2}{\rho}\right)^{3/2} [\bar{Q}_{r2}(x_3, \varepsilon)\Delta_1(x_3, x_5, \varepsilon)\rho^{x_3} - \bar{Q}_{r2}(-x_3, \varepsilon)\Delta_1(-x_3, x_5, \varepsilon)\rho^{-x_3} + \bar{Q}_{r2}(x_5, \varepsilon)\Delta_3(x_3, x_5)\rho^{x_5} - \bar{Q}_{r2}(-x_5, \varepsilon)\Delta_3(x_3, -x_5, \varepsilon)\rho^{-x_5}]$$
$$\bar{\sigma}_\varphi^{(2)} = \Delta^{-1} P \left(\frac{\rho_2}{\rho}\right)^{3/2} [\bar{Q}_{\varphi 2}(x_3, \varepsilon)\Delta_1(x_3, x_5, \varepsilon)\rho^{x_3} - \bar{Q}_{\varphi 2}(-x_3, \varepsilon)\Delta_1(-x_3, x_5, \varepsilon)\rho^{-x_3} + \bar{Q}_{\varphi 2}(x_5, \varepsilon)\Delta_3(x_3, x_5, \varepsilon)\rho^{x_5} - \bar{Q}_{\varphi 2}(-x_5, \varepsilon)\Delta_3(x_3, -x_5, \varepsilon)\rho^{-x_3}] \tag{4.15.16}$$
$$\bar{\sigma}_\theta^{(2)} = \Delta^{-1} P \left(\frac{\rho_2}{\rho}\right)^{3/2} [\bar{Q}_{\theta 2}(x_3, \varepsilon)\Delta_1(x_3, x_5, \varepsilon)\rho^{x_3} - \bar{Q}_{\theta 2}(-x_3, \varepsilon)\Delta_1(-x_3, x_5, \varepsilon)\rho^{-x_3} + \bar{Q}_{\theta 2}(x_5, \varepsilon)\Delta_3(x_3, x_5, \varepsilon)\rho^{x_5} - \bar{Q}_{\theta 2}(-x_5, \varepsilon)\Delta_3(x_3, -x_5, \varepsilon)\rho^{-x_5}]$$

Taking into account the expansions (4.14.4) (if $m = 0, m = 2$) we expand all the magnitudes included in formulas (4.15.15) and (4.15.6) in a series in ε to give

$$\bar{Q}_{j0} = \bar{Q}_{j0}^{(0)} + \bar{Q}_{j2}^{(2)}\varepsilon^2 + \cdots$$
$$\bar{Q}_{j2} = \bar{Q}_{j2}^{(0)} + \bar{Q}_{j2}^{(2)}\varepsilon^2 + \cdots \quad (j = r, \theta, \varphi)$$
$$\Delta = \Delta_0 + \Delta_2\varepsilon^2 + \cdots$$
$$\Delta_i = \Delta_{i0} + \Delta_{i2}\varepsilon^2 + \cdots \tag{4.15.17}$$

$$\gamma(x, y, \varepsilon) = \gamma_0(x, y) + \gamma_2(x, y)\varepsilon^2 + \cdots$$
$$\Phi(x, y, \varepsilon) = \Phi_0(x, y) + \Phi_2(x, y)\varepsilon^2 + \cdots \tag{4.15.18}$$

Substituting (4.15.17), (4.15.18) into (4.15.15) and (4.15.16) and restricting ourselves with the first term of the expansions we obtain the following expressions for the normal stresses on the contour $r = r_1$

$$\begin{aligned}
\bar{\sigma}_{\theta 0}^{(0)} &= 8/3\alpha_1 \rho_2^{3/2}(1-v^2)^{-1}(6\alpha_1\beta_1 + v^2 - 1)(\rho_2^{\alpha_1} - \rho_2^{-\alpha_1})^{-1} P\varepsilon^2 \\
\bar{\sigma}_{\varphi 0}^{(2)} &= \Delta_0^{-1}\rho_2^{3/2} P\Big[\bar{Q}_{\varphi 2}^{(0)}(\alpha_3)\Delta_{10}(\alpha_3,\alpha_5) - \bar{Q}_{\varphi 2}^{(0)}(-\alpha_3)\Delta_{10}(-\alpha_3,\alpha_5) \\
&\quad + \bar{Q}_{\varphi 2}^{(0)}(\alpha_5)\Delta_{30}(\alpha_3,\alpha_5) - \bar{Q}_{\varphi 2}^{(0)}(-\alpha_3)\Delta_{30}(\alpha_3,-\alpha_5)\Big] \qquad (4.15.19)\\
\bar{\sigma}_{\theta 0}^{(2)} &= \Delta_0^{-1}\rho_2^{3/2} P\Big[\bar{Q}_{\theta 2}^{(0)}(\alpha_3)\Delta_{10}(\alpha_3,\alpha_5) - \bar{Q}_{\theta 2}^{(0)}(-\alpha_3)\Delta_{10}(-\alpha_3,\alpha_5) \\
&\quad + \bar{Q}_{\theta 2}^{(0)}\Delta_{30}(\alpha_3,\alpha_5) - \bar{Q}_{\theta 2}^{(0)}(-\alpha_3)\Delta_{30}(\alpha_3,-\alpha_5)\Big]
\end{aligned}$$

where

$$\begin{aligned}
\Delta_0 &= \Phi_0(\alpha_3,\alpha_5) - \Phi_0(-\alpha_3,\alpha_5) - 2\beta_0(\alpha_3,-\alpha_5)\beta_0(\alpha_5,-\alpha_5) \\
\Delta_{10} &= \bar{T}_{r2}^{(0)}(\alpha_5)\beta_0(-\alpha_3,-\alpha_5)\exp(\alpha_5 \ln \rho_2) - \bar{T}_{r2}^{(0)}(-\alpha_5)\beta_0(-\alpha_3,\alpha_5) \qquad (4.15.20)\\
&\quad \times \exp(-\alpha_5 \ln \rho_2) - \bar{T}_{r2}^{(0)}(-\alpha_3)\beta_0(\alpha_5,-\alpha_5)\exp(-\alpha_3 \ln \rho_2)
\end{aligned}$$

$$\begin{aligned}
\Delta_{30} &= \bar{T}_{r2}^{(0)}(-\alpha_3)\beta_0(\alpha_3,-\alpha_5)\exp(-\alpha_3 \ln \rho_2) - \bar{T}_{r2}^{(0)}(\alpha_3)\beta_0(-\alpha_3,-\alpha_5) \\
&\quad \times \exp(\alpha_3 \ln \rho_2) - \bar{T}_{r2}^{(0)}(-\alpha_5)\beta_0(\alpha_3,-\alpha_3)\exp(-\alpha_5 \ln \rho_2) \\
\Phi_0(x,y) &= \beta_0(x,y)\beta_0(-x,-y)ch[(x+y)\ln \rho_2] \\
\bar{T}_{r2}^{(0)} &= 8v\alpha^3 - 12v\alpha^2 - (5v+12)\alpha + 1/2(20-v) \\
\bar{Q}_{r2}^{(0)}(\alpha) &= 4(1+v)\alpha^3 - 2(1+9v)\alpha^2 - (41-7v)\alpha + 1/2(41+97v) \\
\bar{Q}_{\varphi 2}^{(0)} &= 4v\alpha^4 + 2(2-15v)\alpha^2 + 16(1-2v)\alpha + 3\left(5 - \frac{v}{4}\right) \\
\bar{Q}_{\theta 2}^{(0)} &= 2/3\left[4(1+4v-v^2)\alpha^2 + 16(1-2v)\alpha + 15 - 36v - 23v^2\right] \\
\beta_0(x,y) &= (y-x)\left[32v(3v-1)x^2y^2 + 4(19v^2 - 65v + 12)xy(x+y)\right. \\
&\quad - 2(195v^2 + 63v - 20)(x^2 + xy + y^2) + 2(87v^2 - 133v + 12)xy \\
&\quad \left. - (591v^2 + 167v - 20)(x+y) + 239v^2 + 775v - 164\right]
\end{aligned}$$

(4.15.21)

Now we introduce the expressions for the forces related to the midsurface of the plate:

$$\begin{aligned}
T_\varphi\big|_{\rho_1=1} &= T_{\varphi 0}^{(0)} + T_{\varphi 2}^{(0)}\cos 2\varphi \\
T_\theta\big|_{\rho_1=1} &= T_{\theta 0}^{(0)} + T_{\theta 2}^{(0)}\cos 2\varphi
\end{aligned} \qquad (4.15.22)$$

where

$$
\begin{aligned}
T_{\varphi 0}^{(0)} &= 16\alpha_1 \rho_1^{1/2}\left(\rho_2^{\alpha_1} - \rho_2^{-\alpha_1}\right)^{-1} T_0 \\
T_{\theta 0}^{(0)} &= 8/3\left(1-\nu^2\right)^{-1}\alpha_1 \rho_2^{1/2}\left(6\alpha_1\beta_1 + \nu^2 - 1\right)\left(\rho_2^{\alpha_1} - \rho_2^{-\alpha_1}\right)^{-1} T_0 \varepsilon^2 \\
T_{\varphi 2}^{(0)} &= \Delta_0^{-1}\rho_2^{1/2}\Big[\bar{Q}_{\varphi 2}^{(0)}(\alpha_3)\Delta_{10}(\alpha_3,\alpha_5) - \bar{Q}_{\varphi 2}^{(0)}(-\alpha_3)\Delta_{10}(-\alpha_3,\alpha_5) \\
&\quad + \bar{Q}_{\varphi 2}^{(0)}(\alpha_5)\Delta_{30}(\alpha_3,\alpha_5) - \bar{Q}_{\varphi 2}^{(0)}(-\alpha_5)\Delta_{30}(\alpha_3,-\alpha_5)\Big] T_0 \\
T_{\theta 2}^{(0)} &= \Delta_0^{-1}\rho_2^{1/2}\varepsilon^2\Big[\bar{Q}_{\theta 2}^{(0)}(\alpha_3)\Delta_{10}(\alpha_3,\alpha_5) - \bar{Q}_{\theta 2}^{(0)}(-\alpha_3)\Delta_{10}(-\alpha_3,\alpha_5) \\
&\quad + \bar{Q}_{\theta 2}^{(0)}(\alpha_5)\Delta_{30}(\alpha_3,\alpha_5) - \bar{Q}_{\theta 2}^{(0)}(-\alpha_5)\Delta_{30}(\alpha_3,-\alpha_5)\Big] T_0 \\
T_0 &= 2\varepsilon\rho_2 P
\end{aligned}
\tag{4.15.23}
$$

4.16 Torsional Vibrations of a Conical Shell of Variable Thickness

We will consider the torsional vibrations of a truncated hollow cone of variable thickness here. The equation of torsional vibrations in a spherical coordinate system has the form:

$$
\Delta U_\varphi - \frac{1}{r^2 \sin^2\theta} U_\varphi = g G^{-1} \frac{\partial^2 U_\varphi}{\partial t^2} \tag{4.16.1}
$$

1. Assuming $U_\varphi = -\frac{\partial U}{\partial \theta}$ Eq. (4.16.1) can be written in the form:

$$
\Delta U = g G^{-1} \frac{\partial^2 U}{\partial t^2} \tag{4.16.2}
$$

It is assumed that the conical part of the boundary is stress-free, i.e.

$$
\tau_{\theta\varphi} = 0 \quad \text{for } \theta = \theta_n \quad (n = 1,2), r_1 \le r \le r_2, \tag{4.16.3}
$$

and the following boundary conditions are given on the remaining part of the boundary

$$
\tau_{r\varphi} = \tau_s(\theta) e^{i\omega t} \quad \text{for } r = r_s \quad (s = 1,2). \tag{4.16.4}
$$

The solution of Eq. (4.16.2) will be sought in the form

$$U = f(r)\psi(\theta)e^{i\omega t}. \tag{4.16.5}$$

On substituting (4.16.5) into (4.16.2) we obtain the following equations for the functions $f(r)$ and $\psi(\theta)$

$$f'' + \frac{2}{r}f' + \left(\lambda^2 - \frac{\mu^2}{r^2}\right) = 0, \lambda^2 = gG^{-1}\omega^2 \tag{4.16.6}$$

$$\frac{d^2\psi}{d\theta^2} + ctg\theta\frac{d\psi}{d\theta} + \mu^2\psi = 0 \tag{4.16.7}$$

Using Hooke's law, the stresses $\tau_{r\varphi}$ and $\tau_{\theta\varphi}$ can be represented as follows:

$$\tau_{r\varphi} = -G\left[f'(r) - \frac{1}{r}f(r)\right]\frac{d\psi}{d\theta}e^{i\omega t} \tag{4.16.8}$$

$$\tau_{\theta\varphi} = \frac{G}{r}f(r)\left(2ctg\theta\frac{d\psi}{d\theta} + \mu^2\psi\right)e^{i\omega t} \tag{4.16.9}$$

We define the operator T such that

$$T\psi = \left[-\frac{1}{\sin\theta}\frac{d}{d\theta}\left(\sin\theta\frac{d\psi}{d\theta}\right), \left(2ctg\theta\frac{d\psi}{d\theta} + \mu^2\psi\right)_{\theta=\theta_n} = 0\right] \tag{4.16.10}$$

In the space $L_2(\theta_1, \theta_2)$, the problems (4.16.7), (4.16.3) can be given in the operator form as

$$T\psi = \mu^2\psi \tag{4.16.11}$$

where $L_2(\theta_1, \theta_2)$ is a Hilbert space with the weight $\sin\theta$ and the inner product

$$(\psi_1, \psi_2) = \int_{\theta_1}^{\theta_2} \psi_1\psi_2 \sin\theta d\theta \tag{4.16.12}$$

We prove that operator T is self-adjoint. To achieve this, we use the property of operator T given as

$$(T\psi_1, \psi_2) = -\int_{\theta_1}^{\theta_2} \frac{d}{d\theta}\left(\sin\theta \frac{d\psi_1}{d\theta}\right)\psi_2 d\theta = -\sin\theta \frac{d\psi_1}{d\theta}\psi_2\Big|_{\theta_1}^{\theta_2}$$

$$+\int_{\theta_1}^{\theta_2} \sin\theta \frac{d\psi_1}{d\theta}\frac{d\psi_2}{d\theta} d\theta = \frac{\mu^2}{2}\sin\theta\, tg\theta(\psi_1\psi_2 - \psi_1\psi_2)_{\theta_1}^{\theta_2} \quad (4.16.13)$$

$$-\int_{\theta_1}^{\theta_2} \frac{d}{d\theta}\left(\sin\theta \frac{d\psi_2}{d\theta}\right)\psi_1 d\theta = (\psi_1, T\psi_2)$$

Thus, we have shown that

$$(T\psi_1, \psi_2) = (\psi_1, T\psi_2)$$

for any $\psi_1, \psi_2 \in L_2(\theta_1, \theta_2)$. The self-adjointness of the operator T is proved.

So, we see that the spectrum of operator (4.16.11) is real and the eigenvectors are orthogonal and complete.

Now let us consider the problem (4.16.6), (4.16.4)

$$f'' + \frac{2}{r}f' + \left(\lambda^2 - \frac{\mu^2}{r^2}\right) = 0$$

$$\tau_{r\varphi}(r_s, \theta) = \tau_s(\theta).$$

We expand $\tau_s(\theta)$ in the eigenfunctions of the operator (4.16.11):

$$\tau_s(\theta) = \sum_{k=1}^{\infty} a_{ks}\psi_k$$

where

$$a_{ks} = \int_{\theta_1}^{\theta_2} \tau_s\psi_k \sin\theta d\theta = (\tau_s, \psi_k)$$

$$\|\psi_k\|^2 = 1 = \int_{\theta_1}^{\theta_2} \psi_k^2 \sin\theta d\theta, (\psi_k, \psi_n) = \delta_{kn}$$

Then the general solution of (4.16.6), (4.16.4) can be represented as follows:

$$f_k(r) = \frac{1}{\sqrt{r}}[J\mu_k(\lambda r)C_{1k} + Y\mu_k(\lambda r)C_{2k}] \tag{4.16.14}$$

Finally, the function $U(r, \theta, t)$ can be represented in the form

$$U = \sum_{k=1}^{\infty} C_k f_k(r)\psi_k(\theta)e^{i\omega t} \tag{4.16.15}$$

where the constants C_k are determined from the condition

$$\left[f_k'(r) - \frac{1}{r}f_k(r)\right]_{r=r_s} = a_{ks} \tag{4.16.16}$$

As it may be observed from (4.16.7), the torsional wave in the cone spreads out without any dispersion.

2. Let us proceed now to the construction of the asymptotic formulas for the eigenvalues and eigenfunctions of this problem. The general solution of Eq. (4.16.7) has the form:

$$\psi(\theta) = D_1 P_{z-1/2}(\cos\theta) + D_2 Q_{z-1/2}(\cos\theta) \tag{4.16.17}$$

where $P_{z-1/2}(\cos\theta)$, $Q_{z-1/2}(\cos\theta)$ are Legendre functions of the first and second kinds respectively; D_1, D_2 are arbitrary constants.

From condition (4.16.3) we obtain the characteristic equation

$$\begin{aligned}\Delta(z, \theta_1, \theta_2) = {} & \left(z^2 - \frac{1}{4}\right)^2 D^{(0,0)}_{z-1/2}(\theta_1, \theta_2) + 2\left(z^2 - \frac{1}{4}\right) \\ & \times ctg\theta_1 D^{(1,0)}_{z-1/2}(\theta_1, \theta_2) + 2\left(z^2 - \frac{1}{4}\right) ctg\theta_2 D^{(0,1)}_{z-1/2}(\theta_1, \theta_2) \\ & + 4ctg\theta_1 ctg\theta_2 D^{(1,1)}_{z-1/2}(\theta_1, \theta_2) = 0\end{aligned} \tag{4.16.18}$$

where $\mu^2 = \chi(\chi + 1) = z^2 - \frac{1}{4}$. Equation (4.16.18) has a countable number of zeros. They correspond to the following homogeneous solutions

$$U_\varphi = -f_{z_k}(r)\left[2ctg\theta_2 D^{(1,1)}_{z_k-1/2}(\theta,\theta_2) + \left(z^2-\frac{1}{4}\right)D^{(1,0)}_{z_k-1/2}(\theta,\theta_2)\right]e^{i\omega t}$$
$$+\left(z^2-\frac{1}{4}\right)D^{(1,0)}_{z_k-1/2}(\theta,\theta_2)]e^{i\omega t}$$
$$\tau_{\theta\varphi} = G\frac{f_{z_k}(r)}{r}\left[4ctg\theta ctg\theta_2 D^{(1,1)}_{z_k-1/2}(\theta,\theta_2) + 2\left(z^2-\frac{1}{4}\right)\right. \tag{4.16.19}$$
$$\times ctg\theta D^{(1,0)}_{z_k-1/2}(\theta,\theta_2) + 2\left(z^2-\frac{1}{4}\right)ctg\theta_2 D^{(0,1)}_{z_k-1/2}(\theta,\theta_2)$$
$$+\left(z^2-\frac{1}{4}\right)^2 D^{(0,0)}_{z_k-1/2}(\theta,\theta_2)]e^{i\omega t}$$

Equation (4.16.18) has a rather complicated structure. To effectively investigate its roots, we assume that

$$\theta_1 = \theta_0 - \varepsilon, \theta_2 = \theta_0 + \varepsilon \tag{4.16.20}$$

Substituting (4.16.20) into (4.16.18), we obtain

$$D(z,\varepsilon,\theta_0) = \Delta(z,\theta_1,\theta_2) = 0 \tag{4.16.21}$$

Let us formulate our statement: the function $D(z,\varepsilon,\theta_0)$ has two groups of zeros: (a) the first group consists of four zeros which do not depend on the small parameter ε

$$z_{1,2} = \pm 1/2, z_{3,4} = \pm 3/2$$

(b) the second group consists of a countable set of zeros which tend to infinity as $\varepsilon \to 0$.

We outline the proof of the first statement: represent the function $D(z,\varepsilon,\theta_0)$ as follows:

$$D(z,\varepsilon,\theta_0) = \left(z^2-\frac{1}{4}\right)(z^2-9/4)D_0(z,\varepsilon,\theta_0) \tag{4.16.22}$$

Through highly complicated calculations it can be shown that

$$\lim_{z^2\to\frac{1}{4}} D_0(z,\varepsilon,\theta_0) \neq 0, \lim_{z^2\to\frac{9}{4}} D_0(z,\varepsilon,\theta_0) \neq 0$$

which implies the first statement. We can prove that all other zeros $D(z,\varepsilon,\theta_0)$ increase unboundedly as $\varepsilon \to 0$.

To construct the asymptotics of the zeros of the second group we seek them in the form

$$z_n = \delta_n \varepsilon^{-\beta} + O(\varepsilon^{\beta}), \beta \geq 1, (n = k - 4). \quad (4.16.23)$$

Substituting (4.16.23) into the characteristic Eq. (4.16.18) and utilizing the asymptotic expansions for the function (4.2.12) we obtain the following equation for δ_n

$$\begin{aligned} &\sin 2(\delta_n \varepsilon^{1-\beta}) = 0, \delta_n = \frac{n\pi}{2}, (\beta = 1, n = 1, 2, \ldots) \\ &\delta_n = \frac{n\pi}{2} \varepsilon^{1-\beta} [\beta > 1, n = 0(\varepsilon^{1-\beta})]. \end{aligned} \quad (4.16.24)$$

Assuming that ε is a small parameter we present the asymptotic construction of homogeneous solutions corresponding to the two groups of zeros.

By substituting $z_{1,2} = \pm 1/2$ into the expressions (4.16.19) we obtain that these zeros correspond to the trivial solutions.

When the values $z_3 = -\frac{3}{2}, z_4 = 3/2$ are substituted into (4.16.19), we observe that they correspond to the following groups of solutions:

$$\begin{aligned} U_\varphi &= \frac{2}{r} \left[\left(\frac{\cos \lambda r}{\lambda r} + \sin \lambda r \right) C_1 + \left(\frac{\sin \lambda r}{\lambda r} + \cos \lambda r \right) C_2 \right] \sin \theta e^{i\omega t} \\ \tau_{r\varphi} &= \frac{-2G}{r^2} \left[\left(\frac{3 \cos \lambda r}{\lambda r} + 3 \sin \lambda r - \lambda r \cos \lambda r \right) C_1 \right. \\ &\left. + \left(\frac{3 \sin \lambda r}{\lambda r} - 3 \cos \lambda r - \lambda r \sin \lambda r \right) C_2 \right] \sin \theta e^{i\omega t}, \tau_{\theta\varphi} = 0. \end{aligned} \quad (4.16.25)$$

Assuming $\theta = \theta_0 + \varepsilon\eta, -1 \leq \eta \leq 1$ and using the first term of the asymptotic expansion of the Legendre function, the solution corresponding to the second group of zeros can be represented as follows:

$$\begin{aligned} U_\varphi &= -\sum_{n=1}^{\infty} C_n f_{z_n}(r) \cos z_n(\eta - 1) \varepsilon e^{i\omega t} \\ \tau_{r\varphi} &= -G \sum_{n=1}^{\infty} C_n \left[f'_{z_n}(r) - \frac{1}{r} f_{z_n}(r) \right] \cos z_n(\eta - 1) e^{i\omega t} \\ \tau_{\theta\varphi} &= G \sum_{n=1}^{\infty} C_n \frac{f_{z_n(r)}}{r} z_n \sin z_n(\eta - 1) \varepsilon e^{i\omega t} \end{aligned} \quad (4.16.26)$$

Here, we should pay attention to the following fact: although the characteristic equation is independent of frequency, as is clear from (4.16.26), the asymptotic behaviour of the solution depends essentially on the fact whether $\lambda \sim z, \lambda >> z, \lambda << z$.

As noted above, $\theta_0 = \frac{\pi}{2}$ corresponds to a plate of variable thickness. In this case the solution to (4.16.26) remains valid and we should assume $\theta_0 = \frac{\pi}{2}$ in solutions to (4.16.26).

3. It is well-known that when solving problems of forced vibrations for geometries of finite sizes it is important to know the natural frequencies and the corresponding forms of vibrations. The algorithm for solving the problem of forced vibrations of a cone developed above allows us to find the natural frequency with a certain accuracy since the condition of vanishing of the main determinant of system (4.16.16) serves as the equation for determining the natural frequencies of torsional vibration of the cone.

Thus, equating the primary determinant of the system (4.16.16) to zero we obtain the following frequency equation:

$$\begin{aligned} D(\lambda, z) &= 4\lambda^2 L_z^{(1,1)}(\lambda r_1, \lambda r_2) - \frac{6\lambda}{r_2} L_z^{(1,0)}(\lambda r_1, \lambda r_2) \\ &\quad - \frac{6\lambda}{r_2} L_z^{(0,1)}(\lambda r_1, \lambda r_2) + \frac{9}{r_1 r_2} L_z^{(0,0)}(\lambda r_1, \lambda r_2) = 0. \end{aligned} \tag{4.16.27}$$

In the case of $z = \pm 3/2$ the frequency Eq. (4.16.27) takes the form

$$\begin{aligned} D_0(\lambda, \pm 3/2) &= \left[\frac{9}{\lambda^2 r_1 r_2} - \frac{3(r_1^2 + r_2^2)}{r_1 r_2} + 9 + \lambda^2 r_1 r_2\right] \\ &\quad \times \sin \lambda H - \left[3\lambda H + \frac{9H}{\lambda r_1 r_2}\right] \cos \lambda H = 0 \end{aligned} \tag{4.16.28}$$

where $H = r_2 - r_1$ is the height of the cone.

Equation (4.16.28) can be written in the form

$$\begin{aligned} D_0(\lambda, \pm 3/2) &= \left[\lambda^4 r_1^2 r_2^2 - 3\lambda^2 (r_1^2 + r_2^2 - r_1 r_2) + 9\right] \\ &\quad \times \sin \lambda H - 3\lambda H (\lambda^2 r_1 r_1 + 3) \cos \lambda H = 0. \end{aligned} \tag{4.16.29}$$

Assume that for fixed values of H the limiting case $\sin \lambda H \sim \lambda H, \cos \lambda H \sim 1$ is possible. In this case, from (4.16.28) we obtain

$$\omega_0 \sim \sqrt{\frac{3G}{g} \frac{H}{r_1 r_2}} = \sqrt{\frac{3G}{g}\left(\frac{1}{r_1} - \frac{1}{r_2}\right)}. \tag{4.16.30}$$

As noted above, the remaining magnitudes z_n tend to infinity as $\varepsilon \to 0$. Here the following limiting cases are possible:

(a) λ is a finite value, $z \to \infty$ as $\varepsilon \to 0$;
(b) $\lambda \to \infty$ as $\varepsilon \to 0, \lambda > > z$;
(c) $\lambda \to \infty$ as $\varepsilon \to 0, \lambda \sim z$.

In case (a) setting $z = \delta \varepsilon^{-\beta}, \beta \geq 1$ and transforming the frequency equation by using the asymptotic expansions for the Bessel functions [8] given as

$$J_\nu(x) \sim \frac{1}{\sqrt{2\pi\nu}}\left(\frac{ex}{2\nu}\right)^\nu$$
$$Y_\nu(x) \sim \frac{1}{\sqrt{2\pi\nu}}\left(\frac{ex}{2\nu}\right)^{-\nu} \quad (4.16.31)$$

we obtain the following equation

$$\Delta \sim \frac{1}{2\pi}z\left[\left(\frac{r_2}{r_1}\right)^z - \left(\frac{r_2}{r_1}\right)^{-z}\right] = 0 \quad (4.16.32)$$

From (4.16.32) we get that $z = 0$. This cannot be true since $z \to \infty$ as $\varepsilon \to 0$. Thus, in this case the shell cannot perform free vibrations.

In case (b) using the asymptotic behaviour of the Bessel functions for large values of their arguments Eq. (4.16.27) can be written as

$$D = \frac{2}{\pi}\frac{1}{\sqrt{r_1 r_2}}\left[\left(4\lambda + \frac{4z}{r_1 r_2} + \frac{12z}{r_1 r_2} + \frac{9}{\lambda r_1 r_2}\right)\sin\lambda H\right.$$
$$\left. - \frac{2H}{r_1 r_2}(2z+3)\cos\lambda H + O\left(\frac{1}{\lambda^2}\right)\right] = 0 \quad (4.16.33)$$

Assuming $z = z_0\varepsilon^{-\beta}$, we seek λ_n as

$$\lambda_n = \lambda_{n0}\varepsilon^{-q} + O(\varepsilon^q), q > \beta \quad (4.16.34)$$

From (4.16.33) we obtain

$$\lambda_n = \frac{n\pi}{H} \; (q = 1, n = 1, 2, \ldots)$$
$$\lambda_n = \frac{n\pi}{H}\varepsilon^q[q > 1, n = 0(\varepsilon^{-q})] \quad (4.16.35)$$

Finally, in case (c) using the asymptotic behaviour of the functions $J_z(\lambda x), Y_z(\lambda x)$ for large values of their arguments as well as orders, Eq. (4.16.27) can be written in the form:

$$D \sim sh\frac{2}{3}\lambda H = 0. \quad (4.16.36)$$

This implies that $H\lambda = 0$. This cannot hold true for fixed H as by assumption $\lambda \to \infty$.

Thus, we have two series of natural frequencies defined by formulas (4.16.30) and (4.16.35).

References

1. Nuller, B.M.: Solution of the elasticity theory problem of a truncated cone. MTT (5), 102–110 (1967)
2. Naimark, M.A.: Linear differential operators, p. 525. Nauka, Moscow (1969)
3. Bateman, H., Erdeyn, A.: Higher Transcendental Functions, vol. 1, 294 p. Nauka, Moscow (1965)
4. Bazarenko, N.A., Vorovich, I.I.: Analysis of three-dimensional stress and strain states of circular cylindrical shells. Construction of refined applied theories. J. Appl. Math. Mech. **33**(3), 495–510 (1969)
5. Lurie, A.I.: Spatial Problems of Elasticity Theory, p. 491. Gostekhizdat, Moscow (1965)
6. Pao, Y.N., Kaul, R.K.: Waves and vibrations in isotropic and anisotropic plates. In: Mindlin, R. D. (eds.) Applied Mechanics, pp. 149–195. Pergamon Press, New York (1974)
7. Mekhtiev, M.F., Ustinov, Y.A.: Asymptotic behavior of solutions of elasticity theory for plates of variable thickness. In: Proceedings of VIII All-Russia conference on the theory of shells and plates (Rostov-on-Don, 1971). Nauka, Moscow, pp. 58–60 (1973)
8. Babich, V.M., Molotkov, I.A.: Mathematical methods in the theory of elastic waves. Mech Deformable Solids VINITI **10**, 5–62 (1977)

Appendix

1. Study of Stress Concentration in a Plate with a Circular Hole with a Load of a Power Form

In [1], to investigate the influence of the end effect on the stress concentration at a circular hole in a plate of constant thickness the stress distribution on the contour of the hole was analysed numerically when a load of the following type is given:

$$\sigma_r = \chi\eta^p (p = 3, 5), \tau_{rz} = 0 \tag{A.1}$$

A similar analysis for a plate of variable thickness is given below. In this case, since the first terms of the asymptotic expansions for "the boundary layer" solution, as shown above, coincide with the asymptotic solution for a plate of constant thickness, the results of [1] are used in the solution of the infinite system (4.9.3). Additionally, we consider the case $P = 7$.

Here we will consider an axisymmetric problem of bending of an infinite plate of variable thickness with a circular hole of radius r_1. We assume that the following conditions hold true along the holes on the plate:

$$\begin{gathered}\sigma_r = \chi f(\eta) = \chi\left(\eta^p - K_p\eta\right), \tau_{r\varphi} = 0 \\ (p = 3, 5, 7)\end{gathered} \tag{A.2}$$

There is no stress on the surfaces $\theta = \frac{\pi}{2} \mp \varepsilon$. The constant K_p is chosen so that the load is self-balanced.

$$K_p = \int_{-1}^{1} \eta^p \sin 2\varepsilon\eta d\eta \Big/ \int_{-1}^{1} \eta \sin 2\varepsilon\eta d\eta = K_{p0} + \varepsilon^2 K_{p2} + \cdots \tag{A.3}$$

2. $K_{p0} = 3/p + 2, K_{p2} = -4(p-1)/5(p+2)(p+4)$.
3. Let us now consider the construction of the solution. Firstly, we construct a "penetrating" solution which in this case corresponds to the root $z_6 = -\frac{1}{2}\sqrt{13 - 12\nu} + O(\varepsilon^2)$ and is defined by (4.8.7), (4.8.8). (The solution

M. F. Mekhtiev, *Vibrations of Hollow Elastic Bodies*,
Advanced Structured Materials 88,
https://doi.org/10.1007/978-3-319-74354-7

corresponding to $z_2 = -\frac{1}{2}$ is identically equal to zero due to the vanishing of the resultant vector, and the solution corresponding to $z_5 = \frac{1}{2}\sqrt{13-12v} + O(\varepsilon^2)$ is equal to zero due to bounded stress at infinity). Seeking the constant C_6 in the form

$$C_6 = C_{60} + \varepsilon^2 C_{62} + \cdots \tag{A.4}$$

and employing (4.9.5) ($j = 6$) we have:

$$C_{60} = 0,$$
$$C_{62} = \frac{-3(1-v)\left(\sqrt{13-12v}+7\right)}{20\left[7v-8+(v-2)\sqrt{13-12v}\right]\left(1+\sqrt{13-12v}\right)}\frac{p-1}{(p+2)(p+4)}.$$

2. Let us consider now the “boundary layer” solutions which are defined by the asymptotic formulas (4.8.9), (4.8.10). Due to bounded stresses at infinity the inequality $\mathrm{Re}\delta_n > 0$ holds true in these formulas. The unknown constants are sought in the form

$$B_n = B_{n0} + \varepsilon B_{n1} + \cdots. \tag{A.5}$$

At the first stage of asymptotic integration, according to (4.9.7), we obtain the infinite system

$$\sum_{n=1}^{\infty} g_{tn} B_{n0} = H_{t0} (t = 1, 2) \tag{A.6}$$

where the coefficients of the system are given by (4.9.10), and the right-hand sides $\{H_{t0}\}$ are of the form

$$H_{t0} = 2\int_{-1}^{1} \left(\eta^p - K_p\eta\right)\left[(1-v)\delta_t^{-1}F_t'' - v\delta_t F_t(\eta)\right]d\eta. \tag{A.7}$$

The unique solvability of the system and the convergence of the reduction method are justified in [2].

To invert the system (A.6) it is convenient to transform it by separating its real and imaginary parts. Taking into account the relations

$$\delta_{2t} = \overline{\delta_{2t-1}}, B_{2t} = \overline{B_{2t-1}}, H_{2t} = \overline{H_{2t-1}}.$$

the system can be written in the following operator form:

$$MX = f. \tag{A.8}$$

Here $x = \{x_n\}$, $f = \{f_n\}$ are infinite sequences of new unknowns and new right-hand sides respectively satisfying the relations

$$\begin{aligned} x_{2t-1} &= \mathrm{Re}B_{2t-1}, \quad x_{2t} = J_m B_{2t} \\ f_{2s-1} &= \mathrm{Re}H_{2s-1}, \quad f_{2s} = J_m H_{2s-1}, \quad (t, s = 1, 2, \ldots) \end{aligned} \tag{A.9}$$

and M is a symmetric infinite matrix with real entries.

The solution was constructed by means of the inverse matrix M_{20} obtained by O. K. Aksentyan [1] where M_{20} is a truncated matrix corresponding to the 20 complex roots δ_k lying in the right half-plane.

As noted, the system (A.6) is solved by the method of truncation. Consistently the matrices of 2nd, 4th, …, 20th orders are inverted successively. Poisson's ratio $\nu = \frac{1}{3}$ is assumed throughout the calculations.

As we have 14 equations in 20 unknowns in the system (A.6), we determine only the first 14 of the unknowns x_j on using the method of truncation. Calculations were carried out $P = 3$, $P = 5$, $P = 7$. The results are shown in Table A.1.

To illustrate the speed of convergence of determining the unknowns x_j, we give the values of the different approximations for the unknowns x_1, x_2, x_{13}, x_{14}, as these are the most characteristic ones (Table A.2). The superscript indicates the order of the system from which these unknowns are determined.

Table A.1 Values of unknowns for various types of loading

	j	x_j
$p = 3$	1	$0.98433707 \times 10^{-3}$
	2	$0.25800960 \times 10^{-5}$
	3	$0.22866771 \times 10^{-3}$
	4	$-0.10753112 \times 10^{-3}$
	5	$0.63918478 \times 10^{-4}$
	6	$-0.63120704 \times 10^{-4}$
	7	$0.91924586 \times 10^{-5}$
	8	$-0.32099248 \times 10^{-4}$
	9	$-0.10123414 \times 10^{-4}$
	10	$-0.11889187 \times 10^{-4}$
	11	$-0.13507504 \times 10^{-4}$
	12	$0.42514644 \times 10^{-6}$
	13	$-0.90026532 \times 10^{-5}$
	14	$0.57643361 \times 10^{-5}$

(continued)

Table A.1 (continued)

	j	x_j
$p = 5$	1	$0.27190303 \times 10^{-2}$
	2	$-0.50020730 \times 10^{-3}$
	3	$0.37243899 \times 10^{-3}$
	4	$-0.27638508 \times 10^{-3}$
	5	$0.59337646 \times 10^{-4}$
	6	$-0.11704804 \times 10^{-3}$
	7	$-0.19930847 \times 10^{-4}$
	8	$-0.44265384 \times 10^{-4}$
	9	$-0.37241817 \times 10^{-4}$
	10	$-0.62261089 \times 10^{-5}$
	11	$-0.30632563 \times 10^{-4}$
	12	$0.12236160 \times 10^{-4}$
	13	$-0.14972796 \times 10^{-4}$
	14	$0.16300964 \times 10^{-4}$
$p = 7$	1	$0.11856841 \times 10^{-2}$
	2	$-0.54130682 \times 10^{-4}$
	3	$0.20824164 \times 10^{-3}$
	4	$-0.90423852 \times 10^{-4}$
	5	$0.67477444 \times 10^{-4}$
	6	$-0.44920171 \times 10^{-4}$
	7	$0.32501985 \times 10^{-4}$
	8	$-0.28445183 \times 10^{-4}$
	9	$0.15147308 \times 10^{-4}$
	10	$-0.20655407 \times 10^{-4}$
	11	$0.31155192 \times 10^{-5}$
	12	$-0.13789460 \times 10^{-4}$
	13	$-0.43295336 \times 10^{-5}$
	14	$-0.64154449 \times 10^{-5}$

Now according to formula (4.8.10) the stresses can be calculated in the first approximation ($\rho = 1$). The stresses are caclculated at ten distinct points located at the half of the plate height:

$$\eta = 0.1, \eta = 0.2, \eta = 0.3, \eta = 0.4, \eta = 0.5, \eta = 0.6, \eta = 0.7,$$
$$\eta = 0.8, \eta = 0.9, \eta = 1.$$

Table A.2 Values of unknowns for various types of loading

	n	$x_1^{(n)}$	$x_2^{(n)}$
$p = 3$	6	$0.96295472 \times 10^{-3}$	$-0.18843045 \times 10^{-4}$
	8	$0.97806848 \times 10^{-3}$	$-0.50921612 \times 10^{-5}$
	10	$0.97961498 \times 10^{-3}$	$-0.47135238 \times 10^{-6}$
	12	$0.98092693 \times 10^{-3}$	$-0.67465301 \times 10^{-6}$
	14	$0.98246603 \times 10^{-3}$	$0.13836543 \times 10^{-5}$
	16	$0.98330601 \times 10^{-3}$	$0.19730661 \times 10^{-5}$
	18	$0.98375922 \times 10^{-3}$	$0.23209442 \times 10^{-5}$
	20	$0.98433707 \times 10^{-3}$	$0.25800960 \times 10^{-5}$
$p = 5$	6	$0.26282340 \times 10^{-2}$	$-0.52711530 \times 10^{-3}$
	8	$0.26937102 \times 10^{-2}$	$-0.50746516 \times 10^{-3}$
	10	$0.27008288 \times 10^{-2}$	$-0.50354225 \times 10^{-3}$
	12	$0.270740665 \times 10^{-2}$	$-0.50314273 \times 10^{-3}$
	14	$0.27125672 \times 10^{-2}$	$-0.50225484 \times 10^{-3}$
	16	$0.27153705 \times 10^{-2}$	$-0.50133686 \times 10^{-3}$
	18	$0.27171835 \times 10^{-2}$	$-0.50075423 \times 10^{-3}$
	20	$0.27190303 \times 10^{-2}$	$-0.50020730 \times 10^{-3}$
$p = 7$	6	$0.11836799 \times 10^{-2}$	$-0.79069359 \times 10^{-4}$
	8	$0.11955708 \times 10^{-2}$	$-0.64906175 \times 10^{-4}$
	10	$0.11943863 \times 10^{-2}$	$-0.58283341 \times 10^{-4}$
	12	$0.11917097 \times 10^{-2}$	$-0.55804491 \times 10^{-4}$
	14	$0.11899046 \times 10^{-2}$	$-0.54804691 \times 10^{-4}$
	16	$0.11884753 \times 10^{-2}$	$-0.54260934 \times 10^{-4}$
	18	$0.11870118 \times 10^{-2}$	$-0.54048962 \times 10^{-4}$
	20	$0.11856841 \times 10^{-2}$	$-0.54130682 \times 10^{-4}$
	n	$x_{13}^{(n)}$	$x_{14}^{(n)}$
$p = 3$	6		
	8		
	10		
	12		
	14	$0.96878178 \times 10^{-6}$	$0.11228799 \times 10^{-5}$
	16	$-0.43978137 \times 10^{-5}$	$0.55592369 \times 10^{-5}$
	18	$-0.71786692 \times 10^{-5}$	$0.55922628 \times 10^{-5}$
	20	$-0.90026532 \times 10^{-5}$	$0.57643361 \times 10^{-5}$

(continued)

Table A.2 (continued)

$p = 5$	6		
	8		
	10		
	12		
	14	$0.49156253 \times 10^{-5}$	$0.20080921 \times 10^{-5}$
	16	$-0.4692236 \times 10^{-5}$	$0.11736571 \times 10^{-4}$
	18	$-0.10252636 \times 10^{-4}$	$0.14126080 \times 10^{-4}$
	20	$-0.14972796 \times 10^{-4}$	$0.16300964 \times 10^{-4}$
$p = 7$	6		
	8		
	10		
	12		
	14	$-0.31862682 \times 10^{-5}$	$0.84336146 \times 10^{-5}$
	16	$-0.63289110 \times 10^{-5}$	$0.078031042 \times 10^{-5}$
	18	$-0.64948546 \times 10^{-5}$	$-0.30525692 \times 10^{-5}$
	20	$-0.43295336 \times 10^{-5}$	$-0.64154449 \times 10^{-5}$

According to (4.8.10) all stresses on the surface of the hole have the following structure in the first term of the asymptotic expansions:

$$\begin{aligned}
\sigma_r^{(0)} &= 2G\sum_{n=1}^{\infty} B_{n0}F_n''(\eta) \\
\sigma_\varphi^{(0)} &= 2G\sum_{n=1}^{\infty} B_{n0}\left[F_n''(\eta) + \delta_n^2 F_n(\eta)\right] \\
\sigma_\theta^{(0)} &= 2G\sum_{n=1}^{\infty} B_{n0}\delta_n^2 F_n(\eta) \\
\tau_{r\theta}^{(0)} &= 2G\sum_{n=1}^{\infty} B_{n0}\delta_n F_n'(\eta)
\end{aligned} \tag{A.10}$$

The number B_{n0} is determined by the values found in accordance with the notations (A.9).

We define the sum of two adjoint terms of a series in n in formula (A.10) (for odd and even n) as a boundary layer or Saint-Venant's end effect. The values adjacent to n, odd and even, correspond to complex adjoint terms of the series, as δ_{2n} is adjoint to δ_{2n-1} and $B_{2n,0}$ is adjoint to $B_{2n-1,0}$.

To characterize the convergence of the method it is important to study the convergence of $\sigma_\varphi^{(0)}$ by the boundary layers. It is this convergence that determines the quantity of unknowns B_{n0} necessary for calculating the stresses with some accuracy.

Table A.3 shows the first few terms of the asymptotic expansions of stresses $\sigma_{\varphi}^{(0)}$ calculated sequentially for $n = 1, 2, \ldots, 7$.

From Table A.3 it can be seen that the convergence of the point position depends on the plate height as well as the load of the hole $\sigma_r = \chi(\eta p - K_p \eta)$.

As it may be observed readily, the convergence by the boundary layers becomes somewhat worse with the increasing of the index of loading P. The stress $\sigma_{\varphi}^{(0)}$ on

Table A.3 $\frac{1}{\chi}\sigma_{\varphi}^{(0)}$

	η					
	n	0.1	0.2	0.3	0.4	0.5
$p = 3$	2	−0.0186	−0.0352	−0.0479	−0.0548	−0.0541
	4	−0.0030	−0.0133	−0.0341	−0.0608	−0.0822
	6	−0.0089	−0.0154	−0.0303	−0.0477	−0.0813
	8	−0.0108	−0.0198	−0.0300	−0.0407	−0.0727
	10	−0.0101	−0.0187	−0.0305	−0.0426	−0.0625
	12	−0.0108	−0.0148	−0.0320	−0.0500	−0.0610
	14	−0.0103	−0.0181	−0.0303	−0.0523	−0.0604
$p = 5$	2	−0.0526	−0.0987	−0.1318	−0.1463	−0.1378
	4	−0.0291	−0.0678	−0.1233	−0.1659	−0.1911
	6	−0.0310	−0.0626	−0.1077	−0.1534	−0.2019
	8	−0.0429	−0.0714	−0.1006	−0.1364	−0.1994
	10	−0.0288	−0.0762	−0.1161	−0.1254	−0.1812
	12	−0.0369	−0.0640	−0.1207	−0.1395	−0.1596
	14	−0.0376	−0.0671	−0.1106	−0.1325	−0.1534
$p = 7$	2	−0.0210	−0.0397	−0.0538	−0.0610	−0.0595
	4	−0.0066	−0.0406	−0.0406	−0.0658	−0.0846
	6	−0.0145	−0.0453	−0.0333	−0.0510	−0.0793
	8	−0.0102	−0.0487	−0.0420	−0.0489	−0.0634
	10	−0.0107	−0.0337	−0.0427	−0.0583	−0.0599
	12	−0.0139	−0.0227	−0.0369	−0.0621	−0.0667
	14	−0.0100	−0.0277	−0.0350	−0.0582	−0.0751
	η	0.6	0.7	0.8	0.9	1
	n					
$p = 3$	2	−0.0447	−0.0261	0.00108	0.0347	0.0712
	4	−0.0842	−0.0559	0.00332	0.0800	0.1489
	6	−0.1039	−0.0820	0.00029	0.1152	0.1981
	8	−0.1096	−0.1110	0.00192	0.1407	0.2169
	10	−0.1070	−0.1078	0.00490	0.1548	0.2194
	12	−0.1081	−0.1190	0.00780	0.1599	0.1946
	14	−0.1076	−0.1350	0.00184	0.1676	0.1634

(continued)

Table A.3 (continued)

	η					
	n	0.1	0.2	0.3	0.4	0.5
$p = 5$	2	−0.1038	−0.0444	0.0366	0.1314	0.2278
	4	−0.2110	−0.0813	0.0620	0.2293	0.3679
	6	−0.2036	−0.1116	0.0773	0.2949	0.4300
	8	−0.2261	−0.1301	0.0985	0.3352	0.4333
	10	−0.2462	−0.1518	0.1328	0.3607	0.3770
	12	−0.2490	−0.1866	0.1717	0.3773	0.3005
	14	−0.2333	−0.2200	0.2013	0.3824	0.2446
$p = 7$	2	−0.0480	−0.0262	0.0050	0.0429	0.0835
	4	−0.0840	−0.0542	0.0055	0.0824	0.1534
	6	−0.1002	−0.0817	−0.0044	0.1104	0.2033
	8	−0.0956	−0.1044	−0.0242	0.1338	0.2459
	10	−0.0802	−0.1132	−0.0470	0.1518	0.2790
	12	−0.0681	−0.1131	−0.0663	0.1694	0.2928
	14	−0.0610	−0.1115	−0.0795	0.1880	0.2835

the boundary calculated from (A.10) should be equal to zero if the boundary conditions are satisfied. Table A.4 shows how well the boundary conditions hold true.

For more clarity, these results are shown in the graphs given in Figs. A.1 and A.2. Here, taking into account the seven boundary layers, the stress is calculated for the three types of loading.

2. Vibrations of a Hollow Cylinder with Mixed End Conditions

To illustrate the methods of solving dynamic problems introduced above we consider vibrations of a cylindrical shell which is described in Chap. 1. As an example, we consider the first option of mixed end conditions and for simplicity we assume them to be symmetric with respect to the plane $\xi = 0$.

Table A.4 $\varepsilon = 0.1$ $l_0 = 1$ $\lambda_0 = 2$ Applied theory

ξ	U_r	σ_z
0.1	-0.2799×10^{-12}	−0.4943
0.2	-0.2656×10^{-12}	−0.4689
0.3	-0.2422×10^{-12}	−0.4276
0.4	-0.2106×10^{-12}	−0.3718
0.5	-0.1718×10^{-12}	−0.3034
0.6	-0.1272×10^{-12}	−0.2246
0.7	-0.7827×10^{-13}	−0.1382
0.8	-0.2667×10^{-13}	-0.4710×10^{-1}
0.9	0.2583×10^{-13}	0.4560×10^{-1}
1	0.7745×10^{-13}	0.1968

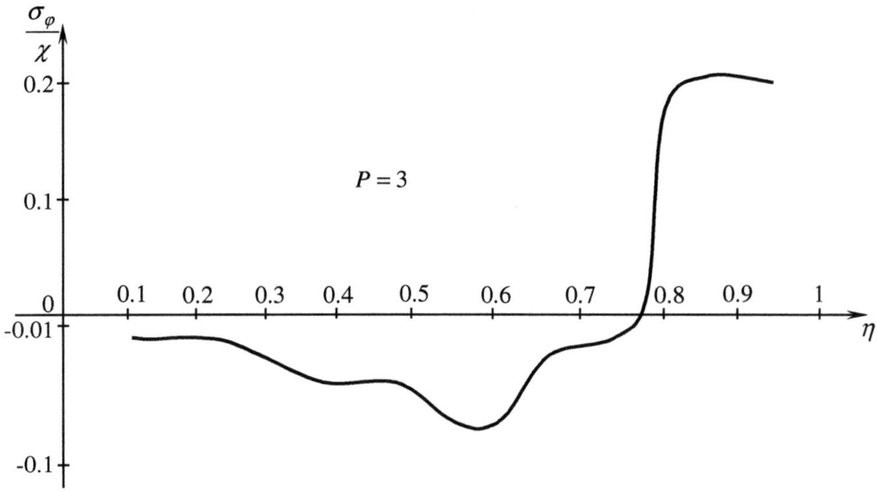

A.1 Variation of ring stress for the three types of loading

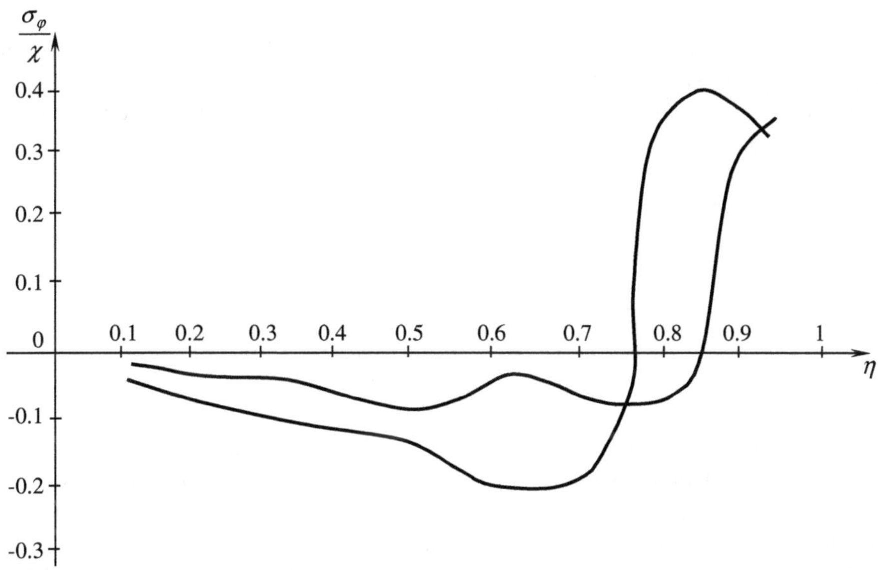

A.2 Variation of ring stress for the three types of loading

Thus, let there be given the following conditions

$$\sigma_z = \chi(1 - c\eta^2)e^{i\omega t}, U_r = 0 \text{ at } \xi = \pm l_0 \tag{A.11}$$

$$\sigma_r = 0, \tau_{rz} = 0 \text{ at } \rho = \rho_s \ (s = 1, 2) \tag{A.12}$$

To satisfy the boundary conditions (A.11) we use generalized orthogonality condition (1.4.5). Numerical calculations were performed for the frequencies $\lambda_0 < 2,5$ for different geometrical parameters of the shell. In this case, using the results of the first chapter, the first term in the asymptotic behaviour of U_r, σ_z, may be represented as follows:

$$\begin{aligned} U_r &= R_0\left\{\sum_{k=1}^{2} -C_k \nu m_k'(\xi) - 2\varepsilon^2 \sum_{k=3}^{\infty} C_k \right. \\ &\quad \times \left[(2-\nu)F_k'(\eta) + (1-\nu)\delta_k^{-2}F_n''(\eta)\right] m_k'(\xi)\Big\} e^{i\omega t} \\ \sigma_z &= 2G\left\{\sum_{k=1}^{2} C_k(1+\nu)(1-\lambda_0^2)m_k'(\xi) + \varepsilon \sum_{k=3}^{\infty} C_k \right. \\ &\quad \times F_n''(\eta)m_k'(\xi)\Big\} e^{i\omega t}. \end{aligned} \tag{A.13}$$

In the series (A.13) the summation is carried out on the roots μ_k located in the upper half-plane ($J_m\mu_k > 0$).

In view of the relations of the generalized orthogonality the desired constants C_k have the form:

$$C_k = -\chi \Delta_k^{-1} ch\mu_k l_0 \int_{-1}^{1} (1 - C_0\eta^2) W_k(\eta) d\eta$$

$$\Delta_k = \int_{-1}^{1} [U_k(\eta)\tau_k(\eta) - Q_{zk}(\eta)W_k(\eta)] d\eta.$$

In the calculations, the values $\nu = \frac{1}{3}$, $C_0 = \frac{1}{3}$, $\chi = 1$ are taken.

Numerical calculations were carried out taking into account the five boundary layers corresponding to the first 10 roots of equation $\sin 2\delta_k + 2\delta_k = 0$. The roots of equation $\sin 2\delta_k + 2\delta_k = 0$ were taken from [3].

The error of satisfying the boundary conditions for stresses does not exceed 5%.

Table A.4 shows the values of U_r, σ_z for the chosen values of the parameters λ_0, ε, l_0, obtained by Kirchhoff-Love theory.

Tables A.5, A.6, A.7 and A.8 shows the values of U_r, σ_z corresponding to the solution of boundary layer type.

Figure A.3 shows the distribution of σ_z along the length of the cylinder.

Table A.5 U_r $\varepsilon = 0.1$ $l_0 = 0.5$ $\lambda_0 = 2$

η					
ξ	0.2	0.4	0.6	0.8	1
0.1	0.3416×10^{-12}	0.4572×10^{-12}	0.2768×10^{-12}	-0.6091×10^{-13}	-0.3060×10^{-12}
0.2	0.7680×10^{-12}	0.1124×10^{-11}	0.9364×10^{-12}	0.3199×10^{-12}	-0.2150×10^{-12}
0.3	0.1436×10^{-11}	0.2189×10^{-11}	0.2351×10^{-11}	0.1789×10^{-11}	0.8163×10^{-12}
0.4	0.2169×10^{-11}	0.3131×10^{-11}	0.4060×10^{-11}	0.6059×10^{-11}	0.5753×10^{-11}
0.5	0.8652×10^{-12}	0.5637×10^{-13}	-0.3513×10^{-11}	-0.1161×10^{-10}	0.5697×10^{-10}

Table A.6 σ_z $\varepsilon = 0.1$ $l_0 = 0.5$ $\lambda_0 = 2$

η					
ξ	0.2	0.4	0.6	0.8	1
0.1	0.1009×10^{-2}	0.1068×10^{-2}	0.1381×10^{-2}	0.1865×10^{-2}	0.1635×10^{-2}
0.2	-0.6518×10^{-4}	-0.8232×10^{-4}	0.5068×10^{-3}	0.2181×10^{-2}	0.3102×10^{-2}
0.3	-0.6577×10^{-2}	-0.6848×10^{-2}	-0.6367×10^{-2}	-0.2095×10^{-2}	0.5410×10^{-2}
0.4	-0.6277×10^{-1}	-0.6307×10^{-1}	-0.623×10^{-1}	-0.6732×10^{-1}	0.1114×10^{-1}
0.5	$-0.3783 \cdot 10^{1}$	$-0.3769 \cdot 10^{1}$	$-0.3718 \cdot 10^{1}$	$-0.3448 \cdot 10^{1}$	-0.9381

Table A.7 U_r $\varepsilon = 0.1$ $l_0 = 1$

η					
ξ	0.2	0.4	0.6	0.8	1
0.1	0.3668×10^{-14}	0.6336×10^{-14}	0.6633×10^{-14}	0.4325×10^{-14}	0.2055×10^{-14}
0.2	0.4106×10^{-14}	0.9591×10^{-14}	0.1399×10^{-13}	0.1377×10^{-13}	0.1184×10^{-13}
0.3	-0.2607×10^{-14}	0.5939×10^{-14}	0.2273×10^{-13}	0.3425×10^{-13}	0.3844×10^{-13}
0.4	-0.3304×10^{-13}	-0.2688×10^{-13}	0.1961×10^{-13}	0.6901×10^{-13}	0.9692×10^{-13}
0.5	-0.1241×10^{-12}	-0.1443×10^{-12}	-0.4075×10^{-13}	0.1038×10^{-12}	0.1981×10^{-12}
0.6	-0.3416×10^{-12}	-0.4572×10^{-12}	-0.2768×10^{-12}	0.6091×10^{-13}	0.3068×10^{-12}
0.7	-0.7680×10^{-12}	-0.1123×10^{-11}	-0.9364×10^{-12}	-0.3199×10^{-12}	0.2150×10^{-12}
0.8	-0.1436×10^{-11}	-0.2189×10^{-11}	-0.2351×10^{-11}	-0.1789×10^{-11}	-0.8163×10^{-12}
0.9	-0.2169×10^{-11}	-0.3131×10^{-11}	-0.4060×10^{-11}	-0.6059×10^{-11}	-0.5753×10^{-11}
1	-0.8652×10^{-12}	-0.5637×10^{-13}	-0.3513×10^{-11}	0.1161×10^{-10}	-0.5697×10^{-10}

Table A.8 σ_z $\varepsilon = 0.1$ $l_0 = 1$ $\lambda_0 = 2$

η					
ξ	0.2	0.4	0.6	0.8	1
0.1	0.1791×10^{-4}	0.1903×10^{-4}	0.1621×10^{-4}	0.5658×10^{-5}	-0.8102×10^{-5}
0.2	0.7045×10^{-4}	0.7484×10^{-4}	0.7274×10^{-4}	0.5095×10^{-4}	0.7174×10^{-5}
0.3	0.1980×10^{-3}	0.2104×10^{-3}	0.2168×10^{-3}	0.1826×10^{-3}	0.7367×10^{-4}
0.4	0.4486×10^{-3}	0.4765×10^{-3}	0.5149×10^{-3}	0.4893×10^{-3}	0.2695×10^{-3}
0.5	0.8125×10^{-3}	0.8627×10^{-3}	0.9881×10^{-3}	0.1067×10^{-2}	0.7319×10^{-3}
0.6	0.1009×10^{-2}	0.1068×10^{-2}	0.1381×10^{-2}	0.1865×10^{-2}	0.1635×10^{-2}
0.7	-0.6517×10^{-4}	-0.8231×10^{-4}	0.5068×10^{-3}	0.2181×10^{-2}	0.3102×10^{-2}
0.8	-0.6577×10^{-2}	-0.6848×10^{-2}	-0.6367×10^{-2}	-0.2095×10^{-2}	0.5410×10^{-2}
0.9	-0.6277×10^{-1}	-0.6307×10^{-1}	-0.6443×10^{-1}	-0.6732×10^{-1}	-0.1114×10^{-1}
1	-0.3783×10^{-1}	-0.3769×10^{-1}	-0.3718×10^{-1}	-0.3448×10^{-1}	-0.9380

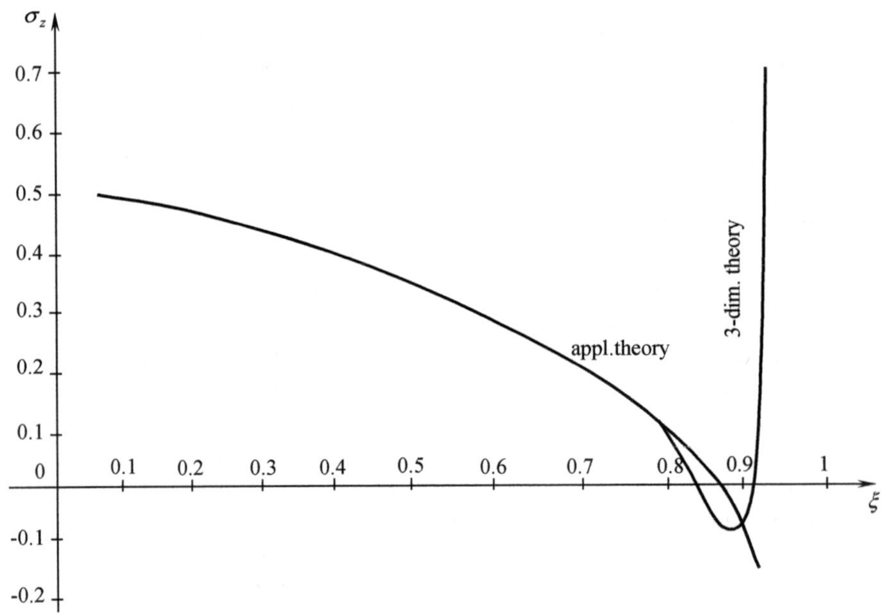

A.3 Distribution of axial stress along the length of the cylinder

3. Torsion of a Spherical Belt by Efforts Distributed Over the Surface of Tapered Cut

We investigate the spatial dynamic stress state of a spherical belt when its faces are free of stresses, but on the conical sections $\theta = \theta_j (j = 1, 2)$ the following boundary conditions are given by

$$\begin{aligned}\tau_{\theta\varphi}(\rho,\theta_1,t) &= \tau(\rho)e^{i\omega t}\text{at} \quad \theta=\theta_1,\\ U_\varphi(\rho,\theta_2,t) &= 0 \text{ at} \quad \theta=\theta_2.\end{aligned} \tag{A.14}$$

Using the results of the second chapter, the solution can be represented in the form

$$\begin{aligned}U_\varphi &= R_0\sum_{k=1}^{\infty}U_k(\rho)\frac{dm_k}{d\theta}e^{i\omega t},\\ \tau_{r\varphi} &= G\sum_{k=1}^{\infty}T_{rk}(\rho)\frac{dm_k}{d\theta}e^{i\omega t},\\ \tau_{\theta\varphi} &= G\sum_{k=1}^{\infty}T_{\theta k}(\rho)\left[2ctg\theta d\frac{m_k}{d\theta}+\left(z_k^2-\frac{1}{4}\right)m_k\right]e^{i\omega t},\\ m_k(\theta) &= C_{1k}P_{z_k-\frac{1}{2}}(\cos\theta)+C_{2k}Q_{z_k-\frac{1}{2}}(\cos\theta).\end{aligned} \tag{A.15}$$

Let us determine the constants C_{1k}, C_{2k} from the condition (2.6.4). To this end we expand $\tau(\rho)$ by the eigenfunctions of the boundary problem (2.6.5):

$$\begin{aligned}&(\tau(\rho) = \sum_{k=1}^{\infty}a_kU_k(\rho)\\ &a_k\int_{\rho_1}^{\rho_2}\tau(\rho)\bar{U}_k(\rho)d\rho\left[\int_{\rho_1}^{\rho_2}U_k(\rho)\bar{U}_k(\rho)d\rho\right]^{-1}\end{aligned} \tag{A.16}$$

Condition (A.14) then gives us

$$\begin{aligned}C_{1k} &= a_k\Delta^{-1}Q^{(1)}_{z_k-1/2}(\cos\theta_2),\\ C_{2k} &= a_k\Delta^{-1}P^{(1)}_{z_k-1/2}(\cos\theta_2).\end{aligned} \tag{A.17}$$

Substituting (A.17) into (A.15), we obtain

$$\begin{aligned}U_\varphi &= R_0\sum_{k=1}^{\infty}a_k\Delta^{-1}U_k(\rho)D^{(1,1)}_{z_k-1/2}(\theta,\theta_2)e^{i\omega t}\\ \tau_{r\varphi} &= G\sum_{k=1}^{\infty}a_k\Delta^{-1}T_{rk}(\rho)D^{(1,1)}_{z_k-1/2}(\theta,\theta_2)e^{i\omega t}\\ \tau_{\theta\varphi} &= G\sum_{k=1}^{\infty}a_k\Delta^{-1}T_{\theta k}(\rho)\Big[2ctg\theta D^{(1,1)}_{z_k-1/2}(\theta,\theta_2)+\\ &\quad+\left(z_k^2-\frac{1}{4}\right)D^{(1,0)}_{z_k-1/2}(\theta,\theta_2)\Big]e^{i\omega t}\\ \Delta &= 2ctg\theta_1D^{(1,1)}_{z_k-1/2}(\theta_1,\theta_2)+\left(z_k^2-\frac{1}{4}\right)D^{(1,0)}_{z_k-1/2}(\theta_1,\theta_2).\end{aligned} \tag{A.18}$$

In the series (A.18) the summation is taken over the roots z_k located in the upper half-plane.

The function $\tau(\eta)$ is written as

$$\tau(\eta) = q_1 + q_2\eta.$$

Figures A.4, A.5 and A.6 show the curves of dependence of stresses $\tau_{\theta\varphi}$ on θ. In these calculations, it is assumed $\theta_1 = 3°$, $\theta_2 = 30°$, $v = \frac{1}{3}$.

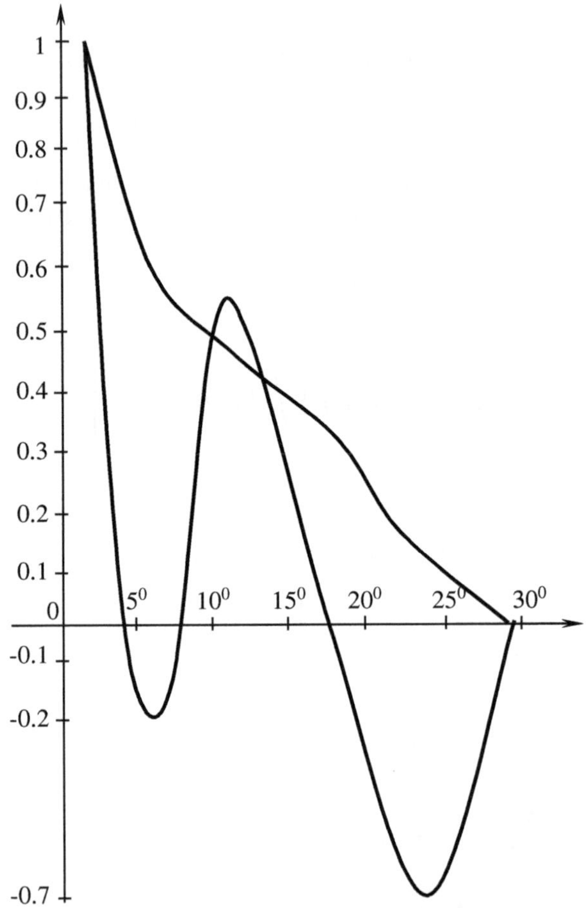

A.4 Variation of tangential stress with respect to spherical coordinate

A.5 Variation of tangential stress by applied and 3-dimensional theories

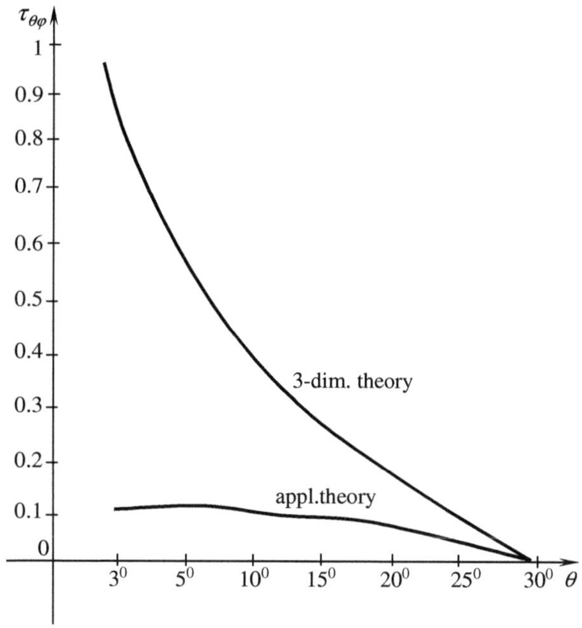

In Fig. A.4, curves 2 were constructed on the basis of a three-dimensional solution with regard to the two expansions when $q_1 = 1, q_2 = 0, \varepsilon = 0, 1, \lambda = 0, 5, \eta = 1$ However, curves 1 correspond to the case where $\lambda \sim \frac{1}{\sqrt{\varepsilon}}, \varepsilon = 0, 1, q_1 = 0, q_2 = 1, \eta = -1$ taking into account 10 terms of the expansion.

Figure A.5 shows the results obtained for $\lambda = 0, 5, \varepsilon = 0, 1, q_1 = 0, q_2 = 1, \eta = 1$ for the applied and three-dimensional theories.

The curves in Fig. A.6 correspond to the case of microwave vibrations $\lambda \sim \frac{1}{\varepsilon}, \varepsilon = 0.01, q_1 = 0, q_2 = 1, \eta = 1$.

Numerical calculations allow to determine the contribution of each solution in the state of stress at different parameters of a shell. The calculations have shown that, depending on the load, the addition of the boundary layer to the solution obtained through the applied theories reaches a considerable value not only in the marginal zone but also in the depth of the domain.

A.6 Curves of tangential stress in case of microwave vibrations

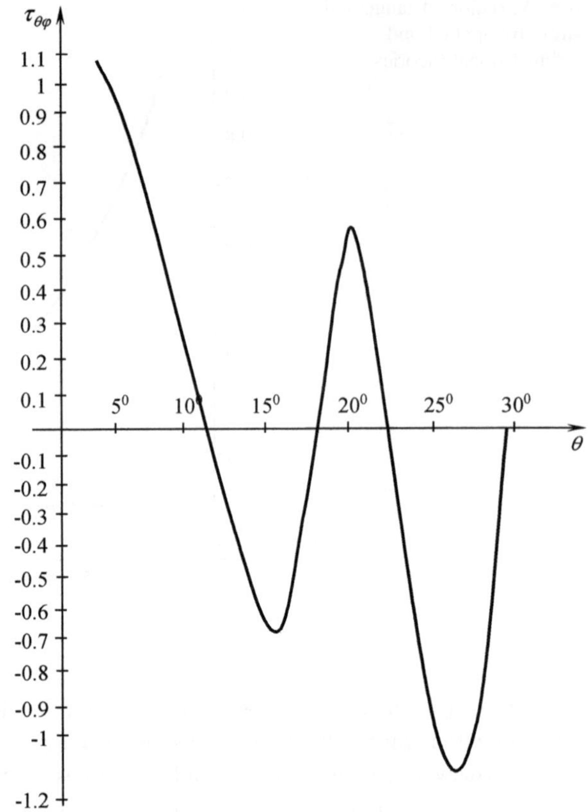

References

1. Aksentyan, O.C.: On stress concentration in thick plates. J. Appl. Math. Mech. **30**(5), 963–970 (1966)

2. Ustinov, Y.A., Yudovich, V.I.: On the completeness of the system of elementary solutions of the biharmonic equation in a half-band. J. Appl. Math. Mech. **37**(4), 706–714 (1973)

3. Malkina O.P.: Stress-strain state of a thick plate under loading symmetric with respect to the median plane, 171p. Dissertion of Candidate of Sciences. Rostov-on-Don (1968)

MIX
Papier aus verantwortungsvollen Quellen
Paper from responsible sources
FSC® C105338

If you have any concerns about our products,
you can contact us on
ProductSafety@springernature.com

In case Publisher is established outside the EU,
the EU authorized representative is:
Springer Nature Customer Service Center GmbH
Europaplatz 3, 69115 Heidelberg, Germany

Printed by Libri Plureos GmbH
in Hamburg, Germany